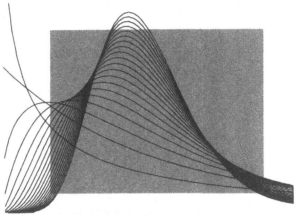

Markov Chain Monte Carlo Simulations and Their Statistical Analysis

With Web-Based Fortran Code

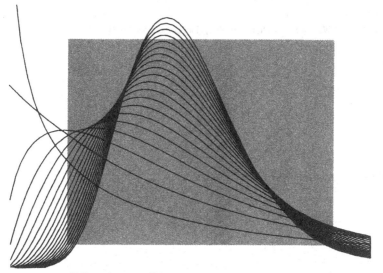

Markov Chain Monte Carlo Simulations and Their Statistical Analysis

With Web-Based Fortran Code

Bernd A. Berg

Florida State University, USA

 **World Scientific**

NEW JERSEY • LONDON • SINGAPORE • BEIJING • SHANGHAI • HONG KONG • TAIPEI • CHENNAI

Published by

World Scientific Publishing Co. Pte. Ltd.
5 Toh Tuck Link, Singapore 596224
USA office: 27 Warren Street, Suite 401–402, Hackensack, NJ 07601
UK office: 57 Shelton Street, Covent Garden, London WC2H 9HE

British Library Cataloguing-in-Publication Data
A catalogue record for this book is available from the British Library.

MARKOV CHAIN MONTE CARLO SIMULATIONS AND THEIR
STATISTICAL ANALYSIS
With Web-Based Fortran Code

ISBN 981-238-935-0

Printed in Singapore.

To Uschi

Preface

This book presents Markov Chain **Monte Carlo (MC)** simulations techniques in a self-contained way. The material should be accessible to advanced undergraduate students and is suitable for a course. In contrast to most other treatments of the subject, we relate the theory directly to computer code. It is my experience that writing suitable computer programs is by far more time consuming than understanding the theoretical concepts. Therefore, a symbiosis of theory and code should allow the beginner to reach the goal of own, original applications in a shorter than usual time. The code is written in Fortran 77[1] and can be downloaded from the Web by following the instructions given in chapter 1. Fortran is chosen, because the code started off in Fortran and it is desirable to present a homogeneous package of programs and subroutines. Also Fortran is still a language which is well-suited for scientific applications and the one which withstood the tooth of time against all odds.

Under Computer Revolution one understands the sustained increase of computational speed by approximately one order of magnitude (a factor of ten) every five years since about 1950. This is sometimes called Moore's law in reference to an article [Moore (1965)] by the former Intel chairman Golden Moore. The, by the year 2000, resulting factor of 10^{10} has led to qualitatively new features, which are most obvious in modern communication technology and by the use of computers in the office environment as well as at home. In physics and other disciplines of science the computer revolution has led to an enormous increase in the use of computational techniques. In particular MC calculations, whose convergence is slow, $\sim 1/\sqrt{t}$, where t is the elapsed computer time, have gained in importance.

---

[1]A C or C++ version of this book may follow. Note that some Fortran compiler allow to create C code.

The theory behind MC simulations is based on statistics and the analysis of the MC generated data is applied statistics. Nowadays abundance of computational power implies also a paradigm shift with respect to statistics: Computationally intensive, but conceptually simple, methods belong at the forefront, whereas traditional analytical simplifications loose in importance. MC simulations are not only relevant for simulating models of interest, but they constitute also a valuable tool for approaching statistics. Our first two chapters provide a self-contained **Introduction to Statistics**. Subsequently, our presentation of Markov chain MC simulations is build on this foundation. Our approach to statistics is strictly from the sampling perspective and illustrated by many MC tests. The statistical methods are later at hand, when data analysis techniques are needed. In particular, we present the important jackknife method in chapter 2 after the bias problem is encountered. In chapters 1 and 2 the focus is on independent random variables. Autocorrelations are treated in chapter 4, after Markov chain MC methods have been introduced. In particular, we investigate the integrated autocorrelation time for various algorithms.

In chapter 3 the conventional Markov Chain MC algorithms are treated after summarizing some statistical physics background. The **impatient reader** may consider to start right away with this chapter and to fill in material from earlier chapters whenever it is needed. The point of departure in chapter 3 is a treatment of the 1953 Metropolis algorithm for simulations of the Gibbs canonical ensemble. The heat bath algorithm follows. To illustrate these methods, our systems of choice are discrete Potts and continuous $O(n)$ models. Both classes of models are programmed for asymmetric (hyper) cubic lattices in arbitrary dimensions ($d = 1, 2, 3, 4, \ldots$).

Advanced MC simulations are the topic of chapters 5 and 6, dealing with multicanonical methods, event driven simulations, cluster algorithms, large scale simulations and parallel computing. Multicanonical simulations cover an entire temperature range and allow for calculations of the entropy and free energy. Event driven updates speed up simulations at low temperatures. Cluster algorithms can almost or even completely eliminate the critical slowing down of MC simulations near second order phase transitions. As a relevant example for parallel algorithms, parallel tempering is treated. Using MPI (Message Passing Interface, which is a popular implementation for parallel computing in Fortran or C) parallel tempering allows for an efficient use of inexpensive PC clusters.

My target group are readers who like to proceed quickly to applications without bothering too much about theoretical details. Therefore, instead of trying to be exhaustive, the focus is on major points. Fortran **assignments** are provided at the end of each section. Their goal is to make practically all numerical results presented in this book fully **reproducible**. Although such reproducibility is not an accepted standard in computational science, it is certainly of importance. To a large extent it is reproducibility which distinguishes scientific insight from hearsay.

To get the assignments up and running on your computer is considered to be a **must**, if you want to master the material of this course. A general introduction to the code is given after the second set of assignments (chapter 1.3). Within their own section the assignments are referred to by their number. Otherwise, chapter and section number of the assignment are also given. For readers, who like to gain a further understanding of some of the introduced concepts, more exercises are collected in appendix B.

This is the first book of a planned series of two. It constitutes an **Introduction** *to MC Simulations and Their Statistical Analysis*. For the sake of getting to simulations after a reasonable amount of preparation, some important statistical concepts as well as interesting physical systems to which simulations are applied, are moved to the second volume. For the first volume, I have in a subjective way picked those methods, which were frequently encountered in my own research practice. There is no claim that the solutions presented are de facto the best. References are also kept to a minimum with the exception of the outlook and conclusions, where some of the missing issues are touched. Otherwise, mainly books and papers whose results are actually presented are quoted. (My own own papers are overrepresented because they provide reference values needed to test the Fortran code.)

The second volume of this book is intended to fill in additional material about statistics and to extend significantly the scope of physical systems considered. Some of the statistical material of the second volume will be fairly basic and is normally treated at the beginning of a statistics course. It is re-shuffled with the intention to limit the statistics treatment of the first book to methods needed for the analysis of MC data in the later chapters. So it comes that, for instance, a treatment of the elementary binomial distribution, as well as of the fundamental Kolmogorov axioms, is relegated to the second book. Other statistical approaches of the second volume are more advanced, like bootstrap data analysis, Bayesian statistics and maximum entropy methods. The additional physical systems, included in the second volume, range from lattice gauge theory to all-atom protein models. What keeps such diverse topics together is that they are all well suited for studies by MC simulations.

There exists already an extensive and excellent literature on Markov chain MC methods, for example the books[2] by [Allen and Tildesley (1990)], [Frenkel and Smit (1996)], [Kalos and Whitlock (1986)], [Gould and Tobochnik (1996)], [Landau and Binder (2000)], [Liu (2001)], [Newman and Barkena (1999)]. But a treatment which includes step by step computer code and corresponding numerical assignments appears to be missing. The translation of the general concepts into code can be quite tedious and a major obstacle for the learning student. Further, it is not such a good idea that a newcomer starts to write her or his own code before studying examples which lay out some main ideas. At the best, the wheel gets invented over and over again.

Let me emphasize one more point: This book is primarily about **computational techniques** and only occasionally about physics. What about all the great physics applications waiting out there? Certainly they are important. However, it is unlikely that your physics applications of MC simulations will be of high quality if you do not learn to master the techniques first.

Parts of this book are based on lectures on the subject which I have given at the Florida State University, at the Technical University Vienna, at the European Center for Theoretical Studies in Nuclear Physics in Trento, at Leipzig University, at the Okazaki National Research Institutes, and at the National University of Singapore. Further, I benefitted from numerous discussions with faculty and students. I would like to thank all institutions for their support and the involved students and faculty for their interest. In particular, my thanks go to Alain Billoire, Kurt Binder, Peter Grassberger, Ulrich Hansmann, Hsiao-Ping Hsu, Wolfhard Janke, Harald Markum, Yuko Okamoto, Robert Swendsen, Alexander Velytsky, and Jian-Sheng Wang.

Bernd A. Berg (2004)

---

[2][Allen and Tildesley (1990)] and [Frenkel and Smit (1996)] focus on *molecular dynamics*, which is not a subject of our presentation (see our summary and conclusions for more information). Markov chain MC methods found their way also into computer science, mathematics and statistics. These disciplines were creative in inventing their own terminology. The book by [Liu (2001)] is written in their notations and bridges to some extent the language barrier to physics.

Contents

Chapter 1

Sampling, Statistics and Computer Code

This chapter combines a treatment of **basic statistical concepts** with our numerical approach and a presentation of the **structure of our Fortran code**. The notion of probability is introduced via **sampling** and followed by a discussion of probability distributions and other standard concepts of **descriptive statistics**, like q-tiles and confidence intervals, functions, expectation values and sums of independent random variables, characteristic functions and the central limit theorem. The numerical generation of arbitrary probability distributions is achieved through the use of **uniformly distributed pseudo random numbers**. **Sorting**, implemented by a heapsort routine, is seen to be at the heart of dealing computationally with confidence intervals.

1.1 Probability Distributions and Sampling

We give an operational definition of **probability**, based on **sampling** due to **chance**. But, chance is one of those basic concepts in science which elude a derivation from more fundamental principles.

A **sample space** is a set of **events**, also called **measurements** or **observations**, whose occurrence depends on chance.

Examples:

(i) The sample space defined by a die is $\{1, 2, 3, 4, 5, 6\}$ and an event is obtained by throwing the die.

(ii) We may define as sample space the set of all intervals $[a, b]$ with $0 \leq a \leq b \leq 1$, where an event specified by a and b is said to occur if $x \in [a, b]$. Here $x \in (-\infty, +\infty)$ is a real number obtained by chance in a yet unspecified way.

Let us assume, for the moment, that events are outcomes of experiments which can be replicated arbitrarily often under conditions assumed to be *identical*. In particular this ensures that subsequent events are statistically independent. Carrying out such repetitions of the same experiment is called **sampling**. The outcome of each experiment provides an event called **data point**. In N such experiments, for instance throwing the die again and again, we may find the event A to occur with **frequency** n, $0 \leq n \leq N$. The **probability** assigned to the event A is a number $P(A)$, $0 \leq P(A) \leq 1$, so that

$$P(A) \; = \; \lim_{N \to \infty} \frac{n}{N}. \tag{1.1}$$

This equation is sometimes called the **frequency definition of probability**. However, as one can never carry out an infinite number of experiments, it is better to rely on an *a-priori* understanding of probability and to require (1.1) to happen for events which occur with probability $P(A)$. For finite N this allows one to predict a likely range for n/N if the error, due to the finite N in (1.1), can be controlled in the sense of probability. Kolmogorov has 1933 formulated axioms for probabilities which can be found in statistics texts like [Van der Waerden (1969)] or [Brandt (1983)]. The book *Mathematical Statistics* by van der Waerden is recommended, because it is clear and well suited to be read by physicists. In the following we embark on an intuitive approach, exploiting equation (1.1) further.

Let A be an element of a sample space. The probability of A is a number $P(A)$, $0 \leq P(A) \leq 1$, assigned to each element of the sample space. $P(A) = 0$ means the event occurs never and $P(A) = 1$ means the event occurs with certainty.

Example: For a perfect die all probabilities are identical, $P(i) = 1/6$ for $i \in \{1, 2, 3, 4, 5, 6\}$.

The situation is more involved for the other sample space introduced above as example (ii). Let us denote by $P(a, b)$ the probability that $x^r \in [a, b]$ where x^r is a **random variable** (indicated by the superscript ^{r}) drawn in the interval $(-\infty, +\infty)$ with the **probability density** $f(x)$. A probability density is a positive function $f(x) \geq 0$, which fulfills the normalization condition

$$1 = \int_{-\infty}^{+\infty} f(x)\,dx \tag{1.2}$$

and is defined so that $P(a,b)$ is given by

$$P(a,b) = \int_a^b f(x)\,dx. \tag{1.3}$$

The other way round, knowledge of all probabilities $P(a,b)$ implies knowledge of the probability density $f(x)$. For $x > y$:

$$f(x) = \lim_{y \to x} \frac{P(y,x)}{x-y} \geq 0. \tag{1.4}$$

The **cumulative distribution function** of the random variable x^r is defined as the probability that the random variable is smaller or equal than some given real number x

$$F(x) = P(x^r \leq x) = \int_{-\infty}^{x} f(x)\,dx. \tag{1.5}$$

$F(x)$ is monotonically increasing with x and strictly monotone in all regions where $f(x)$ is non-zero. The word *cumulative* in front of *distribution function* will often be omitted. Physicists tend to use the notation **probability distribution** (or just distribution) for the probability density, whereas in statistics it denotes the distribution function. Here probability distribution will be used to denote both, followed by more precise explanations if necessary.

Using the Dirac delta function, **discrete probabilities** can be accommodated into continuous mathematics. Their probability density is

$$f(x) = \sum_i P_i\,\delta(x - x_i) \tag{1.6}$$

where

$$\delta(x - x_i) = 0 \text{ for } x \neq x_i \quad \text{and} \quad \int_{-\infty}^{+\infty} dx\,\delta(x - x_i) = 1. \tag{1.7}$$

Let us turn to truly continuous distributions. A particularly important case is the **uniform probability distribution** in the interval $[0,1)$, because this is what most random number generators implement. The uniform probability density $f(x) = u(x)$ is defined by

$$u(x) = \begin{cases} 1 & \text{for } 0 \le x < 1; \\ 0 & \text{elsewhere.} \end{cases} \tag{1.8}$$

The corresponding distribution function is

$$U(x) = \int_{-\infty}^{x} u(x)\,dx = \begin{cases} 0 & \text{for } x < 0; \\ x & \text{for } 0 \le x \le 1; \\ 1 & \text{for } x > 1. \end{cases} \tag{1.9}$$

Remarkably, the uniform distribution allows for the construction of general probability distributions. Let

$$y = F(x) = \int_{-\infty}^{x} f(x')\,dx'$$

and assume that a real number $\epsilon > 0$ exist so that $f(x') > 0$ for $x' \in (x - \epsilon, x]$. Then $F(x')$ is strictly monotone in this interval and, therefore, the inverse $x = F^{-1}(y)$ exist. For y^r being a uniformly distributed random variable in the range $[0, 1)$ it follows that

$$x^r = F^{-1}(y^r) \tag{1.10}$$

is distributed according to the probability density $f(x)$. An example is the **Cauchy distribution** for which the mapping (1.10) is illustrated in figure 1.1. The probability density and cumulative distribution function of the Cauchy distribution are given by

$$f_c(x) = \frac{\alpha}{\pi\,(\alpha^2 + x^2)} \tag{1.11}$$

and

$$F_c(x) = \int_{-\infty}^{x} f_c(x')\,dx' = \frac{1}{2} + \frac{1}{\pi}\tan^{-1}\left(\frac{x}{\alpha}\right), \quad \alpha > 0. \tag{1.12}$$

Hence, the Cauchy distributed random variable x^r is generated from the uniform $y^r \in [0, 1)$ through

$$x^r = \alpha\,\tan(\pi y^r - \pi/2)\,. \tag{1.13}$$

To gain computational speed in a Cauchy random number generator this can be simplified to

$$x^r = \alpha\,\tan(2\pi y^r)\,, \tag{1.14}$$

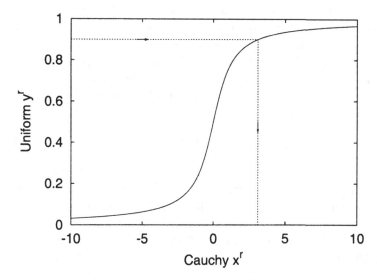

Fig. 1.1 Illustration of equation (1.10): Mapping (1.13) of the uniform to the Cauchy distribution. Let y^r be a uniform random number, the arrows lead to the corresponding Cauchy random number.

where it is used that one is not interested in the individual values of the random events x^r but only in generating its correct distribution.

1.1.1 *Assignments for section 1.1*

(1) Calculate the probability density for the distribution of the scalar product of two unit vectors which are uniformly distributed on the circle, *i.e.*, for $x = \cos(\phi)$, where the angle ϕ between the vectors is uniformly distributed in the range $0 \leq \phi \leq \pi$. Solution: $f(x) = \pi^{-1}/\sqrt{1-x^2}$ for $-1 \leq x \leq 1$.

(2) Repeat the previous exercise for the scalar product of two unit vectors, which are uniformly distributed on the sphere. Solution: $f(x) = 1 - |x|$ for $-1 \leq x = \cos(\theta) \leq 1$, where θ is the angle between the vectors.

1.2   Random Numbers

We will develop applications of probability theory by simulating chance. For this purpose we need a reliable source of random numbers. Some processes in nature, like nuclear decay and other quantum events in physics, are supposed to be truly random and could be used to provide uniformly distributed random numbers. However, for practical purposes this would be clumsy: Although one could consider to devote a hard drive for the exclusive purpose of storing true random numbers, the period would still be rather limited and to get fast access times would be a hassle. It is more convenient to rely on **pseudo-random number generators**. The aim is to provide numbers x in the range

$$0 \leq x < 1 \qquad (1.15)$$

by a deterministic procedure so that, for almost all practical purposes, the obtained numbers can be used as uniformly distributed random numbers. Obviously, one application is excluded: the numbers cannot be random when tested against their own generating procedure. Hence the notion **pseudo, which is dropped in the following**. According to [Marsaglia *et al.* (1990)] a list of desirable properties for random number generators is:

(i)   *Randomness.* The generator should provide a sequence of uniformly distributed random numbers, suitable for all reasonable applications. In particular, the generator should pass stringent tests for randomness.

(ii) *Long period.* One requests the ability to produce, without repeating the initial sequence, all of the random variables of a huge sample that current computer speeds allow for.

(iii) *Computational efficiency.* The execution should be rapid, with modest memory requirements.

(iv) *Repeatability.* Initial conditions (seed values) completely determine the resulting sequence of random variables.

(v)   *Portability.* Identical sequences of random variables may be produced on a wide variety of computers (for given seed values).

(vi) *Homogeneity.* All subsets of bits of the numbers are random, from the most- to the least-significant bits.

Marsaglia *et al.* used statistical distributions to test random number generators for randomness. Physicists have added a number of their applications as new tests. In particular exact results for the Ising model allow for tests which give many generators difficulties. The work by [Vattulainen *et al.* (1995)] is one of the papers on this subject and many earlier references can be found therein, see also pp.176 of [Landau and Binder (2000)].

Most Fortran compilers implement a default generator, often called RAN or RANF. Normally it is a **congruential random number generator** based on the recursion

$$I_n = a\,I_{n-1} + b \pmod{m}, \qquad (1.16)$$

where I_n, a, b and m are all are integers. Uniformly distributed random numbers are then given by

$$x_n = \frac{\text{FLOAT}\,(I_n)}{\text{FLOAT}\,(m)} . \qquad (1.17)$$

Theorems, for proofs see the book by [Knuth (1981)], state that good choices of the integers a, b and m exist so that the period of the series becomes m, *i.e.* a permutation

$$\pi_0, \pi_1, \ldots \pi_{m-2}, \pi_{m-1}$$

of the numbers $0, 1, \ldots m - 2, m - 1$ is generated by the recursion (1.16). In implementations the period m is limited by the available computer word size, so that typical maximum periods are 2^{32} or 2^{64}. The generator is reasonably random for most applications, but has the problem that k-dimensional coordinates lie on $\leq m^{1/k}$ subspaces of dimension $(k - 1)$. For $k = 3$ this amounts to $2^{32/3} \approx 1600$ $2d$ subspaces. In addition: Some system implementations relied on faulty values of a and b, and the generator is normally not portable as the maximum word-size can only be exploited by implementations on the system level. In conclusion, the system generator is suitable for playing around, but for serious applications the reader is recommended to rely on his or her own carefully selected random number generator.

Here Fortran code for the random number generator by [Marsaglia *et al.* (1990)] is provided. It relies on a combination of two generators:

x_n from a lagged Fibonacci series $I_n = I_{n-r} - I_{n-s} \bmod 2^{24}$, $r = 97$, $s = 33$ with a period of $(2^{24} - 1) \times 2^{94} \approx 2^{110}$, which on its own is already a reasonably good generator (but fails the statistical birthday spacing test).

y_n from the arithmetic series $I - k, I - 2k, I - 3k, \ldots$, $\bmod\,[2^{24} - 3]$, where $2^{24} - 3 = 16777213$ is a prime number and $k = 7654321$. On its own this series is too regular for a random number generator.

The final random numbers u_n are obtained by the combination

$$u_n = x_n \circ y_n$$

where $\circ$ denotes the *binary* operation

$$x \circ y = \begin{cases} x - y & \text{for } x \geq y; \\ x - y + 1 & \text{otherwise.} \end{cases}$$

For most applications this generator is a good compromise. It fulfills the statistical criteria for randomness [Marsaglia *et al.* (1990)] and does reasonably well in the physics tests [Vattulainen *et al.* (1995)]. It has a long period (ii) of 2^{144} and fulfills criteria (iv) to (vi). Further study of this generator reveals an underlying chaotic process [Lüscher (1994)].

Portability of the Marsaglia random number generator is a convenience of major practical importance. It is achieved by using numbers for which 32 bit `real*4` floating point (fp) arithmetic is exact, employing the IEEE[1] standard representation of 32 bit fp numbers, which is

$$\pm 2^p \left(1 + b_1 2^{-1} + b_2 2^{-2} + \cdots + b_{23} 2^{-23}\right) \tag{1.18}$$

where $b_i = 0$ or 1 and p is an integer $-64 \leq p \leq 64$.

Concerning the computational efficiency (iii), other generators, like already the congruential one, are faster. However, this is only an important issue for special large scale applications for which the generation of random numbers eats up a major fraction of the CPU time. Then one has to think about implementing a faster generator, which for the purpose of speed will sacrifice some of the nice properties of the Marsaglia generator. The shift-register random number generator, for references see [Vattulainen *et al.* (1995)], would be an example. Additional, extensive information about random number generators is easily found by searching the Web. For instance, type *random numbers* into the **google** search machine at google.com. In particular, the *Scalable Parallel Pseudo Random Number Generator Library* (SPRNG) allows downloads of computer code from a website at the Florida State University [Mascagni and Srinivasan (2000)].

---

[1]IEEE stands for *Institute of Electrical and Electronics Engineers*. The standards specified by IEEE are obeyed by most computers.

Table 1.1 The program `mar.f`, which illustrates the use of the routines `rmaset.f`, `ranmar.f` and `rmasave.f`, is depicted. If the file `ranmar.d` does not exist, the default seeds `iseed1=1` and `iseed2=0` of the Marsaglia random number generator are used.

```
      program marsaglia ! Bernd Berg, June 2 1999.
C IMPLEMENTATION OF MARSAGLIA'S RANDOM NUMBER GENERATOR.
C (MARSAGLIA, ZAMAN AND TSANG, STAT.& PROB. LETTERS 8 (1990) 35.)
C - The seed parameters are integers, -1801 <= iseed1 <= 29527,
C   -9373 <= iseed2 <= 20708. Distinct pairs give independent sequences.
C - When ranmar.d does not exist, iseed1=1 and iseed2=0 reproduces
C   table 4 of the Marsaglia et al. paper.
C - Whenever the file ranmar.d exists: Continuation using the seeds
C   saved by RMASAVE is ensured. Compare the last number of the previous
C   run with the first of the continuation!
      include '../../ForLib/implicit.sta'
      parameter(iuo=6,iud=10,iseed1=1,iseed2=0,ndat=004)
c
      inquire(file='ranmar.d',exist=lexist)
      if(.not.lexist .and. iseed1.eq.1 .and. iseed2.eq.0) then
        write(iuo,*) ' '
        call rmaset(iuo,iud,iseed1,iseed2,'ranmar.d')
        write(iuo,*) 'Table of Marsaglia et al.:'
        DO II=1,20005
          CALL RANMAR(XR)
          IF(II.GT.20000)
     &      WRITE(IUO,'(2X,7I3)') (MOD(INT(XR*16.**I),16),I=1,7)
        END DO
      end if
C
      write(iuo,*) ' '
      call rmaset(iuo,iud,iseed1,iseed2,'ranmar.d')
      do idat=1,ndat
        call ranmar(xr)
        write(iuo,*) 'idat, xr =',idat,xr
      end do
c
      call rmasave(iud,'ranmar.d')
      call ranmar(xr)
      write(iuo,*) 'extra xr =  ',xr
      stop
      end

      include '../../ForLib/rmaset.f'
      include '../../ForLib/ranmar.f'
      include '../../ForLib/rmasave.f'
```

Table 1.2 Output produced by running `mar.f` with its default settings, `iseed1=1` and `iseed2=0`. The first run is done without a data file `ranmar.d` present in the run directory. The second (`CONTINUATION`) run is performed with `ranmar.d` as obtained from the first run.

```
RANMAR INITIALIZED.
Table of Marsaglia et al.:
    6  3 11  3  0  4  0
   13  8 15 11 11 14  0
    6 15  0  2  3 11  0
    5 14  2 14  4  8  0
    7 15  7 10 12  2  0
```

RANMAR INITIALIZED.		MARSAGLIA CONTINUATION.	
idat, xr = 1	0.116391063	idat, xr = 1	0.495856345
idat, xr = 2	0.96484679	idat, xr = 2	0.577386141
idat, xr = 3	0.882970393	idat, xr = 3	0.942340136
idat, xr = 4	0.420486867	idat, xr = 4	0.243162394
extra xr =	0.495856345	extra xr =	0.550126791

Our Fortran implementation of Marsaglia random numbers consists of three primary subroutines: `rmaset.f` to set the initial state of the random number generator, `ranmar.f` which provides one random number per call, and `rmasave.f` to save the final state of the generator. In addition `rmafun.f` is a function version of `ranmar.f` and calls to these two routines are freely interchangeable. Related is also the subroutine `rmacau.f`, which uses equation (1.14) to generate Cauchy random numbers (1.14) with $\alpha = 1$. **Instructions about using the Fortran code** are given in section 1.3. Read those before using any of the code to this book! All the subroutines mentioned so far are part of the directory `ForLib` of STMC (see figure 1.3).

The subroutine `rmaset.f` initializes the generator to mutually independent sequences of random numbers for distinct pairs of seed values

$$-1801 \leq \text{iseed1} \leq 29527 \quad \text{and} \quad -9373 \leq \text{iseed2} \leq 20708 . \quad (1.19)$$

This property makes the generator quite useful for parallel processing. Note that the quadruple (I,J,K,M) in `rmaset.f` corresponds to (I,J,K,L) in the paper of [Marsaglia *et al.* (1990)]. The range (1.19) of our two seeds generates precisely their range 1 to 178, not all 1, for I,J,K, and 0 to 168 for L. The particular implementation of their paper is reproduced for $\text{iseed1} = 1$ and $\text{iseed2} = 0$.

The use of the subroutines `rmaset.f`, `ranmar.f` and `rmasave.f` is illustrated by the program `mar.f`, see assignment 2. The program is depicted in table 1.1 and its source code is contained in the folder `ForProb/Marsaglia` of STMC (see figure 1.3). The results of assignment 2 are given in table 1.2, which illustrates a start and a continuation run of the generator.

Let us comment on **discretization**. Due to the finite number of bits of a computer word, we cannot generate a continuous distribution of numbers $x \in [0, 1)$. Instead, we choose from a discrete number of possibilities with a small finite stepsize Δx. Similarly, all numbers recorded in real experiments are *a-priori* limited to a finite number of digits, as defined by the resolution of the recording device. In other words, computer simulations as well as real sampling create always a discrete number of possibilities. In contrast to that, the mathematical calculus is most efficient when formulated for continuous variables. Discrete random numbers can only give satisfactory sampling results when the scale on which we sample is sufficiently large compared to the discretization Δx. For instance, we would never converge towards a meaningful histogram when we would choose the range of a single histogram entry smaller than Δx. Most important, it follows that concepts of calculus which make a difference on a scale smaller than Δx have to be irrelevant for the final results, or our sampling cannot describe them correctly.

To illustrate this point: The discretization stepsize of the Marsaglia random number generator is

$$\epsilon_{\text{Marsaglia}} = 2^{-24} . \tag{1.20}$$

This relatively large stepsize makes the series reproducible even on computers with rather limited precision, but implies that there are only $2^{24} = 16,777,216$ distinct random numbers available. Therefore, the lowest allowed value, zero (1.15), will with almost certainty show up in a lengthy MC simulation, whereas it is of measure zero in the continuum. In assignment 5 we count how often the smallest and the largest Marsaglia random numbers show up when 10^{10} numbers are generated with the default seed. Note that the largest Marsaglia random number is $1 - 2^{-24}$. One can refine the discretization by combining two Marsaglia random numbers into one. Compare equation (3.123) of chapter 3.

1.2.1   *Assignments for section 1.2*

(1) **Before proceeding with any of the assignments: Read the introduction to the Fortran code given in the next section!** Fortran based solutions to most assignments are provided in the folders of the directory Assignments (see figure 1.3), which normally contain readme.txt files.

(2) **Marsaglia random numbers:** Run the program mar.f to reproduce the results of table 1.2. Understand how to re-start the random number generator as well as how to perform different starts when the continuation data file ranmar.d does not exist. Note: You find mar.f in ForProg/Marsaglia and it includes subroutines from ForLib. To compile properly, mar.f has to be located two levels down from a root directory (our STMC), which contains ForLib, ForProg, and so on. For instance, mar.f may be run in the folder a0102_02 of figure 1.3. See the next section for more details.

(3) **Frequency definition (1.1) of a probability:** Use Marsaglia random numbers with the default seed, iseed1=1 and iseed2=0, to generate consecutive series of $N = 2^{2k-1}$ random numbers with $k = 1, ..., 11$. For each series count the frequency n of events $x \in [0, 0.4)$, calculate the frequency ratio $R = n/N$ and its error $|R - 0.4|$. Result: Upon correct installation of the random number generator you get the numbers depicted in table 1.3. As expected, qualitatively sampling converges towards 0.4, but we have not yet quantitative control over the error.

(4) **Histogramming of the uniform probability density:** Use the default seeds to generate consecutive series of N=100 and N=10 000 random numbers. For each series build and plot its histogram of ten entries, which cover the equally spaced intervals $[0, 0.1], [0.1, 0.2], \ldots$. Normalize the integral over the histogram curve to one. For gnuplot users the call to the subroutine hist_gnu.f in the program hist.f generates the gnuplot script his.plt, which provides the plot of this assignment. Otherwise, use your favorite graphical package. In any case, if everything is done correctly, you obtain the result of figure 1.2. Visible is a tremendous improvement in uniformity for the large statistics of 10 000 random numbers over the small statistics of 100 random numbers. Note that with each histogram we check ten events for the convergence of sampling, whereas in assignment 3 only a single event is monitored.

Table 1.3 Sampling illustration of the frequency definition (1.1), see assignment 3.

K,N,R,ERROR:	1	2	0.5000000	0.1000000E+00
K,N,R,ERROR:	2	8	0.1250000	0.2750000
K,N,R,ERROR:	3	32	0.4375000	0.3750000E-01
K,N,R,ERROR:	4	128	0.3906250	0.9375000E-02
K,N,R,ERROR:	5	512	0.4003906	0.3906250E-03
K,N,R,ERROR:	6	2048	0.4047852	0.4785156E-02
K,N,R,ERROR:	7	8192	0.3952637	0.4736328E-02
K,N,R,ERROR:	8	32768	0.4010620	0.1062012E-02
K,N,R,ERROR:	9	131072	0.3975983	0.2401733E-02
K,N,R,ERROR:	10	524288	0.4002132	0.2132416E-03
K,N,R,ERROR:	11	2097152	0.4000158	0.1583099E-04

(5) Use the Marsaglia random number generator with its default seed to generate 10^{10} random numbers and count how often the smallest and the largest value show up. Answer: The zero shows up 563 times and the largest random number 561 times. Measure the user CPU time, which this run takes on your computer. (Depending on the speed of your computer you will have to wait some while for the results. Maybe, you better run this assignment in the background and/or over night. There are remarks about CPU time measurements at the end of the next section.)

1.3   About the Fortran Code

Our code is designed to run on platforms which support a Fortran 77 (f77) compiler. Freely use is made of some Fortran 77 extensions, like for instance a wordlength exceeding six characters. None of the tested f77 compilers had difficulties with the extensions. Some peculiarities have to be understood before running the code and for operating systems other than Linux slight modifications may be necessary. Under Linux the code has been extensively tested with the freely available (but not very good) gnu g77 compiler. Under Microsoft (MS) Windows the g77 compiler is part of the **Cygwin** package, which emulates many Linux features under Windows. Cygwin can be freely downloaded from many websites (go to http://cygwin.com and choose *Install or update now!*) of which the author recommends the one maintained by NASA. A fast internet connection is necessary, because the Cygwin

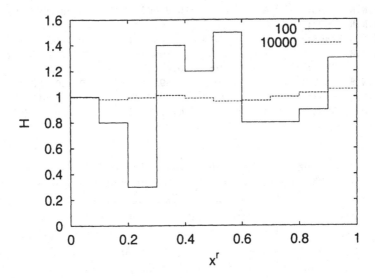

Fig. 1.2 Normalized histograms for the 100 and the 10 000 uniformly distributed random numbers of assignment 4.

package has the size of several gigabytes.

Occasionally checks of our Fortran code were performed under Windows with the (no longer available) MS fl32 Fortran 77 compiler, under true Unix on Alpha workstations with the DIGITAL Fortran 77 compiler (formerly DEC Fortran) and on Sun workstations.

To **download** the Fortran code for this book you may either visit the online bookshop of *World Scientific* at

http://www.wspc.com.sg/books

and find the website of this book by searching for the author, or you may visit the author's homepage, which is presently at

http://www.hep.fsu.edu/˜berg

and follow from there the link to the STMC code. Once you find the code, follow the download instructions given there.

The Fortran routines are provided and prepared to run in a tree structure of folders depicted in figure 1.3, which unfolds from the downloaded

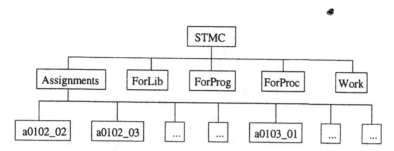

Fig. 1.3 Statistics and Monte Carlo (STMC): Tree of Fortran program folders.

file. STMC is our parent directory and the Fortran code is contained in the directories ForLib, ForProg and ForProc. ForLib contains a library of functions and subroutines which are listed in table 1.4. The library is closed in the sense that no reference to non-standard functions or subroutines outside the library is ever made. Fortran programs are contained in the folder ForProg and procedures for interactive use in ForProc. Details are given later. The Fortran programs and subroutines are also listed in the index.

To the extent that is is feasible within reasonable programming efforts, computer code should not only be efficient and clever, but also clear. For instance, instructions which are self-evident to the average programmer ought to be preferred over scriptic coding, even if the self-evident code is a few lines longer. Consistent rules should be used to indicate whether a variable is **real, integer, logical** and so on, instead of hiding this in endless declaration lists. The most important goal of **structured programming** is to make the control structure manifestly apparent to the programmer (the computer does not care). Therefore, one should limit the code to using the standard control structures, which are the plain do, the do-while, the do-until and the do-break[2] for the **iteration structure** and the **if structure**. The control flows should be emphasized by a systematic use of indentation.

Modularization and encapsulation are further important concepts of structured programming. For instance, the control structure can hardly be

_____

[2]In Fortran the latter three require goto statements of the "tame" type, see the first chapter of *Numerical Recipes* [Press *et al.* (1992)] for a more detailed discussion.

Table 1.4 Contents of ForLib. The *.f files are Fortran functions and subroutines. In addition ForLib contains the implicit.sta statement file and its alternate versions, the definitions of constants in constants.par and its alternates, and a number of common block definitions in *.com files.

F_df.	bino_nk_df.f	gau_df.f	p_mu_a.f	rwght_ts_zln.f
F_pd.f	bino_nk_xq.f	gau_metro.f	p_ts_z0ln.f	rwght_ts_zlnj.f
F_qdf.f	bino_pd.f	gau_pd.f	p_ts_z0lnj.f	sebar_e.f
F_xq.f	bino_qdf.f	gau_qdf.f	p_ts_zln.f	sebar_e_as.f
Ftest.f	cau_df.f	gau_xq.f	p_ts_zlnj.f	steb0.f
ac_int.f	cau_qdf.f	gaudif.f	parabola.f	steb1.f
ac_intj.f	cau_xq.f	heap_iper.f	perm.f	steb2.f
addln.f	chi2_df.f	heap_per.f	potts1_wght.f	steb_rat.f
addln2.f	chi2_pd.f	heapisort.f	potts_act.f	stebj0.f
addln_cut.f	chi2_qdf.f	heapsort.f	potts_act_tab.f	stebj1.f
autcora.f	chi2_xq.f	hist_gnu.f	potts_actm.f	stebjj1.f
autcorf.f	chi2pdf_df.f	icpointer.f	potts_actm2.f	stmean.f
autcorj.f	chi2pdf_pd.f	ipointer.f	potts_order.f	stop_mpi.f
autcorja.f	chi2pdf_qdf.f	isfun.f	potts_ran.f	stud_df.f
autoj_tau.f	chi2pdf_xq.f	ixcor.f	potts_rwght.f	stud_pd.f
bbi1_nk_df.f	datjack.f	kolm1.f	potts_wghb.f	stud_qdf.f
bbi1_nk_xq.f	datjack2.f	kolm1_as.f	potts_wght.f	stud_xq.f
bbi_df.f	df_gnu.f	kolm2_as.f	potts_z0ln.f	studdif.f
bbi_nk_df.f	eigen_2x2.f	kolm2_as2.f	potts_zln.f	tun_cnt.f
bbi_nk_xq.f	ellipse.f	kolm2_del2.f	pt_rec0.f	tun_cnt2.f
bbi_qdf.f	error_f.f	lat_init.f	qdf_gnu.f	wrat_to_b.f
beta.f	f_interpol.f	latc_init.f	qtiles.f	write_mpi_i.f
beta_i.f	fct_ln.f	lpt_ex_ia.f	ranmar.f	write_progress.f
bias.f	fct_ln_init.f	mat_gau.f	razero.f	xy_act.f
bin_inta.f	fi1.f	nsfun.f	read_steb0.f	xy_act1.f
bining.f	fit_g.f	nsum.f	rmacau.f	xy_ran.f
bino1_nk_df.f	fit_ggnu.f	o3_act.f	rmafun.f	xy_ran1.f
bino1_nk_xq.f	fit_l.f	o3_ran.f	rmag_steb0.f	
bino_cln.f	fit_lgnu.f	on_act.f	rmagau.f	
bino_coef.f	gamma_ln.f	on_ran.f	rmasave.f	
bino_df.f	gamma_p.f	p_e_compare.f	rmaset.f	

implicit.04	implicit.08	implicit.16	implicit.sta
constants.04	constants.08	constants.16	constants.par

binom.com	latc.com	on.com	p_pt_mpi.com	potts1.com	potts_muca.com
lat.com	muca.com	p_eds.com	potts.com	potts_hb.com	xy1.com

transparent when an if statement depends on a control variable which is passed through a common block and can be assigned anywhere in a (large) number of routines which share the common block. By such reasons some people advocate to abandon the use of common blocks all together. In our context this is impractical as outlined in the following, but we shall limit the use of common blocks to a few well-defined situations.

(1) In typical MC simulations various subroutines perform distinct tasks on the same set of arrays and variables. Allocating such arrays and variables in shared common blocks, which are passed via `include` statements into the programs, makes sure that their definitions are consistent in all routines. A change of a common block, like adding an array or variable, propagates then automatically into all affected routines. Changing the argument structure in each affected routine separately would be more tedious and error prone.

(2) The dynamical variables of a MC simulation are stored in arrays, which can become quite large. For efficient performance these arrays have to stay at fixed physical storage locations, unless moving them is required by the logic of the calculations. Common blocks force the compiler to allocate a fixed storage space. For some older Fortran compilers use of subroutine arguments led to a considerable slowing down. In program languages other than Fortran "dynamical storage allocation" has sometimes severely downgraded the performance of a code.

(3) Some control variables simply reflect the status of certain dynamical variables at runtime, for instance the internal energy. It is natural to include them in the common block of the dynamical variables. Typically, such control variables need an initialization in some ..._init.f routine and are updated together with the dynamical variables.

The structure discussed in our remarks 1–3 becomes only relevant from chapter 3 on.

(4) Groups of routines may have to exchange information with one another, which is better hidden away from the main program and all other routines, to decrease the likelihood to overwrite this information incidentally. The transfer of the relevant variables is then best done by letting the group of routines share a designated common block. The common block, discussed below, for the Marsaglia random number generator fulfills this purpose.

Together with the common blocks go parameter statements which dimension the arrays in question. We distinguish two situations. Either a common block and/or its associated parameter statement will most likely never be modified (as it is the case for the Marsaglia random number generator), or (frequent) changes are expected. In the first case the parameter statements and common blocks become simply part of the code. In the second case we introduce a **hyperstructure:** Those common blocks and their parameter statements are kept in separate files and **include statements** transfer them into the code.

We deal with implicit variable declarations in a similar way. All our routines of the ForLib library use the statement

$$\text{include 'implicit.sta'} \tag{1.21}$$

where the file implicit.sta contains by default the lines

```
implicit real*8 (a-h,o-z)
implicit logical (l)
```

which define the declaration of a-h,o-z and l for our entire code. It is used together with the Fortran default declaration that i,j,k,m,n are integers and changes the Fortran default for l from integer to logical. On machines with 64 bit words (*e.g.* alpha processors) real*8 is the Fortran default for a-h,o-z, whereas for other architectures (*e.g.* PCs) the default is real*4. Our implicit.sta statement makes real*8 the platform independent default, while allowing to switch easily to real*4 or real*16. For this purpose we keep together with implicit.sta the files implicit.04, implicit.08 and implicit.16 in ForLib, see table 1.4. We use real*8 as our default, because for most of our practical applications the degrading in processing speed on processors with 32 bit words is irrelevant and normally more than offset by the comfort of less worries about rounding errors. Should speed improvements due to real*4 arithmetic really matter, like in some MC applications, exchange of the implicit.sta file propagates through the entire code. In the same way, the change to implicit.16 can be done, if accuracy beyond real*8 matters.

Parameters have to be defined with the precision used in implicit.sta. This is done in the file

$$\text{constants.par} \tag{1.22}$$

which is propagated into the code through include 'constants.par' statements. Definitions with the precisions real*4, real*8 and real*16 are, respectively, stored in the files constants.04, constants.08 and constants.16. Consistent changes of the precision are achieved by using implicit.sta and constants.par files which match. Initialization of non-integer constants in functions and subroutines is avoided through use of the constants.par file. For instance, the number one is defined in the constants.* files as ONE = 1.E0 for real*4, ONE = 1.D0 for real*8 and ONE = 1.Q0 for real*16. In the same way other frequently used numbers are defined in these files. Additional fp numbers of the desired precision can be obtained by multiplication with an integer. For instance,

$$ONE * I$$

converts the integer I to its corresponding fp number. As another example, X=ONE/(ONE*2) assigns to X the fp number 0.5 in the precision declared for ONE and X in implicit.sta (this particular number is also directly defined as HALF in constants.par). Do **not** use FLOAT(I), because the precision will be real*4 only.

Main programs and closely related specialized subroutines are contained in the folders of ForProg and listed in table 1.5. These programs are only **ready to compile, when they are located two levels down from the parent directory** STMC. Otherwise, the include statements discussed below (1.23) need to be changed. For instance, mar.f has been copied to the folder

$$Assignments/a0102\_02,$$

where it can be run to solve assignment 2. For our **assignments** the **subdirectory notation** aiijj_kk is as follows: ii labels the chapter number, jj the section number and kk the number of the assignment within the section. For a new project it is recommended to create a subdirectory in the directory Work of figure 1.3.

The reason why our main programs compile only two levels down from our parent directory is the decision to allow for a tree structure of folders, while enforcing an include hyperstructure which propagates certain changes to all affected programs. The tree has to be sufficiently branched to allow for many projects. Therefore, our programs include Fortran functions, subroutines, the implicit.sta and other files through statements of the form

$$include '../../ForLib/fln.ext' \qquad (1.23)$$

where fln stands for the filename and .ext for its extension. For example our main program mar.f contains include '../../ForLib/rmaset.f' . Only on their correct level, the programs are ready to run.

Under Windows the MS fl32 compiler finds also the files named in the include statements (1.23), despite the fact that the proper MS DOS nota-

Table 1.5 Fortran programs and subroutines of the subdirectories of ForProg.

AutoC/auto_gau.f	MC_Potts/p_e_cases.f	MPI_Potts/p_ptmpi0.f
AutoC/bauto.f	MC_Potts/p_e_hist.f	MPI_Potts/p_ptmpi1.f
AutoC/uncor_gau.f	MC_Potts/p_e_init.f	MPI_Potts/p_ptmpi2.f
Ferdinand/ferdinand.f	MC_Potts/p_e_mc.f	MPI_Potts/potts_mchb.f
MC_On/ana_ts1_on.f	MC_Potts/p_hb_ts.f	MPI_Potts/pt_rec0.f
MC_On/ana_ts_on.f	MC_Potts/p_met_ts.f	MPI_Potts/pt_rec1.f
MC_On/ana_ts_onm.f	MC_Potts/p_met_tsm.f	MPI_Potts/pt_rec2.f
MC_On/o2_met.f	MC_Potts/p_metr_ts.f	Marsaglia/mar.f
MC_On/o2_ts.f	MC_Potts/p_mu.f	Marsaglia/ran_range.f
MC_On/o3_init.f	MC_Potts/p_mu_init.f	Test/addln_test.f
MC_On/o3_mchb.f	MC_Potts/p_mu_rec.f	Test/hello_world.f
MC_On/o3_tshb.f	MC_Potts/p_mu_ts.f	Test/isort_test.f
MC_On/o3_tshbm.f	MC_Potts/potts1_hist.f	Test/lran.f
MC_On/on_init.f	MC_Potts/potts1_histn.f	Test/permutations.f
MC_On/on_met.f	MC_Potts/potts1_init.f	chi2/chi2_tabs.f
MC_On/on_ts.f	MC_Potts/potts1_met.f	fit_g/gfit.f
MC_On/on_tsm.f	MC_Potts/potts1_metn.f	fit_g/gfitj.f
MC_On/sphere_test.f	MC_Potts/potts_cl.f	fit_g/subg_1ox.f
MC_On/xy_init.f	MC_Potts/potts_clw.f	fit_g/subg_exp.f
MC_On/xy_init1.f	MC_Potts/potts_hihb.f	fit_g/subg_exp2.f
MC_On/xy_met.f	MC_Potts/potts_hist.f	fit_g/subg_linear.f
MC_On/xy_met0.f	MC_Potts/potts_histcl.f	fit_g/subg_power.f
MC_On/xy_met1.f	MC_Potts/potts_histw.f	fit_g/subg_power2.f
MC_On/xy_ts.f	MC_Potts/potts_init.f	fit_g/subg_power2f.f
MC_On/xy_ts0.f	MC_Potts/potts_inithb.f	fit_g/subg_power2p.f
MC_On/xy_ts1.f	MC_Potts/potts_mchb.f	fit_l/lfit.f
MC_Potts/ana_hist.f	MC_Potts/potts_met.f	fit_l/lfitj.f
MC_Potts/ana_kol.f	MC_Potts/potts_met_r.f	fit_l/subl_1ox.f
MC_Potts/ana_mag.f	MC_Potts/potts_metn.f	fit_l/subl_exp.f
MC_Potts/ana_pmu.f	MC_Potts/potts_ts.f	fit_l/subl_linear.f
MC_Potts/ana_pmu_ts.f	MC_Potts/potts_ts_r.f	fit_l/subl_power.f
MC_Potts/ana_pmuh.f	MC_Potts/pottshb_ts.f	histogram/hist.f
MC_Potts/ana_ts1_p.f	MC_Potts/pottsn_ts.f	lat/lat3d.f
MC_Potts/ana_ts_p.f	MC_Potts/prod_mu.f	lat/lat3d_ckb.f
MC_Potts/ana_ts_pm.f	MPI_Potts/ana_pt.f	lat/lat4d.f
MC_Potts/ana_ts_pr.f	MPI_Potts/lpt_ex_ia.f	lat/lat4d_ckb.f
MC_Potts/ana_tshb_p.f	MPI_Potts/p_hbtest_mpi.f	student/stud_tabs.f
MC_Potts/p_clsw_ts.f	MPI_Potts/p_inithb_mpi.f	
MC_Potts/p_clw_ts.f	MPI_Potts/p_pt_mpi.f	

tion is \, *i.e.* include '..\..\ForLib\fln.ext', in contrast to the / Linux and true Unix notation of (1.23). A slight complication occurs nevertheless when using the MS fl32 compiler. It concerns the routines in ForLib which are, like the just mentioned subroutine rmaset.f, by themselves transferred via include into the main programs. These routines use also the implicit.sta statement and other files of ForLib. Under the Linux and true Unix Fortran compilers this works by using include statements of the form

$$\text{include 'fln.ext'} \tag{1.24}$$

in these subroutines. However, the MS fl32 compiler searches then in the folder where the main program is compiled. The simplest and **recommended solution** is to create copies of the *.com, *.par and *.sta files of ForLib in the actual working directory. Another solution is to change the include statements (1.24) in all functions and subroutines to the form (1.23), but then the code does no longer compile under Linux and true Unix.

In general, when difficulties are encountered with some platform, a save solution is to change all include statement from the form (1.23) to the form (1.24) and to copy whatever files are needed into the working directory. Of course, one may also compile the routines of ForLib into a run time library and eliminate all include statements for functions and subroutines. However, then one has to learn how to create such libraries and one has to recompile the library whenever changes to ForLib routines are made. Another advantage of using include statements is that dependencies of the code become exhibited.

Fortran **procedures for interactive use** are provided in the folders of ForProc and are listed in table 1.6. They are ready to compile. Running the executables results in an interactive on-screen dialogue with the user.

Table 1.6 Fortran procedures *.f of the directory ForProc.

F_stud/F_stud.f	Gau_dif/gau_dif.f	ratio_eb/ratio_eb.f
F_test/F_test.f	Stud_dif/stud_dif.f	

An example of routines which share a named common blocks are the subroutines ranmar.f, rmaset.f and rmasave.f. The Marsaglia random number generator keeps its status in the named common block

$$\text{COMMON/RASET1/U(97), C, CD, CM, I, J}$$

which is directly part of the code. When changes are anticipated the common blocks are not directly written into the program, but included via statements

$$\text{include 'fln.com' or include '../../ForLib/fln.com'}. \qquad (1.25)$$

Finally, if data assignments have to be made they may also be included via

$$\text{include 'fln.dat' or include '../../ForLib/fln.dat'} \qquad (1.26)$$

statements. Our definitions of statements, parameters, common blocks and data assignments keep the program package together beyond a collection of individual pieces. When developing the code further, it is recommended to leave the include structure intact, although some programs and subroutines can easily be isolated when required.

Whenever graphical output is generated by our code, it is in a form suitable for **gnuplot**. Gnuplot is a freely available, easy to use, graphical package which is installed on practically all Unix and Linux platforms. A Windows version exists too. Some of our programs, for instance hist.f, generate fln.plt files which are used by simply typing gnuplot fln.plt on Unix based platforms. Under Windows the appropriate mouse clicks have to be made in the gnuplot menu. Our fln.plt files plot data from files which are generated by our Fortran programs. To find out more about gnuplot, search the Web.

1.3.1 *CPU time measurements under Linux*

Often one would like to measure the **CPU time**, which an application takes. Sophisticated software tools exist for this purpose. For instance under Linux **gprof** (gnuplot profile) provides a routine by routine analysis of the performance of the code, where the g77 compiler has to be used with the -pg option. Use the Linux manual command man (here man gpof and man g77) to find out details.

Often one likes to know nothing more than the CPU time used for an entire run. Interactively this is easily achieved in Linux. If a.exe is the executable, the command

$$\text{time ./a.exe} \qquad (1.27)$$

(simply time a.exe in some implementations of Linux) returns the CPU time once the run is finished. However, for longer runs one would like to write the timing result on a file and this is a bit tricky. Running the time command in the background with

$$\texttt{time ./a.exe > a.txt \&}$$

does not work. The Fortran unit 6 output will be on `a.txt`, but the CPU time information is simply lost. **Under the tcshell** (enter with the command `tcsh`) the solution to the dilemma is to create an executable file with, say, the name `CPUtime`, which contains the line (1.27). Run this file with the command

$$\texttt{CPUtime > \& a.txt \&} \tag{1.28}$$

in the background. Then the file `a.txt` contains after completion of the task all the desired information. In the tcshell the first **&** of the command enforces to write also system information on the output file `a.txt`. We are now ready for the CPU time measurement of assignment 5 of the previous section. Note, that the same trick can be used to force the gnu g77 compiler to write its error messages to an output file, *i.e.*

$$\texttt{g77 -0 fln.f > \& fln.txt \&} \tag{1.29}$$

does the trick. Here -O is the optimization level recommended by the author to be used with the g77 compiler.

1.4 Gaussian Distribution

The **Gaussian** or **normal distribution** is of major importance in statistics. Its probability density $f(x) = g(x)$ is given by

$$g(x) = \frac{1}{\sigma\sqrt{2\pi}} e^{-x^2/(2\sigma^2)} \tag{1.30}$$

where σ^2 is the **variance** of the Gaussian distribution and $\sigma > 0$ its **standard deviation**. The general definitions of variance and standard deviation are given in section 1.7. The Gaussian distribution function $G(x)$ is related to that of variance $\sigma^2 = 1$ by

$$G(x) = \int_{-\infty}^{x} g(x')\, dx' = \frac{1}{\sqrt{2\pi}} \int_{-\infty}^{x/\sigma} e^{-(x'')^2/2}\, dx''$$
$$= \frac{1}{2} + \frac{1}{2}\operatorname{erf}\left(\frac{x}{\sigma\sqrt{2}}\right) \tag{1.31}$$

where

$$\text{erf}(x) = \frac{2}{\sqrt{\pi}} \int_0^x e^{-(x')^2} dx' \tag{1.32}$$

defines the **error function**. It is related to the **incomplete gamma function** $P(a, x)$ by $\text{erf}(x) = \text{sgn}(x) \, P(1/2, x^2)$, where $\text{sgn}(x) = 1$ for $x > 1$ and $\text{sgn}(x) = -1$ for $x < 1$. The numerical calculation of the error function and other special functions is the topic of appendix A.1.

In principle we could now generate **Gaussian random numbers** by means of equation (1.10). However, the numerical calculation of the inverse error function is slow and makes this an impractical procedure. Much faster is the method discussed in the following. First, we express the joint (product) probability density of two independent Gaussian distributions in polar coordinates

$$\frac{1}{2\pi \, \sigma^2} e^{-x^2/(2\sigma^2)} \, e^{-y^2/(2\sigma^2)} \, dx \, dy = \frac{1}{2\pi \, \sigma^2} e^{-r^2/(2\sigma^2)} \, d\phi \, r dr \, .$$

We have the relations

$$x^r = r^r \cos \phi^r \quad \text{and} \quad y^r = r^r \sin \phi^r \tag{1.33}$$

between the corresponding random variables. It is now easy to generate ϕ^r and r^r: ϕ^r is simply uniform in $[0, 2\pi)$ and the distribution function for r^r in $[0, \infty)$ is found to be

$$F(r) = \int_0^r dr' \, \frac{d}{dr'} \left(e^{-(r')^2/(2\sigma^2)} \right) = 1 - e^{-r^2/(2\sigma^2)} \, ,$$

where it is obvious from the result that the normalization is correct. Consequently, for u^r being a uniformly distributed random variable in the $[0, 1)$:

$$r^r = \sigma \sqrt{-2 \ln(1 - u^r)} \, .$$

In this way a generation of two uniformly distributed random numbers translates via (1.33) into two Gaussian distributed random numbers x^r and y^r. It is notable that x^r and y^r are statistically independent (*i.e.*, both can be used in sequence). This is seen as follows: Assume y^r is given to be y_0, then

$$\text{const} \int_0^{2\pi} d\phi \int_0^\infty r \, dr \, \delta(y_0 - r \, \sin \phi) \, e^{-r^2/(2\sigma^2)} = \text{const} \int_{-\infty}^{+\infty} dx \, e^{-x^2/(2\sigma^2)} \, ,$$

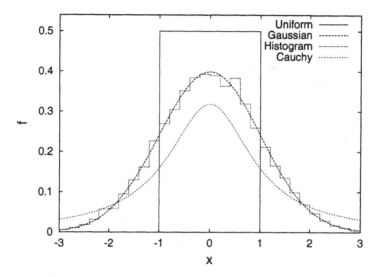

Fig. 1.4 Probability densities for a uniform, a Gaussian and a Cauchy distribution. Further, the normalized histogram for the initial 10 000 random numbers generated by `rmagau.f` is plotted. See assignment 1.

where the constant normalizes the integrals to one. Our subroutine `rmagau.f` generates Gaussian random numbers with variance $\sigma^2 = 1$ along these lines.

1.4.1 *Assignments for section 1.4*

(1) **Probability densities:** Use gnuplot or your favorite plotting package to draw in the range $[-3, +3]$ the uniform probability density with support in $[-1, +1]$, the $\sigma = 1$ Gaussian and the $\alpha = 1$ Cauchy probability densities. Generate the initial (`iseed1=1` and `iseed2=0`) 10 000 normally distributed random numbers with `rmagau.f` and histogram them with respect to thirty equally spaced intervals in the same graph. The results are depicted in figure 1.4.

(2) **Distribution functions:** Again in the range $[-3, +3]$, draw a picture of the cumulative distribution functions for the probability densities of

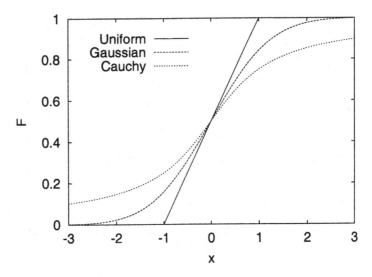

Fig. 1.5 Cumulative distribution functions for a uniform, a Gaussian and a Cauchy distribution, see assignment 2.

the previous assignments. For the Gaussian case error_f.f and other subroutines, all discussed in appendix A.1, are needed. The results are shown in figure 1.5.

1.5   Confidence Intervals

The cumulative distribution function (1.5) allows for a number of useful definitions to describe the statistical distribution of its random variable x^r. For simplicity, we assume in the following a strictly monotone distribution function $F(x)$. This condition is easily relaxed by focusing on piecewise monotone behavior. Patching such pieces together is straightforward but tedious. The essential concepts are already caught by discussing distribution functions which are strictly monotone.

Let $0 \leq q \leq 1$ be given, one defines **q-tiles** x_q by means of

$$F(x_q) = q . \qquad (1.34)$$

The x_q are also called **quantiles** or **fractiles** [Brandt (1983)]. The **median** $x_{\frac{1}{2}}$ is often (but not always) the **typical** value of the random variable x^r.

To calculate q-tiles x_q we have to invert the cumulative distribution function (1.5), a problem which we already encountered in the context of generating general distributions from the uniform distribution (1.10). The q-tiles of the uniform distribution (1.9) are trivial: $x_q = q$. For the Cauchy distribution the quantiles x_q follows immediately from equation (1.12). Nevertheless, for the sake of a consistent notation we have devoted the Fortran function cau_xq.f of our library to this task. For a given q it returns the Cauchy q-tile

$$x_q = \text{CAU\_XQ(q)} . \qquad (1.35)$$

The inversion of the Gaussian distribution function (1.31) is less trivial and requires a root finder. The task is eased by the fact that the distribution function is strictly monotone. On an initial interval, which contains the result, the bisection method converges and its speed is for most practical applications (other than random numbers generation) satisfactory. Our Fortran function fi1.f (this notation stands for function inverse, routine 1) implements the bisection method:

$$\text{FI1}(F, Y, X1, X2) \qquad (1.36)$$

returns X=FI1 so that Y=F(X) holds, where F(X) is a user supplied Fortran function. An appropriate initial range [X1, X2], known to contain the answer, has to be supplied on input and the function has to be monotone in this range (increasing or decreasing). The search for Y=F(X) is done by bisection for X1 $\leq$ X $\leq$ X2. The use of the function FI1 is illustrated in gau_xq.f which returns the Gaussian q-tile

$$x_q = \text{GAU\_XQ(q)} . \qquad (1.37)$$

For further information about the bisection method and other root finders see, for instance, *Numerical Recipes* [Press et al. (1992)].

The q-tiles are of immediate statistical relevance, because the probability to find the random variable x^r in the range $[x_{q_1}, x_{q_2}]$, $q_1 < q_2$ is given by

$$P(x_{q_1} \leq x^r \leq x_{q_2}) = q_2 - q_1 . \qquad (1.38)$$

The probability $q_2 - q_1$ is called **confidence** and $[x_{q_1}, x_{q_2}]$ **confidence interval**. Given a probability p, a corresponding **symmetric confidence interval** exists. Namely,

$$x^r \in [x_q, x_{1-q}] \text{ with probability } p \quad \Longrightarrow \quad q = \frac{1}{2}(1-p). \quad (1.39)$$

One may also be interested in the smallest, *i.e.* $(x_{q_2} - x_{q_1}) = $ Minimum, confidence interval so that $P(x_{q_1} \le x^r \le x_{q_2}) = p$ holds. For asymmetric distributions the smallest confidence interval will normally not fulfill the relation $q_2 = 1 - q_1$. To follow up on this is left to the reader.

Typical choices for symmetric confidence intervals are

$$
\begin{array}{rcccc}
p = & 0.70\,, & 0.90\,, & 0.950\,, & \text{etc.} \\
q = & 0.15\,, & 0.05\,, & 0.025\,, & (1.40) \\
\text{normal} \approx & 1\sigma\,, & & 2\sigma\,. &
\end{array}
$$

The last row refers to multiples of the standard deviation σ of the normal distribution (1.31), which correspond (approximately) to the p and q values of their columns. The precise probability content of the confidence intervals

$$[x_q, x_{1-q}] = [-n\sigma, n\sigma] \text{ for } n = 1, 2$$

of the normal distribution is given by $p = 68.27\%$ for one σ and $p = 95.45\%$ for two σ.

Examples of x_q values are discussed in the following.

(1) Consider the uniform distribution and take $p = 0.95$. The corresponding lower and upper q-tiles are $x_{0.025} = 0.025$ and $x_{0.975} = 0.975$.

(2) Consider the Cauchy distribution (1.12): For $p = 0.95$ we find $x_{0.975} = 12.706$ and $x_{0.025} = -12.706$, *i.e.*, $x_{0.975} - x_{0.025} = 25.412$. For $p = 0.70$ we get $x_{0.850} = 1.9626$ and $x_{0.150} = -1.9626$, *i.e.*, $x_{0.850} - x_{0.150} = 3.9252$. The calculation can be done with cau_xq.f, see assignment 1.

(3) Consider the Gaussian distribution (1.31): For $p = 0.95$ we find $x_{0.975} = 1.96$ and $x_{0.025} = -1.96$, *i.e.*, $x_{0.975} - x_{0.025} = 3.92$. For $p = 0.70$ we get $x_{0.850} = 1.0364$ and $x_{0.150} = -1.03634$, *i.e.*, $x_{0.850} - x_{0.150} = 2.07$. The calculation can be done with gau_xq.f, see assignment 1.

Dividing the length of the 95% confidence interval by that of the 70% confidence interval, we find a factor 1.9 for the normal distribution and a factor 6.5 for the Cauchy distribution. This is a consequence of the long

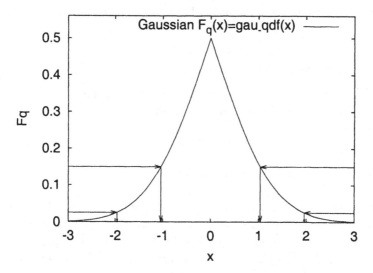

Fig. 1.6 Gaussian peaked distribution function (1.41) and estimates of x_q for the 70% (approximately $1\,\sigma$) and 95% (approximately $2\,\sigma$) confidence intervals, compare (1.40).

tails of the Cauchy distribution versus the fast fall off of the Gaussian distribution.

Let me introduce a **peaked distribution function** $F_q(x)$ which, although not standard notation, provides a useful way to visualize probability intervals of a distribution. F_q is defined by

$$F_q(x) = \begin{cases} F(x) \text{ for } F(x) \le \frac{1}{2}\,; \\ 1 - F(x) \text{ for } F(x) > \frac{1}{2}\,. \end{cases} \tag{1.41}$$

In figure 1.6 $F_q(x)$ is depicted for the Gaussian distribution. Fortran implementations of peaked distribution functions are cau_qdf.f for the Cauchy and gau_qdf.f for the Gaussian case. By construction the maximum of the peaked distribution function $F_q(x)$ is at the median $x = x_{\frac{1}{2}}$. If one likes to cut out a certain probability content of the tails, one has to pick the F_q graph at these values and to project on the corresponding q-tiles x_q. For the 70% and 95% confidence intervals this is illustrated in the figure.

Of course, one can use the distribution function $F(x)$ itself for the same purpose, but $F_q(x)$ has two advantages: The median is clearly exhibited and the accuracy of the ordinate is doubled. Note that many (but not all!) probability densities take their maximum at the median or close to it. In this situation F_q plots are similarly useful as histograms and avoid for continuous distributions ambiguities due to choosing a discretization of the abscissa.

1.5.1   *Assignment for section 1.5*

(1) Reproduce the results for the Cauchy and Gaussian q-tiles given in the examples after equation (1.40). Rely on cau_xq.f and gau_xq.f from ForLib.

1.6   Order Statistics and HeapSort

Sampling provides us with an empirical distribution function and in practice the problem is to estimate confidence intervals from the empirical data. The general theory of **confidence or fiducial limits** was developed by [Fisher (1930-1933)] and [Neyman and Pearson (1933)]. In the following we introduce some of the basic concepts and combine them with appropriate numerical tools.

Assume we generate n random numbers $x_1, ..., x_n$ according to a probability distribution $F(x)$. The n random numbers constitute then a **sample**. We may re-arrange the x_i in increasing order. Denoting the smallest value by x_{π_1}, the next smallest by x_{π_2}, etc., we arrive at

$$x_{\pi_1} \leq x_{\pi_2} \leq \cdots \leq x_{\pi_n} \qquad (1.42)$$

where $\pi_1, \ldots, \pi_n$ is a permutation of $1, \ldots, n$. Each of the x_{π_i} is then called an **order statistic**. More generally, the word **statistic** is occasionally used in the literature to denote **functions of random variables**.

As long as the data are not yet created, an order statistic x_{π_i} is a random variable and should properly be denoted by

$$x_{\pi_i}^r . \qquad (1.43)$$

As the distinction between random variables and data can be quite tedious, we shall sometimes drop the superscript r and simply keep in mind that certain quantities may be random variables.

An order statistic provides **estimators** for q-tiles. For instance, let n be odd, $n = 2m - 1$, $(m = 1, 2, ...)$. Then there is a middle value of the data

$$\overline{x}_{\frac{1}{2}} = x_{\pi_m} \tag{1.44}$$

which is an estimator for the median $x_{\frac{1}{2}}$ and therefore called the **sample median**. Here and in the forthcoming we use the overline symbol to distinguish estimators (which before being created are random variables) from exact results: $\overline{x}_{\frac{1}{2}}$ is an estimator of the exact $x_{\frac{1}{2}}$. The estimator is obtained by sampling random variables x^r, see assignment 1 for the case of Cauchy random numbers.

At the moment we are most interested in an estimator for the distribution function $F(x)$ (1.5) itself. Such an estimator is the **empirical distribution function** $\overline{F}(x)$ defined by

$$\overline{F}(x) = \frac{i}{n} \quad \text{for} \quad x_{\pi_i} \leq x < x_{\pi_{i+1}}, \ i = 0, 1, \ldots, n - 1, n \tag{1.45}$$

with the definitions $x_{\pi_0} = -\infty$ and $x_{\pi_{n+1}} = +\infty$. The **empirical peaked distribution function** is then defined in the same way as for the exact case (1.41)

$$\overline{F}_q(x) = \begin{cases} \overline{F}(x) \text{ for } \overline{F}(x) \leq \frac{1}{2} \, ; \\ 1 - \overline{F}(x) \text{ for } \overline{F}(x) > \frac{1}{2} \, . \end{cases} \tag{1.46}$$

To calculate $\overline{F}(x)$ and $\overline{F}_q(x)$ one needs an efficient way to **sort** n data values in ascending (or descending) order. ForLib contains the Fortran subroutine heapsort.f which is depicted in table 1.7. An input array X(N) of n real numbers has to provided. A call to HEAPSORT(N,A) returns X(N) in the order

$$X(1) \leq X(2) \leq \cdots \leq X(N) \, .$$

The routine heapsort.f relies on two steps: First the data $x_1, \ldots, x_n$ are arranged in a heap, then the heap is sorted.

A **heap** is a partial ordering so that the number at the top is larger or equal than the two numbers in the second row, provided at least three numbers x_i exist. Each number of the second row has under it two smaller or equal numbers in the third row and so on, until all numbers $x_i, \ldots, x_n$ are exhausted. The heap for Marsaglia's first ten random numbers is depicted in table 1.8. It is obtained from our subroutine heapsort.f by printing

Table 1.7 Heapsort Fortran subroutine `heapsort.f`.

```
      SUBROUTINE HEAPSORT(N,X)
C COPYRIGHT, BERND BERG, JUNE 16, 1999.
      include 'implicit.sta'
      DIMENSION X(N)
C
C BUILDING THE HEAP:
C ==================
      NHALF=N/2
      DO M=NHALF,1,-1
        TEMP=X(M)
        I1=M
        I2=M+M
1       CONTINUE
C         INDEX THE LARGER OF X(I2) AND X(I2+1) AS I2:
          IF(I2.LT.N .AND. X(I2+1).GT.X(I2)) I2=I2+1
          IF(X(I2).GT.TEMP) THEN
            X(I1)=X(I2)
            I1=I2
            I2=I1+I1
          ELSE
            I2=N+1
          ENDIF
        IF(I2.LE.N) GO TO 1
        X(I1)=TEMP ! FINAL POSITION FOR TEMP=X(M).
      END DO
C
C SORTING THE HEAP:
C =================
      DO I=N,3,-1 ! I IS THE NUMBER OF STILL COMPETING ELEMENTS.
        TEMP=X(I)
        X(I)=X(1) ! STORE TOP OF THE HEAP.
        I1=1
        I2=2
2       CONTINUE
C         INDEX THE LARGER OF X(I2) AND X(I2+1) AS I2:
          IF((I2+1).LT.I .AND. X(I2+1).GT.X(I2)) I2=I2+1
          IF(X(I2).GT.TEMP) THEN
            X(I1)=X(I2)
            I1=I2
            I2=I1+I1
          ELSE
            I2=I
          ENDIF
        IF(I2.LT.I) GO TO 2
        X(I1)=TEMP ! FINAL POSITION FOR TEMP=X(I).
      END DO
      TEMP=X(2)
      X(2)=X(1)
      X(1)=TEMP
      RETURN
      END
```

Table 1.8 The heap, generated by the upper half of the **heapsort.f** routine from Marsaglia's first ten random numbers.

			.9648			
		.6892			.9423	
	.5501		.4959	.5774		.8830
.2432		.4205	.1164			

the array X after the first DO loop finished, *i.e*, just before the comment SORTING THE HEAP. The top element of the heap is X(1), in the second row X(2) and X(3) follow, in the third row we have X(4), X(5), X(6) and X(7). The remaining three array elements constitute the forth and last row. Let M = [N/2] be defined by Fortran integer division, *i.e.*, [N/2] is the integer of largest magnitude so that $|[N/2]| \leq |N/2|$ holds. The recursion of heapsort.f starts by comparing the array element at M with those at I2 = 2M or I2+1 if I2+1 = N. Then M is replaced by M-1 and the procedure is repeated until M = 1 is completed.

In the second part of the routine the heap is sorted. Due to the partial order of the heap, we know that X(1) is the largest element of the array. Hence, X(1) is moved to X(N). Before doing so, X(N) is stored in a temporary location TEMP. From the numbers under X(1), at positions X(2) and X(3), the larger one is then moved into position X(1). Its vacant spot is filled by the larger of its underlings, if they exist, and so on. The algorithm for this section is pretty much the same as the one used when building the heap. The computer time needed to succeed with the sorting process grows like $n \log_2 n$ because there are $\log_2 n$ levels in the heap. See [Knuth (1968)] for a detailed discussion of the heapsort and other sorting algorithms.

One may add a second array of length n in heapsort.f to keep track of the permutations $\pi_1, \ldots, \pi_2$ in equation (1.42). This is done in our subroutine heap_per.f. A call to HEAP_PER(N,A,IPER) returns besides the sorted array X the permutation array IPER. If one wants to keep the original order of the array X, the solution is simply to copy X to a workspace array W and to call HEAP_PER(N,W,IPER). One may also use heap_per.f to generate a random permutation of the numbers $1, \ldots, n$ by feeding it with an array of uniformly distributed random numbers. (In passing, let us note that the routine perm.f of ForLib generates one permutation out of the $n!$ possibilities in n instead of $n \log_2 n$ steps. But n is limited to $1, 2, \ldots, 12$ and extension beyond $n = 12$ is in Fortran tedious. The working of the routine

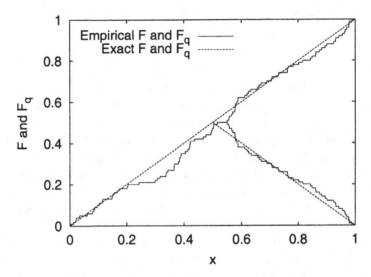

Fig. 1.7 Cumulative (1.45) and peaked (1.46) distribution functions for Marsaglia's first 100 uniform random numbers, see assignment 2.

perm.f is illustrated by the program permutations.f which is located in the folder ForProg/Tests.)

Figure 1.7 displays the empirical cumulative and peaked distribution functions obtained by sorting Marsaglia's first 100 random numbers, see assignment 2. Once the empirical distribution function is known, it can be used to estimate q-tiles x_q without relying on a constraint to the number of data as in equation (1.44). The subroutine qtiles.f is provided for this purpose. It relies on the definition (1.45) of the empirical distribution function $\overline{F}(x)$ and estimates x_q by linear interpolation between the two support points which are closest to the given $q = \overline{F}$ value ($q \geq 1/n$ is required).

1.6.1 *Assignments for section 1.6*

(1) Use Marsaglia's default seeds and generate eight subsequent sequences of $n = 2m - 1$ Cauchy distributed random numbers with $m = 4^k$,

Table 1.9 Median, quartiles and standard confidence q-tiles from sampling of sequences of $n = 2m - 1$, $m = 4^k$, $k = 2, \ldots, 8$ Cauchy random numbers, see assignment 1.

```
VARIOUS QTILES FOR CAUCHY RANDOM NUMBERS:
```

K	NDAT	X_0.5	X_0.25	X_0.75	X_0.15	X_0.85
2	31	-.1311	-.7801	.8973	-1.1979	1.4936
3	127	.1038	-.6728	.6411	-1.4110	1.3584
4	511	-.0625	-1.1535	.8680	-2.0433	1.7800
5	2047	-.0684	-1.1294	.9531	-2.0686	1.8786
6	8191	-.0178	-.9758	.9824	-1.9685	1.9418
7	32767	.0028	-.9828	.9921	-1.9651	1.9518
8	131071	-.0036	-.9988	.9945	-1.9594	1.9573
	EXACT:	.0000	-1.0000	1.0000	-1.9626	1.9626

$k = 2, \ldots, 8$. Use `heapsort.f` and `qtiles.f` to calculate for each sequence the sample median, lower and upper quartiles, and finally $x_{0.15}$ and $x_{0.85}$ for the 70% confidence interval (1.40). Compare the thus obtained estimates of table 1.9 with the exact values and the results for the estimator (1.44).

(2) Calculate the empirical cumulative and peaked distribution functions for our usual first 100 uniformly distributed random numbers and plot them in comparison with the exact functions. Figure 1.7 should result. Remark: The random numbers are the same as those used for the noisy histogram of figure 1.2.

1.7 Functions and Expectation Values

Another way to describe the statistical distribution of a random variable x^r is by means of **expectation values**. The entire distribution may eventually be recovered from an appropriate set of expectation values. We first introduce **functions of random variables**. The function of a random variable is itself a random variable: $\phi^r = \phi(x^r)$. The expectation value of such a function is defined by

$$\widehat{\phi} = \langle \phi^r \rangle = \int_{-\infty}^{+\infty} \phi(x)\, f(x)\, dx \,. \tag{1.47}$$

In physics the notation $\langle \ldots \rangle$ is often used to denote expectation values, whereas in the statistics literature one finds notations like $E[\phi^r]$. Note that in contrast to ϕ^r the expectation value $\widehat{\phi}$ is not a random variable, but a real or complex number (depending on whether the function $\phi(x)$ is real or complex).

1.7.1 *Moments and Tchebychev's inequality*

Frequently used statistical functions are the **moments**

$$\lambda_n = \widehat{x^n} = \langle (x^r)^n \rangle = \int_{-\infty}^{+\infty} x^n\, f(x)\, dx, \ \ n = 1, 2, 3, \ldots . \tag{1.48}$$

The notation $\widehat{x^n}$ is needed because $\widehat{x}^n$ is the nth power of the first moment and in general distinct from the nth moment. The first moment is the expectation value of the random variable x^r,

$$\widehat{x} = \langle x^r \rangle \tag{1.49}$$

and called the mean value or just the **mean**. **Reduced moments** are defined as moments about the mean

$$\mu_n = \langle (x^r - \widehat{x})^n \rangle = \int_{-\infty}^{+\infty} (x - \widehat{x})^n\, f(x)\, dx, \ \ n = 1, 2, 3, \ldots . \tag{1.50}$$

More generally, one can define moments $\mu_n(c)$ about an arbitrary point c. When the argument c is omitted $c = \widehat{x}$ is implied. The reduced second moment μ_2 is called the **variance** and the notations

$$\mu_2 = \sigma_x^2 = \sigma^2(x^r) = \mathrm{var}(x^r) \tag{1.51}$$

are used. The square root of the variance

$$\sigma = \sigma_x = \sigma(x^r) \ = \ \sqrt{\sigma^2(x^r)} \tag{1.52}$$

is called **standard deviation**. When no confusion is possible, we suppress the random variable x^r in the arguments of the variance and the standard deviation. Following equation (1.40) we already discussed confidence intervals for the standard deviation of the normal distribution. When "normal" confidence limits apply in estimates, the standard deviation is often (in particular in physics) quoted as the **statistical error** or **error bar** of the estimate. In detail error estimates are discussed in the next chapter.

The **skewness** is defined by

$$\gamma = \frac{\mu_3}{\sigma^3} \tag{1.53}$$

and used to analyze a probability distribution for asymmetry around the mean $\hat{x}$. The skewness is the reduced third moment in units of the third power σ^3 of the standard deviation. Of course, the reduced third moment μ_3 itself can be used for the same purpose.

The importance of the variance and the standard deviation comes from the fact that they allow **estimates of confidence intervals** when the **Central Limit Theorem** of section 1.8 applies, which covers many cases of importance. Although these estimates are less direct than those from the order statistic, they have often computational advantages. Therefore, these quantities dominate the traditional statistical analysis, in particular the recipes used by physicists. As todays powerful PCs handle order statistics methods with ease one can doubt whether the traditional view is still appropriate. Anyway, the traditional approach plays an important role and we introduce a number of its results.

Before coming to the central limit theorem, let us prove **Tchebychev's inequality**. Provided the variance σ exists (*i.e.*, the defining integral (1.50) is convergent)

$$P(|x^r - \hat{x}| > a\sigma) < a^{-2} \quad \text{holds for any real number } a > 0 . \tag{1.54}$$

Tchebychev's inequality gives an upper limit for the probability content of the confidence interval covered by $\pm a\sigma$ standard deviations around the mean. However, the limit is too crude to be of practical importance.

Proof of the Tchebychev inequality: The probability of equation (1.54) can be re-written as

$$P = P(|x^r - \hat{x}| > a\sigma) = P(t^r > a^2\sigma^2)$$

where $t^r = (x^r - \hat{x})^2 > 0$ is a new random variable. Let us denote by $g(t)$ the probability density of t^r. Then

$$P = \int_{a^2\sigma^2}^{\infty} g(t) \, dt \quad \text{and} \quad \sigma^2 = \int_0^{\infty} t \, g(t) \, dt.$$

We may estimate the integral for σ^2 as follows:

$$\sigma^2 = \int_0^{a^2\sigma^2} t \, g(t) \, dt + \int_{a^2\sigma^2}^{\infty} t \, g(t) \, dt \geq$$

$$\int_{a^2\sigma^2}^{\infty} t\, g(t)\, dt \; > \; a^2\sigma^2 \int_{a^2\sigma^2}^{\infty} g(t)\, dt = a^2\sigma^2\, P$$

and solving for P completes the proof.

Examples of moments

(1) The moments and reduced moments of the uniform distribution in $[0,1)$ are given by

$$\widehat{x^n} \; = \; \int_0^1 x^n\, dx \; = \; \frac{1}{n+1} \qquad (1.55)$$

and

$$\mu_n \; = \; \int_{-0.5}^{+0.5} x^n\, dx \; = \; \begin{cases} 2^{-n}/(n+1) \text{ for } n \text{ even,} \\ 0 \text{ for } n \text{ odd.} \end{cases} \qquad (1.56)$$

In particular, the mean is $\widehat{x} = 0.5$ and the variance is $\sigma^2 = 1/12$.

(2) The mean of the normal distribution (1.30) is $\widehat{x} = 0$. Moments and reduced moments for n even are given by

$$\widehat{x^n} = \mu_n \; = \; \frac{1}{\sigma\sqrt{2\pi}} \int_{-\infty}^{+\infty} dx\, x^n\, \exp\left(-\frac{x^2}{2\sigma^2}\right) = \frac{(\sigma\sqrt{2})^n}{\sqrt{\pi}} \Gamma\left(\frac{n+1}{2}\right). \qquad (1.57)$$

In particular, the variance is $\sigma^2(x^r) = \sigma^2$ and the fourth moment is $\mu_4 = 3\sigma^4$. For efficient calculations of the Gamma function use gamma_ln.f, see appendix A.1. For n odd the reduced moments are zero due to the symmetry of the probability density about its mean.

(3) For the Cauchy distribution moments are undefined because of the slow fall-off of the Cauchy probability density (1.11).

(4) For a **bimodal distribution**, which generates the number 1 with probability p, $0 < p < 1$ and the number 0 with probability $q = 1 - p$, the variance is

$$\sigma^2 = pq \qquad (1.58)$$

Proof: The expectation value is $\widehat{x} = p$ and the reduced second moment becomes $\sigma^2 = \langle(p-x)^2\rangle = p^2 - 2p\langle x\rangle + \langle x^2\rangle = -p^2 + p = p(1-p) = pq$.

Similar as for the q-tiles, we like to **estimate expectation values by sampling**. The estimator of the nth moment λ_n from a sample of N events is

$$\overline{x^n} = \frac{1}{N} \sum_{i=1}^{N} x_i^n .$$ (1.59)

Similarly as before (1.44), we use the overline symbol to distinguish the estimator $\overline{x^n}$ from the exact result $\widehat{x^n}$. Let us recall that, before being measured, estimators are random variables. In full random variable notation equation (1.59) becomes

$$\overline{x^n}^r = \frac{1}{N} \sum_{i=1}^{N} (x_i^r)^n \text{ with } f_1(x) = f_2(x) = \cdots = f_N(x) = f(x)$$ (1.60)

where we stress explicitly that the defining probability densities of all the random variables x_i^r, $(i = 1, \ldots N)$ are equal. This is not obvious, if one just writes down the sum of the random variables, because the subscripts i on the random variables may refer to different distributions as well as to different drawings from the same distribution. We shall normally drop the explicit presentation of $f_1(x) = f_2(x) = \cdots = f_N(x)$ when it is anyway clear that this identity is valid. Care is needed because we shall also use equations for which $f_1(x) \neq f_2(x)$ and so on holds, for instance equation (1.63) in the next subsection.

The introduction of estimators for reduced moments (with unknown exact mean values) and other more complicated functions of random variables is postponed to the next chapter, because bias complications are encountered.

For $n = 1$ equation (1.60) becomes the estimator of the **arithmetic mean**

$$\overline{x}^r = \frac{1}{N} \sum_{i=1}^{N} x_i^r .$$ (1.61)

Like any random variable $\overline{x}^r$ depends on chance and is defined by its probability distribution. We are led to the important question: **What is the probability distribution for a sum of random variables?** Once we know the probability distribution of $\overline{x}^r$ we can calculate its confidence intervals and understand quantitatively how and why sampling leads to statistically improved estimates.

1.7.2   *The sum of two independent random variables*

We consider first the case of two **random variables**, x_1^r and x_2^r, which are described by the probability densities $f_1(x)$ and $f_2(x)$, respectively. We assume that x_1^r and x_2^r are **independent**. This means the occurrence of an event corresponding to $f_1(x)$ does not influence the occurrence of an event corresponding to $f_2(x)$ and vice versa. The joint probability density of x_1^r and x_2^r is simply the product

$$f(x_1, x_2) \;=\; f_1(x_1)\, f_2(x_2) \;. \tag{1.62}$$

The other way round, this equation is taken as the mathematical definition of the independence of two random variables. Let us now consider the sum

$$y^r \;=\; x_1^r + x_2^r \;. \tag{1.63}$$

We denote the probability density of y^r by $g(y)$. The corresponding cumulative distribution function is given by

$$G(y) \;=\; \int_{x_1 + x_2 \le y} f_1(x_1)\, f_2(x_2)\, dx_1\, dx_2 \;. \tag{1.64}$$

This may be rewritten as

$$G(y) \;=\; \int_{-\infty}^{+\infty} f_1(x_1)\, dx_1 \int_{-\infty}^{y - x_1} f_2(x_2)\, dx_2 \;=\; \int_{-\infty}^{+\infty} f_1(x)\, F_2(y - x)\, dx \;,$$

where $F_2(x)$ is the distribution function of the random variable x_2^r. We take the derivative with respect to y and obtain the probability density of y^r

$$g(y) \;=\; \frac{dG(y)}{dy} \;=\; \int_{-\infty}^{+\infty} f_1(x)\, f_2(y - x)\, dx \;. \tag{1.65}$$

The probability density of a sum of two independent random variables is the **convolution of the probability densities** of these random variables.

Examples:

(1) Sums of uniformly distributed random numbers $x^r \in (0, 1]$:
   (a) Let $y^r = x^r + x^r$, then

$$g_2(y) \;=\; \begin{cases} y & \text{for } 0 \le y \le 1, \\ 2 - y & \text{for } 1 \le y \le 2, \\ 0 & \text{elsewhere.} \end{cases} \tag{1.66}$$

(b) Let $y^r = x^r + x^r + x^r$, then

$$
g_3(y) = \begin{cases}
y^2/2 & \text{for } 0 \le y \le 1, \\
(-2y^2 + 6y - 3)/2 & \text{for } 1 \le y \le 2, \\
(y-3)^2/2 & \text{for } 2 \le y \le 3, \\
0 & \text{elsewhere.}
\end{cases} \tag{1.67}
$$

(2) The sum of two normally distributed random variables $y^r = x_1^r + x_2^r$ with

$$
f_1(x) = \frac{1}{\sigma_1 \sqrt{2\pi}} \exp\left(-\frac{1}{2}\frac{x^2}{\sigma_1^2}\right) \quad \text{and} \quad f_2(x) = \frac{1}{\sigma_2 \sqrt{2\pi}} \exp\left(-\frac{1}{2}\frac{x^2}{\sigma_2^2}\right).
$$

Their convolution (1.65) is a Gaussian integral, which is solved by the appropriate variable substitution (completion of the square). The result is

$$
g(y) = \frac{1}{\sqrt{2\pi(\sigma_1^2 + \sigma_2^2)}} \exp\left(-\frac{1}{2}\frac{y^2}{(\sigma_1^2 + \sigma_2^2)}\right), \tag{1.68}
$$

i.e., we find a Gaussian distribution with variance $\sigma_1^2 + \sigma_2^2$.

1.7.3 *Characteristic functions and sums of N independent random variables*

The convolution (1.65) takes on a simple form in **Fourier space**. In statistics the **Fourier transformation** of the probability density is known as **characteristic function**. Complex random functions are defined by

$$
f^r = f(x^r) = v(x^r) + i\,w(x^r) = v^r + i\,w^r \tag{1.69}
$$

and their expectation values are

$$
\langle f^r \rangle = \langle v^r \rangle + i\,\langle w^r \rangle. \tag{1.70}
$$

The characteristic function $\phi(t)$ of the random variable x^r is defined as the expectation value of e^{itx^r}:

$$
\phi(t) = \langle e^{itx^r} \rangle = \int_{-\infty}^{+\infty} e^{itx} f(x)\,dx. \tag{1.71}
$$

Alias, it is the Fourier transformation of the probability density $f(x)$. In the discrete case (1.6) we obtain the Fourier series

$$\phi(t) = \sum_i p_i\, e^{itx_i} \, . \tag{1.72}$$

Knowledge of the characteristic function $\phi(t)$ is equivalent to knowledge of the probability density $f(x)$, as we obtain $f(x)$ back by means of the inverse Fourier transformation

$$f(x) = \frac{1}{2\pi} \int_{-\infty}^{+\infty} dt\, e^{-itx}\, \phi(t) \, . \tag{1.73}$$

The derivatives of the characteristic function are

$$\phi^{(n)}(t) = \frac{d^n \phi(t)}{dt^n} = i^n \int_{-\infty}^{+\infty} dx\, x^n\, e^{itx}\, f(x) \, . \tag{1.74}$$

Therefore,

$$\phi^{(n)}(0) = i^n \lambda_n \quad \text{where} \quad \lambda_n = \int_{-\infty}^{+\infty} dx\, x^n\, f(x) \tag{1.75}$$

is the nth moment (1.48) of $f(x)$. If $f(x)$ falls off faster than any power law, all moments exist and the characteristic function is analytic around $t = 0$

$$\phi(t) = \sum_{n=0}^{\infty} \frac{\lambda_n}{n!} (it)^n \, . \tag{1.76}$$

In this case the probability density $f(x)$ is completely determined by its moments. One may expand around other points than $x_0 = 0$. For instance, when the first moment λ_1 is non-zero, $x_0 = \lambda_1$ is a frequent choice of the expansion point and, if the conditions are right, the characteristic function is determined by the reduced moments (1.50).

The characteristic function is particularly useful for investigating sums of random variables. Let us return to equation (1.63), $y^r = x_1^r + x_2^r$, and denote the characteristic functions of y^r, x_1^r and x_2^r by $\phi_y(t)$, $\phi_{x_1}(t)$ and $\phi_{x_2}(t)$, respectively. We find

$$\phi_y(t) = \langle e^{(itx_1^r + itx_2^r)} \rangle = \int_{-\infty}^{+\infty} \int_{-\infty}^{+\infty} e^{itx}\, e^{ity}\, f_1(x_1)\, f_2(x_2)\, dx_1\, dx_2$$

$$= \phi_{x_1}(t)\, \phi_{x_2}(t) \, . \tag{1.77}$$

The characteristic function of a sum of random variables is the product of their characteristic functions. The results for two random variables generalize immediately to sums of N independent random variables

$$y^r = x_1^r + \cdots + x_N^r . \tag{1.78}$$

The characteristic function of y^r is now

$$\phi_y(t) = \prod_{i=1}^{N} \phi_{x_i}(t) \tag{1.79}$$

and the probability density of y^r is the Fourier back-transformation of this characteristic function

$$g(y) = \frac{1}{2\pi} \int_{-\infty}^{+\infty} dt \, e^{-ity} \, \phi_y(t) . \tag{1.80}$$

1.7.4 *Linear transformations, error propagation and covariance*

In practice one deals often with linear transformations of random

$$y_i^r = a_{i0} + \sum_{j=1}^{N} a_{ij} x_j^r \quad \text{for } i = 1, \ldots, M . \tag{1.81}$$

Their characteristic functions are

$$\phi_{y_i}(t) = \left\langle e^{it\left(a_{i0} + \sum_{j=1}^{N} a_{ij} x_j^r\right)} \right\rangle \tag{1.82}$$

$$= e^{ita_{i0}} \int \prod_{j=1}^{N} dx_j \, e^{it \, x_j} \, f_j(x_j) = e^{ita_{i0}} \prod_{j=1}^{N} \phi_{x_i}(a_{ij} t) .$$

The probability densities $g_i(y)$ of the random variables y_i^r are then given by the inverse Fourier transformation (1.80).

Of particular importance are the mean and the variance of the random variable y_i^r, which are best obtained directly from the defining equation (1.81). Using linearity the mean is

$$\widehat{y}_i = \langle y_i^r \rangle = a_{i0} + \sum_{j=1}^{N} a_{ij} \langle x_j^r \rangle = a_{i0} + \sum_{j=1}^{N} a_{ij} \widehat{x}_j . \tag{1.83}$$

The variance is bit more involved

$$\sigma^2(y_i^r) = \langle (y_i^r - \widehat{y}_i)^2 \rangle = \left\langle \left[\sum_{j=1}^{N} a_{ij} (x_j^r - \widehat{x}_j) \right]^2 \right\rangle =$$

$$\left\langle \sum_{j=1}^{N} a_{ij} (x_j^r - \widehat{x}_j) \sum_{k=1}^{N} a_{ik} (x_k^r - \widehat{x}_k) \right\rangle = \sum_{j=1}^{N} a_{ij}^2 \langle (x_j^r - \widehat{x}_j)^2 \rangle$$

$$= \sum_{j=1}^{N} a_{ij}^2 \, \sigma^2(x_j^r) \qquad (1.84)$$

where we used that

$$\langle (x_j^r - \widehat{x}_j)(x_k^r - \widehat{x}_k) \rangle = \langle (x_j^r - \widehat{x}_j) \rangle \langle (x_k^r - \widehat{x}_k) \rangle = 0 \text{ holds for } j \neq k \quad (1.85)$$

because the random variables x_j^r and x_k^r are independent (1.62) for $j \neq k$. Equation (1.84) is often quoted as **error propagation** and written as

$$\sigma^2(y_i^r) = \sum_{j=1}^{N} \sigma^2(x_j^r) \, a_{ij}^2 \quad \text{with} \quad a_{ij}^2 = \left(\frac{\partial y_i^r}{\partial x_j^r} \right)^2 . \qquad (1.86)$$

The second equality is quite formal and an abuse of notation, because random variables are treated as if the were ordinary variables. Equation (1.86) simply works, because it reproduces the linear variance (1.84).

It is notable that the new random variables y_i^r are **correlated**. The **covariance matrix** $C = (c_{ij})$ of a set of random variables y_i^r, $i = 1, \ldots, M$ is a $M \times M$ matrix defined by

$$c_{ij} = \text{cov}(y_i^r, y_j^r) = \langle (y_i^r - \widehat{y}_i)(y_j^r - \widehat{y}_j) \rangle . \qquad (1.87)$$

For the particular case of the linear relationship (1.81) we obtain

$$\text{cov}(y_i^r, y_j^r) = \sum_{k=1}^{N} \sum_{l=1}^{N} a_{ik} \, a_{jl} \, \langle (x_k^r - \widehat{x}_k)(x_l^r - \widehat{x}_l) \rangle = \sum_{k=1}^{N} a_{ik} \, a_{jk} \, \sigma^2(x_k^r) . \qquad (1.88)$$

Again, quite formally, we can rewrite this equation as

$$c_{ij} = \text{cov}(y_i^r, y_j^r) = \sum_{k=1}^{N} \sum_{l=1}^{N} \frac{\partial y_i^r}{\partial x_k^r} \frac{\partial y_j^r}{\partial x_l^r} \, \text{cov}(x_k^r, x_l^r) = \sum_{k=1}^{N} \frac{\partial y_i^r}{\partial x_k^r} \frac{\partial y_j^r}{\partial x_k^r} \, \sigma^2(x_k^r) \qquad (1.89)$$

because this reproduces the previous equation.

Let us now give up the linear relationship (1.81) and turn to **arbitrary correlated random variables** y_i^r. The **covariance** matrix $C = (c_{ij})$ is still defined by equation (1.87) and a real, symmetric $M \times M$ matrix. The eigenvectors of the matrix C define new random variables which have a diagonal covariance matrix and the eigenvalues of C as their variances. We have already seen (1.85) that the covariance matrix of independent random variables is diagonal. However, the inverse of this statement is in general not true, because there can still be correlations between random variables in their higher moments.

We may solve the eigenvalue equation

$$C \vec{x} = \lambda \vec{b} \tag{1.90}$$

by standard methods of linear algebra, for instance see *Numerical Recipes* [Press *et al.* (1992)]. Assuming that there are no linear dependencies in the matrix C, we obtain M eigenvectors $\vec{b}_i$ with their corresponding (possibly degenerate) eigenvalues λ_i. The eigenvectors are normalized to be orthonormal

$$\vec{b}_i^\dagger \cdot \vec{b}_j = \delta_{ij} \tag{1.91}$$

where $\vec{b}_i$ is a column vector, $\vec{b}_i^\dagger$ the corresponding row vector and δ_{ij} the Kronecker delta symbol

$$\delta_{ij} = 1 \text{ for } i = j \text{ and } 0 \text{ otherwise.} \tag{1.92}$$

The elements of the vector $\vec{b}_i$ are b_{ji}, $j = 1, \ldots, M$ and the elements of $\vec{b}_i^\dagger$ are $b_{ij}^\dagger = b_{ji}$, $j = 1, \ldots, M$. We define

$$B = (b_{ji}) \tag{1.93}$$

to be the matrix of the column vectors $\vec{b}_i$. Its transpose $B^\dagger = (b_{ij}^\dagger) = (b_{ji})$ is the matrix of the row vectors $\vec{b}_i^\dagger$. Further, we define D_λ to be the *diagonal* matrix of the eigenvalues $\lambda_1, \ldots, \lambda_N$ and find that C is diagonalized by the transformation

$$B^\dagger C B = B^\dagger D_\lambda B = D_\lambda . \tag{1.94}$$

We rewrite the definition (1.87) of the matrix elements c_{ij} as

$$c_{ij} = \langle \Delta y_i^r \Delta y_j^r \rangle \text{ with } \Delta y_i^r = y_i^r - \widehat{y}_i \tag{1.95}$$

and the components of equation (1.94) are

$$\sum_{j=1}^{M}\sum_{k=1}^{M} b_{ij} \langle \Delta y_j^r \, \Delta y_k^r \rangle b_{kl} = \left\langle \sum_{j=1}^{M} b_{ij} \Delta y_j^r \sum_{k=1}^{M} \Delta y_k^r b_{kl} \right\rangle = \lambda_i \, \delta_{il} \, .$$

$$(1.96)$$

These are precisely the variances $\sigma^2(\Delta x_i^r)$ of the random variables Δx_i^r defined by

$$\Delta x_i^r = \sum_{j=1}^{M} b_{ji} \, \Delta y_i^r \tag{1.97}$$

and equation (1.96) reveals that their covariances vanish. As the variances and covariances of $\Delta x_i^r = x_i^r - \widehat{x}_i$ do not depend on the constants $\widehat{x}_i = \sum_{j=1}^{M} b_{ji}\widehat{y}_i$, the random variables

$$x_i^r = \sum_{j=1}^{M} b_{ji} \, y_i^r \, . \tag{1.98}$$

are "uncorrelated" in the sense of vanishing covariances, but they may exhibit correlations between their higher moments. We call the x_i^r of equation (1.98) **diagonal random variables**.

1.7.5   *Assignments for section 1.7*

(1) Generalize the Gaussian probability density (1.30) and its cumulative distribution function (1.31) to an arbitrary mean value $\widehat{x} = a$ with, as before, variance σ^2. Find the reduced moments (1.57) for this case.

(2) Use Marsaglia's default seeds and generate ten consecutive sequences of $n = 2^{2k-1}$, $k = 1, \dots, 10$ uniformly distributed random numbers and, with the same initial seeds, Cauchy distributed random numbers. For the uniform random numbers calculate the moment estimators $\overline{x}$, $\overline{x^2}$, $\overline{x^4}$ and for the Cauchy random numbers the moment estimators $\overline{x}$, $\overline{x^2}$. You should obtain table 1.10. Compare this with exact results.

(3) Use the same seeds as in the previous assignment and `rmagau.f` to generate ten consecutive sequences of $n = 2^{2k-1}$, $k = 1, \dots, 10$ normally distributed random numbers. For each sequence calculate the moment estimators $\overline{x}$, $\overline{x^2}$, $\overline{x^4}$, $\overline{x^6}$ and $\overline{x^8}$. Show that table 1.11 results. Use equation (1.57) and `gamma_ln.f` to obtain the exact results of this table.

Table 1.10 Moments of uniform and Cauchy random numbers from sampling of sequences of length 2^{2k-1}, $k = 1, \ldots, 10$, see assignment 2.

MOMENTS 1-4 FOR UNIFORM RANDOM NUMBERS AND
1-2 FOR CAUCHY RANDOM NUMBERS:

K	NDAT	U1	U2	U3	U4	C1	C2
1	2	.5406	.4722	.4499	.4334	.3364	.4
2	8	.6002	.4076	.3028	.2400	3.0939	68.6
3	32	.5172	.3575	.2775	.2295	-.5185	9.5
4	128	.4700	.2990	.2152	.1668	.7962	52.5
5	512	.5042	.3380	.2553	.2056	-.2253	831.3
6	2048	.5043	.3381	.2550	.2049	.1321	3535.2
7	8192	.5049	.3391	.2559	.2058	-4.5898	119659.2
8	32768	.4989	.3321	.2490	.1993	.0661	43545.2
9	131072	.5015	.3348	.2514	.2013	-2.8896	623582.7
10	524288	.5000	.3334	.2501	.2001	6.1029	13823172.6

1.8 Sample Mean and the Central Limit Theorem

1.8.1 *Probability density of the sample mean*

Let $y^r = x^r + x^r$ for probability densities $f_1(x) = f_2(x)$. The arithmetic mean of the sum of the two random variables is $\overline{x}^r = y^r/2$. We denote the probability density (1.65) of y^r by $g_2(y)$ and the probability density of the arithmetic mean by $\widehat{g}_2(\overline{x})$. They are related by the equation

$$\widehat{g}_2(\overline{x}) = 2 g_2(2\overline{x}), \tag{1.99}$$

which follows by substituting $y = 2\overline{x}$ into $g_2(y)\, dy$, *i.e.*,

$$1 = \int_{-\infty}^{+\infty} g_2(y)\, dy = \int_{-\infty}^{+\infty} g_2(2\overline{x})\, 2d\overline{x} = \int_{-\infty}^{+\infty} \widehat{g}_2(\overline{x})\, d\overline{x} \ .$$

Similarly, we obtain for the probability density of the arithmetic mean (1.61) of N identical random variables the relationship

$$\widehat{g}_N(\overline{x}) = N g_N(N\overline{x}) \tag{1.100}$$

where g_N is defined by equation (1.80).

Examples:

Table 1.11 Moments of $\sigma = 1$ Gaussian random numbers from sampling of sequences of length 2^{2k-1}, $k = 1, \ldots, 10$, see assignment 3.

CALCULATION OF NORMAL GAUSSIAN MOMENTS 1, 2, 4, 6 AND 8:

K	NDAT	G1	G2	G4	G6	G8
1	2	.1882	.1237	.0278	.0065	.0015
2	8	-.1953	1.6206	5.9399	28.2180	147.2287
3	32	-.3076	.9731	2.6462	8.9705	32.7718
4	128	-.3730E-01	.9325	2.3153	9.2695	47.7155
5	512	-.6930E-02	.9794	2.8333	12.6530	70.9580
6	2048	.3361E-01	.9676	2.7283	12.6446	80.6311
7	8192	-.1088E-01	1.0175	3.0542	14.6266	93.0394
8	32768	.2158E-03	1.0015	3.0497	15.5478	110.8962
9	131072	-.3590E-04	1.0087	3.0434	15.2814	107.8183
10	524288	.6201E-03	.9989	3.0020	15.1051	106.9634
	EXACT:	.0000	1.0000	3.0000	15.0000	105.0000

(1) The probability densities for the sums of two and three uniformly distributed random numbers are given by equations (1.66) and (1.67), respectively. Equation (1.100) gives the corresponding distribution functions for the arithmetic means $\widehat{g}_2(\overline{x})$ and $\widehat{g}_3(\overline{x})$, which are depicted in figure 1.8.

(2) The probability density for the sum of two Gaussian distributed random variables is given by equation (1.68). Equation (1.100) yields for the corresponding arithmetic average, $\overline{x}^r = (x_1^r + x_2^r)/2$, a Gaussian distribution with variance $(\sigma_1^2 + \sigma_2^2)/4$. For the special case $\sigma = \sigma_1 = \sigma_2$ the variance is reduced by a factor of two, the standard deviation by a factor of $\sqrt{2}$:

$$\widehat{g}_2(\overline{x}) = \frac{1}{(\sigma/\sqrt{2})\sqrt{2\pi}} \exp\left(-\frac{1}{2}\frac{\overline{x}^2}{(\sigma^2/2)}\right) . \tag{1.101}$$

These examples suggest that sampling leads to convergence of the mean by reducing its variance. Equation (1.101) makes this explicit for the normal distribution. For the uniform distribution figure 1.8 shows that the distribution of the mean gets narrower by increasing the number of random variables from two to three. The variance is reduced and for three random variables the probability density looks a bit like a Gaussian.

We use the characteristic function (1.71) to understand the general be-

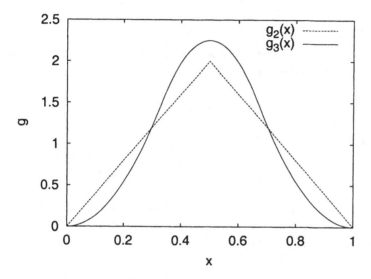

Fig. 1.8 Probability densities for the arithmetic means of two and three uniform random variables, $\hat{g}_2(\overline{x})$ and $\hat{g}_3(\overline{x})$, respectively. Compare equations (1.100), (1.66) and (1.67) .

havior. The characteristic function of a sum of independent random variables is the product of their individual characteristic functions (1.79). Obviously, then

$$\phi_y(t) \;=\; [\phi_x(t)]^N \tag{1.102}$$

for a sum of N independent random variables (1.78). The characteristic function for the corresponding arithmetic average (1.61) is obtained by using equation (1.100)

$$\phi_{\overline{x}}(t) \;=\; \int_{-\infty}^{+\infty} d\overline{x}\, e^{it\overline{x}}\, \hat{g}_N(\overline{x}) \;=\; \int_{-\infty}^{+\infty} N d\overline{x}\, e^{it\overline{x}}\, g_N(N\overline{x})$$

$$\;=\; \int_{-\infty}^{+\infty} dy\, \exp\left(i\,\frac{t}{N}\,y\right)\, g_N(y) \; .$$

Hence, using (1.79) we find

$$\phi_{\bar{x}}(t) = \phi_y\left(\frac{t}{N}\right) = \left[\phi_x\left(\frac{t}{N}\right)\right]^N \qquad (1.103)$$

and this equation is central for the problem of sampling.

Example: The normal distribution.

For the probability density (1.30) the characteristic function is obtained by Gaussian integration to be

$$\phi(t) = \exp\left(-\frac{1}{2}\sigma^2 t^2\right). \qquad (1.104)$$

Defining $y^r = x^r + x^r$ we have

$$\phi_y(t) = [\phi(t)]^2 = \exp\left(-\frac{1}{2}2\sigma^2 t^2\right). \qquad (1.105)$$

As expected this is the characteristic function of a Gaussian with variance $2\sigma^2$. We obtain the characteristic function of the arithmetic average $\bar{x}^r = y^r/2$ by the substitution $t \to t/2$:

$$\phi_{\bar{x}}(t) = \exp\left(-\frac{1}{2}\frac{\sigma^2}{2}t^2\right). \qquad (1.106)$$

In agreement with equation (1.101) the variance is reduced by a factor of two.

1.8.2   *The central limit theorem*

The central limit theorem is the most important consequence of equation (1.103). Let us consider a probability density $f(x)$ and assume that its reduced second moment exists. It is sufficient to consider $\hat{x} = 0$. Otherwise one can define a characteristic function by Fourier transformation around $x_0 = \hat{x}$. The existence of the second moment implies that the characteristic function is at least two times differentiable and we obtain the expansion

$$\phi_x(t) = 1 - \frac{\sigma_x^2}{2}t^2 + \mathcal{O}(t^3). \qquad (1.107)$$

The first term is a consequence of the normalization (1.2) of the probability density, which makes the zeroth moment equal to one. The first moment is $\phi'(0) = i\hat{x} = 0$, so that there is not contribution linear in t. Inserting the approximate characteristic function (1.107) into equation (1.103) for the characteristic function of the mean, we obtain the **result**

$$\phi_{\overline{x}}(t) = \left[1 - \frac{\sigma_x^2}{2N^2}t^2 + \mathcal{O}\left(\frac{t^3}{N^3}\right) \right]^N = \exp\left[-\frac{1}{2}\frac{\sigma_x^2}{N}t^2 \right] + \mathcal{O}\left(\frac{t^3}{N^2}\right).$$
(1.108)

The probability density of the arithmetic mean $\overline{x}^r$ converges towards the Gaussian probability density with variance

$$\sigma^2(\overline{x}^r) = \frac{\sigma^2(x^r)}{N}.$$
(1.109)

Recall the role of the random variables as argument of the variance (1.51).

1.8.2.1 *Counter example*

The Cauchy distribution (1.11) provides an untypical, but instructive, case for which the central limit theorem does not work. This is expected as its second moment does not exist. Nevertheless, the characteristic function of the Cauchy distribution exists. For simplicity we take $\alpha = 1$ and get

$$\phi(t) = \int_{-\infty}^{+\infty} dx \, \frac{e^{itx}}{\pi(1+x^2)} = \exp(-|t|).$$
(1.110)

The integration is left as a exercise, which involves the residue theorem (B.1), to the reader. Using equation (1.103) for the characteristic function of the mean of N random variables, we find

$$\phi_{\overline{x}}(t) = \left[\exp\left(-\frac{|t|}{N} \right) \right]^N = \exp(-|t|).$$
(1.111)

The surprisingly simple result is that the probability distribution for the mean values of N independent Cauchy random variables agrees with the probability distribution of a single Cauchy random variable. Estimates of the Cauchy mean cannot be obtained. Empirical consequences are already exhibited in table 1.10. Indeed, the mean does not exist. This is untypical, because in most cases of practical importance the mean exists and sampling improves, at least asymptotically, its variance in accordance with the prediction (1.109) of the central limit theorem.

1.8.3   *Binning*

The notion of **binning introduced here should not be confused with histogramming!** In the literature histogrammed data are often called binned data. The concept of binning used here is of newer origin and was possibly first introduced by Ken Wilson, but more likely independently discovered and re-discovered by many researchers, see [Flyvberg and Peterson (1989)]. Once one is faced with the error analysis of data from MC simulations (discussed later), the concept is quite obvious.

Binning reduces NDAT data to NBINS binned data, where each binned data point is the arithmetic average of

$$\text{NBIN} = [\text{NDAT}/\text{NBINS}] \quad \text{(Fortran integer division.)} \qquad (1.112)$$

original data points. Preferably NDAT is a multiple of NBINS, although this is not absolutely necessary. The NBIN data within one bin are normally used in the order of the times series of the original measurements. The purpose of the binning procedure is twofold:

(1) When the the central limit theorem applies, the binned data will become practically Gaussian when NBIN becomes large. Most error analysis methods (see the next chapter) rely on the assumption that the data are normally distributed. If this is not true for the original data, binned data can fulfill this assumption.
(2) When data are generated by a Markov process, see chapters 3 and 4, subsequent events are correlated. For binned data these correlations are reduced and can be neglected, once NBIN is sufficiently large compared to the autocorrelation time, see chapter 4.1.1.

Binning of NDAT data into NBINS binned data is done by the subroutine

$$\text{BINING}(\text{NDAT}, \text{DATA}, \text{NBINS}, \text{DATAB}) \qquad (1.113)$$

of ForLib. On input the routine receives the array DATA(NDAT) and on output the binned data are returned in the array DATAB(NBINS). Each element of DATAB is the average over NBIN (1.112) original data. A warning is issued if NDAT is not a multiple of NBINS (*i.e.*, NBINS $*$ NBIN $<$ NDAT), but only the first time when this occurs. An example for the distribution of binned data is given in figure 1.9: The 10 000 uniformly distributed random numbers of figure 1.2 are reduced to 500 binned data and compared with the applicable normal distribution, see assignment 2.

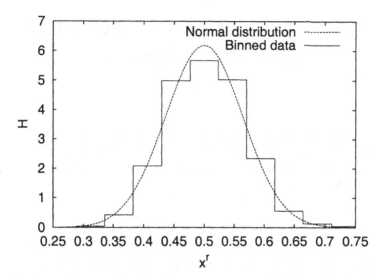

Fig. 1.9 Comparison of a histogram of 500 binned data with the normal distribution $\sqrt{(120/\pi)}\,\exp[-120\,(x-1/2)^2]$, see assignment 2. Each binned data point is the average of 20 uniformly distributed random numbers. The distribution of the original 10 000 uniformly distributed random numbers is depicted in figure 1.2.

1.8.4 *Assignments for section 1.8*

(1) Generalize equations (1.104), (1.105) and (1.106) for the characteristic function of the Gaussian distribution with expectation values $\hat{x} = 0$ to the Gaussian distribution with $\hat{x} = a \neq 0$.

(2) Generate again the 10 000 uniformly distributed random numbers which are used in figure 1.2. Apply the subroutine bining.f of ForLib to generate 500 bins, each averaging 20 data. Subsequently, reproduce figure 1.9 and explain the factors of the normal distribution used for comparison with the histogram. (The histogram shown was created by the routine hist_gnu.f with NHIST=11.)

Chapter 2

Error Analysis for Independent Random Variables

Treated in this chapter are: Gaussian Confidence Intervals, Error Bars and Sampling (Bias, Gaussian Difference Test, χ^2-Distribution and Error of the Error Bar, Student Distribution, Student Difference Test, F-Test), and Parameter Estimation (Fitting).

2.1 Gaussian Confidence Intervals and Error Bars

The central limit theorem (1.108) underlines the importance of the normal distribution. In many applications we are interested in estimating expectation values. Assuming we have a large enough sample, the arithmetic mean of a suitable expectation value becomes normally distributed and the calculation of the confidence intervals is reduced to studying the normal distribution. In particular under physicists, it has become the convention to use the **standard deviation** (1.52) of the sample mean

$$\sigma = \sigma(\overline{x}^r) \quad \text{with} \quad \overline{x}^r = \sum_{i=1}^{N} x_i^r \tag{2.1}$$

to indicate its confidence intervals $[\hat{x} - n\sigma, \hat{x} + n\sigma]$. Note that the important dependence of σ on N is suppressed. The N dependence could be

Table 2.1 Probability content p of Gaussian confidence intervals $[\hat{x} - n\sigma, \hat{x} + n\sigma]$, $n = 1, \ldots, 6$, and $q = (1 - p)/2$, see assignment 1.

n	1	2	3	4	5	6
p	.68	.95	1.0	1.0	1.0	1.0
q	.16	.23E-01	.13E-02	.32E-04	.29E-06	.99E-09

54

emphasized by using the notation σ_N instead of σ. For a Gaussian distribution equation (1.31) yields the probability content p of the confidence intervals (2.1) to be

$$p = p(n) = G(n\sigma) - G(-n\sigma) = \frac{1}{\sqrt{2\pi}} \int_{-n}^{+n} dx\, e^{-\frac{1}{2}x^2} = \mathrm{erf}\left(\frac{n}{\sqrt{2}}\right) . \quad (2.2)$$

Table 2.1 compiles the results for $n = 1, \ldots, 6$ standard deviations ($q = (1-p)/2$ was previously defined (1.39) for symmetric confidence intervals).

Table 2.1 gives the likelihoods that the sample mean $\bar{x}$ will be found in the indicated confidence intervals around the exact expectation value $\hat{x}$. In practice the roles of $\bar{x}$ and $\hat{x}$ are interchanged: The sample mean $\bar{x}$ is measured and one would like to know the likelihood that the **unknown** exact expectation value $\hat{x}$ will be in a certain confidence interval around the measured value. For the moment, let us make the (unrealistic) assumption that we know the exact standard deviation (2.1) of the sample mean. Then, the relationship

$$\bar{x} \in [\hat{x} - n\sigma, \hat{x} + n\sigma] \iff \hat{x} \in [\bar{x} - n\sigma, \bar{x} + n\sigma] \quad (2.3)$$

solves the problem. In the Gaussian approximation the probabilities to find $\hat{x}$ in the confidence intervals $[\bar{x} - n\sigma, \bar{x} + n\sigma]$, ($n = 1, 2, \ldots$) are as well given by the numbers of table 2.1. Conventionally, these estimates are quoted as

$$\hat{x} = \bar{x} \pm \Delta\bar{x} \quad (2.4)$$

where the **error bar** $\Delta\bar{x}$ is either the exact standard deviation (2.1) of the sample mean or an estimator of it. The typical situation, addressed now, is that we have to estimate the error bar from the data.

When we present numerical estimates of error bars, we will in general give two digits and use the **conservative rounding convention** that the second digit is always rounded to the next higher value. The rationale for our convention is that the error of the error bar (to be discussed later) will normally exceed one unit of the second digit and by Murphy's law[1] the real error tends always to be larger than the estimated one. Besides Murphy's law, there is the statistics reason, that in many situations the Gaussian distribution is used an approximation of a somewhat bróader distribution.

---

[1] Murphy was an engineer concerned with safety issues. His law exists in many variants of the same theme, of which the best known appears to be, *anything that can go wrong, will go wrong.*

2.1.1 *Estimator of the variance and bias*

In the Gaussian limit (1.109) $\sigma_{\overline{x}} = \sigma_x / N$ holds. An obvious estimator for the variance σ_x^2 is

$$(s_x'^r)^2 \;=\; \frac{1}{N} \sum_{i=1}^{N} (x_i^r - \overline{x}^r)^2 \tag{2.5}$$

where the prime indicates that we shall not be happy with it and later replace $(s_x'^r)^2$ by an improved estimator $(s_x^r)^2$. With the estimator $(s_x'^r)^2$ we encounter for the first time the **bias** problem, which will be discussed in more details when we introduce the jackknife approach. An estimator is said to be biased when its expectation value does not agree with the exact result. In our case

$$\langle (s_x'^r)^2 \rangle \;\neq\; \sigma_x^2 \;. \tag{2.6}$$

This is immediately recognized for the case $N = 1$: Our estimator (2.5) yields always zero, independently of the exact result. Another problem is that, due to using an estimator instead of the exact σ_x^2, the confidence probabilities of table 2.1 are not exact anymore. Again, $N = 1$ is the obvious example: The probability to find the exact expectation value $\widehat{x}$ in the estimated confidence interval is zero. The Student distribution, discussed in section 2.3, solves this problem for normally distributed data by correcting the confidence probabilities for finite N.

Although we do still expect our estimator (2.5) to converge towards the exact result for $N \to \infty$, equation (2.6) is annoying. For finite, possibly small, values of N an estimator $(s_x^r)^x$ with

$$\langle (s_x^r)^2 \rangle \;=\; \sigma_x^2 \tag{2.7}$$

ought to be preferred. An estimator whose expectation value agrees with the true expectation value is called **unbiased**. For the variance it is rather straightforward to construct an unbiased estimator $(s_x^r)^x$. The bias of the definition (2.5) comes from replacing the exact mean $\widehat{x}$ by its estimator $\overline{x}^r$. The latter is a random variable, whereas the former is just a number. Using linearity of expectation values we find

$$\langle (s_x'^r)^2 \rangle = \frac{1}{N} \sum_{i=1}^{N} \left(\langle (x_i^r)^2 \rangle - 2 \langle x_i^r \, \overline{x}^r \rangle + \langle (\overline{x}^r)^2 \rangle \right) \tag{2.8}$$

and investigate this equation term by term. First, $(x_i^r)^2$ is, independently of i, an unbiased estimator of the second moment $\widehat{x^2}$

$$\langle (x_i^r)^2 \rangle = \widehat{x^2} . \tag{2.9}$$

Next, we encounter a little problem, x_i^r and $\bar{x}^r$ are not independent random variables, because x_i^r enters the definition (1.61) of $\bar{x}^r$. This is resolved by writing the term in question as a sum of independent random variables

$$\langle x_i^r \bar{x}^r \rangle = \frac{1}{N} \left(\langle (x_i^r)^2 \rangle + \sum_{j;j \neq i} \langle x_i^r x_j^r \rangle \right) = \frac{1}{N} \left(\widehat{x^2} - \widehat{x}^2 \right) + \widehat{x}^2 = \frac{1}{N} \sigma_x^2 + \widehat{x}^2 ,$$
$$\tag{2.10}$$

where, for the last equal sign, we have used

$$\sigma_x^2 = \langle (x^r - \widehat{x})^2 \rangle = \left(\langle (x^r)^2 \rangle - 2 \widehat{x} \langle x^r \rangle + \widehat{x}^2 \right) = \left(\widehat{x^2} - \widehat{x}^2 \right) . \tag{2.11}$$

Using this identity also for $\sigma_{\bar{x}}^2$ we find for the last term in (2.8)

$$\langle (\bar{x}^r)^2 \rangle = \widehat{x}^2 + \sigma_{\bar{x}}^2 = \widehat{x}^2 + \frac{1}{N} \sigma_x^2 . \tag{2.12}$$

Inserting (2.9), (2.10) and (2.12) into (2.8) gives

$$\langle (s_x'^r)^2 \rangle = \left(\widehat{x^2} - \widehat{x}^2 \right) - \frac{1}{N} \sigma_x^2 = \left(\frac{N-1}{N} \right) \sigma_x^2 . \tag{2.13}$$

We note that the bias is $-\sigma_x^2/N$ and disappears $\sim 1/N$ for large N. The desired **unbiased estimator of the variance** is given by

$$(s_x^r)^2 = \frac{N}{N-1} (s_x'^r)^2 = \frac{1}{N-1} \sum_{i=1}^{N} (x_i^r - \bar{x}^r)^2 . \tag{2.14}$$

Correspondingly (1.109), the unbiased estimator of the variance of the sample mean is

$$(s_{\bar{x}}^r)^2 = \frac{1}{N(N-1)} \sum_{i=1}^{N} (x_i^r - \bar{x}^r)^2 . \tag{2.15}$$

2.1.2 *Statistical error bar routines* (steb)

The square root $s_{\bar{x}}^r$ of the estimator (2.15) estimates the standard deviation of the sample mean[2] $\sigma_{\bar{x}}$, i.e., the error bar of equation (2.4) is normally

---

[2]This estimator is biased again, because the square root is a non-linear function of the mean.

given by

$$\Delta \bar{x} = s_{\bar{x}}^r .$$
(2.16)

This error bar estimate is implemented by our Fortran subroutine steb0.f. A data array X(N) has to be provided on input. Then a call to

$$\text{STEBO}(N, X, XM, XV, XE)$$
(2.17)

returns the following estimates: The mean XM, the variance XV and the error bar XE, see assignment 2 to practice its use. XV is the variance of x^r (estimated from the data), the variance of $\bar{x}^r$ is XE**2. When only the sample mean is needed, one may alternatively rely on our Fortran function stmean.f:

$$\text{XM} = \text{STMEAN}(N, X) .$$
(2.18)

Often one encounters the problem of **observations with different accuracies**. Let us first assume that the relative weights of the observations are known exactly. This is for instance the case when each observation is the average of N_i independent measurement, all sampled with the same probability density, possibly by different (experimental) groups. In this case one calculates the mean as the weighted average

$$\bar{x}^r = \sum_{i=1}^{N} w_i \bar{x}_i^r \text{ where } \sum_{i=1}^{N} w_i = 1 .$$
(2.19)

The w_i are called **weights**. Our previous sample mean (1.61) is recovered as the case where all weights are equal, $w_i = 1/N$ and $\bar{x}_i^r = x_i^r$. For our example of (otherwise identical) measurements with distinct statistics we have

$$\bar{x}_i^r = \frac{1}{N_i} \sum_{j=1}^{N_i} x_j^r \text{ and } w_i = \frac{N_i}{\sum_{i=1}^{N} N_i} .$$
(2.20)

By similar considerations as in subsection 2.1.1 (see the exercises for section 2.1 of appendix B) one can show that

$$(s_{\bar{x}}^r)^2 = \frac{1}{N-1} \sum_{i=1}^{N} w_i (\bar{x}_i^r - \bar{x}^r)^2$$
(2.21)

is the unbiased estimator of the variance of the mean. Our Fortran subroutine steb1.f uses the weighted, unbiased estimators (2.19) and (2.21) to calculate the sample mean XM and its error bar XE. As before, the variance of the sample mean is XE**2, whereas XV is no longer calculated. Besides the data array X(N) also a weight array W(N) must now be provided on input. A call to

$$\text{STEB1}(\text{N}, \text{X}, \text{W}, \text{XM}, \text{XE}) \tag{2.22}$$

returns then XM and XE. See assignments 3 and 4 to practice its use.

When one does not know the exact relative weights, they may be estimated from the relation

$$w_i = \frac{1}{s_{x_i}^2} \left(\sum_{i=1}^{N} \frac{1}{s_{x_i}^2} \right)^{-1} \tag{2.23}$$

which holds because $(s_{x_i}^r)^2$ is an estimator of $\sigma_{x_i}^2$ and $\sigma_{x_i}^2 = \sigma_x^2/N_i$. For instance, this equation may be applied when one combines measurements from different (experimental) groups, who state their results with different error bars. Estimating the weight via equation (2.23) is at the edge of what can be done, because estimating the weights statistically introduces undesired additional noise, which can lead to instabilities. Use (2.23) only when better estimates of the weights, like $w_i \sim N_i$ in (2.20), are not known.

Instead of using equation (2.21), one can estimate the variance of the sample mean also from

$$(s_{\bar{x}}^r)^2 = \sum_{i=1}^{N} w_i^2 \, (s_{\bar{x}_i}^r)^2 \; . \tag{2.24}$$

This equation relies on $\sigma_{\bar{x}}^2 = w_i \, \sigma_{\bar{x}_i}^2$, so it is the weighted sum (2.19) of different estimators of $\sigma_{\bar{x}}^2$. **It depends on the situation whether it is favorable to use (2.21) or (2.24):** When there are many $\bar{x}_i$ and each one is only an average of very few data, equation (2.21) should be used. On the other side, when there are only few $\bar{x}_i$ values, but each relies on an average of many data, equation (2.24) is the better choice. In practical applications one can determine the optimal choice by a simulation similar to the one of assignment 4.

Equations (2.23) and (2.24) are implemented in our Fortran subroutine steb2.f. On input each data point DAT(I), I=1,...,N has to be supplied together with its error bar EB(I), I=1,...,N. Equation (2.24) is applied when the weight factor array W(N) is also given on input. Otherwise one has to set W(1) < 0 and equation (2.23) will be used. A call to

$$\text{STEB2}(N, DAT, EB, W, XM, XE) \qquad (2.25)$$

returns the sample mean XM and its error bar XE. The weight factors are also available on output. In the case that the input measurements rely on small samples, and their exact relative weights are not known, care has to be exercised as is illustrated by assignment 3 of section 2.4, see table 2.10.

2.1.2.1 *Ratio of two means with error bars*

Frequently, one has two means with (approximately) Gaussian error bars given and likes to know their ratio and the error bar of this ratio. For this purpose we provide a short subroutine, steb_rat.f in ForLib. A call to

$$\text{STEB\_RAT}(XM1, EB1, XM2, EB2, RATM, RATE, REsmall) \qquad (2.26)$$

needs on input the two mean values XM1 and XM2 with their error bars EB1 and EB2. It returns on output the input values unchanged and the ratio

$$\text{RATM} \pm \text{RATE} \; = \; \frac{XM1}{XM2} \pm \text{RATE}$$

with a **conservative** estimate of its error bar RATE: Gaussian error propagation is assumed, but for the denominator the deviation which maximizes $|XM1/(XM2 \pm EB2)|$ is used. For $EB2 \ll XM2$ this makes little difference, but if EB2 becomes so large that asymmetric error bars should be used for the ratio, our choice leads to an overestimate of RATE. For control the smaller error bar estimate is returned as REsmall. To perform this calculation interactively the program

$$\text{ratio\_eb.f} \qquad (2.27)$$

is provided in the folder ForProc/ratio_eb.

2.1.3 *Gaussian difference test*

The final topic of this section is the **Gaussian difference test**. In practice one is often faced with the problem to compare two different empirical estimates (for instance from two experimental physics groups) of some mean. Let us denote these estimates by $\bar{x}$ and $\bar{y}$. In general $\bar{x}$ and $\bar{y}$ will differ and the question arises, does this imply that there is a real difference between the expectations values $\hat{x}$ and $\hat{y}$ or is the observed difference due to chance? How large must $D = \bar{x} - \bar{y}$ be to indicate a real difference between $\hat{x}$ and $\hat{y}$?

To approach this question we assume that both estimates are correct estimates of the same expectation value, that $\bar{x}^r$ is normally distributed with mean $\hat{x}$ and variance $\sigma_{\bar{x}}^2$ and $\bar{y}^r$ normally distributed with mean $\hat{y} = \hat{x}$ and variance $\sigma_{\bar{y}}^2$. It follows (use the characteristic function (1.71)) that the random variable

$$D^r = \bar{x}^r - \bar{y}^r \tag{2.28}$$

is normally distributed with

$$\text{mean } \hat{D} = 0 \text{ and variance } \sigma_D^2 = \sigma_{\bar{x}}^2 + \sigma_{\bar{y}}^2. \tag{2.29}$$

Therefore, the quotient

$$d^r = \frac{D^r}{\sigma_D} \tag{2.30}$$

is normally distributed with expectation zero and variance one (the denominator σ_D is the exact variance of D^r). From equation (1.31) we find the probability

$$P = P(|d^r| \le d) = G_0(d) - G_0(-d) = 1 - 2G_0(-d) = \text{erf}\left(\frac{d}{\sqrt{2}}\right). \tag{2.31}$$

The **likelihood that the observed difference** $|\bar{x} - \bar{y}|$ **is due to chance** is defined to be

$$Q = 1 - P = 2G_0(-d) = 1 - \text{erf}\left(\frac{d}{\sqrt{2}}\right). \tag{2.32}$$

The definition of Q implies: If the assumption is correct, *i.e.* identical distributions are sampled, then Q is a uniformly distributed random variable in the range $(0, 1]$.

The Gaussian difference test allows to impose a cut-off value Q_{cut} on Q. For $Q \le Q_{\text{cut}}$ we say that the two results differ in a **statistically significant** way. Of course, the choice of the cut-off value is somewhat a matter of taste. In this book, as in many applications (biology, medicine, and others), we will mostly use $Q_{\text{cut}} = 0.05$ as a standard. Still, in one out of twenty cases $Q \le 0.05$ emerges when both distributions are identical. Finally, if Q is very close to one, say 0.99, one has an indication that the error bars are overestimated. Generalizations of the test are discussed in subsequent sections.

Table 2.2 Numerical examples for the Gaussian difference test, see assignment 6.

$\bar{x}_1 \pm \sigma_{\bar{x}_1}$	1.0 ± 0.1	1.0 ± 0.05	1.000 ± 0.025	1.0 ± 0.1	1.0 ± 0.1
$\bar{x}_2 \pm \sigma_{\bar{x}_2}$	1.2 ± 0.1	1.2 ± 0.05	1.200 ± 0.025	1.2 ± 0.0	1.2 ± 0.2
Q	0.16	0.0047	0.15×10^{-7}	0.046	0.37

To perform the Gaussian difference test interactively the program

$$\texttt{gau\_dif.f} \tag{2.33}$$

is provided in the folder ForProc/Gau_dif. Compile and run this program! The relevant subroutine call in gau_dif.f is to

$$\texttt{GAUSDIF(XM1, EB1, XM2, EB2, Q)} \tag{2.34}$$

of ForLib. This subroutine compares two means according to the Gaussian difference test. Input: The first mean XM1 and its error bar EB1, the second mean XM2 and its error bar EB2. Output: The probability Q of the Gaussian difference test. **Examples** are given in table 2.2.

2.1.3.1 *Combining more than two data points*

Table 2.3 Measurements of the deflection of light by the sun, as collected in Weinberg's book. The numbers in parenthesis are the error bars, referring to the last two digits of the estimates.

#	Site and year	θ_0 (sec)	Researchers
1	Sobral 1919	1.98 (16)	F.W Dyson, *et al.*
2	Principe 1919	1.61 (40)	F.W Dyson, *et al.*
3	Australia 1922	1.77 (40)	G.F. Dodwell and C.R. Davidson.
4	Australia 1922	1.79 (37)	C.A. Chant and R.K. Young.
5	Australia 1922	1.72 (15)	W.W. Campbell and R. Trumpler.
6	Australia 1922	1.82 (20)	W.W. Campbell and R. Trumpler.
7	Sumatra 1929	2.24 (10)	E.F. Freundlich, *et al.*
8	U.S.S.R. 1936	2.73 (31)	A.A. Mikhailov.
9	Japan 1936	1.70 (43)	T. Matukuma, et al.
10	Brazil 1947	2.01 (27)	G. van Biesbroeck.
11	Sudan 1952	1.70 (10)	G. van Biesbroeck.
12	Owens Valley 1969	1.77 (20)	G.A. Seielstad, *et al.*
13	Goldstone 1969	1.85 (21)	D.O. Muhleman, *el al.*
14	NRAO 1970	1.57 (08)	R.A. Sramek.
15	NRAO 1970	1.87 (30)	J.M. Hill.

The Gaussian difference test is also a powerful tool when more than two data points are considered. To give an example we reproduce in table 2.3 estimates for the deflection of light by the sun, which [Weinberg (1972)] compiled in his book *Gravitation and Cosmology*. The results up to 1952 rely on measurements performed during solar eclipses and the later on measurements of the deflection of radio waves. Meanwhile, more accurate experiments have been performed. Our interest is limited in the statistical combination of the results of table 2.3. The theoretical value is

$$\theta_0(\text{sec}) = 1.75 \qquad (2.35)$$

and we address two questions:

(1) Are the measurements consistent with one another?
(2) Are the combined measurements consistent with the theoretical value?

To address the first question, we combine $N - 1$ of the N data using steb2.f (2.25) with the W(1)<0 option. The Gaussian difference test is then used to compare this average with the remaining data point $\theta_{0i} \pm \Delta\theta_{0i}$ for $i = 1, \ldots, N$. According to a cut-off value Q_{cut} the data point with the smallest Q value, $Q_{\min}$, may then be deleted. However, we cannot take $Q_{\text{cut}} = 0.05$, because we perform now N Gaussian difference tests and the likelihood to produce a $Q \leq Q_{\text{cut}}$ value becomes $Q_{0,\text{cut}}$ given by

$$Q_{0,\text{cut}} = 1 - (1 - Q_{\text{cut}})^N \implies Q_{\text{cut}} = 1 - \exp\left[\ln(1 - Q_{0,\text{cut}}) / N\right] . \quad (2.36)$$

In the usual way, we may choose a reasonably small values of $Q_{0,cut}$ from which Q_{cut} follows. When no data point yields a Q value smaller than Q_{cut}, the final result is the average from all N data. Otherwise, the data point with the smallest Q value, $Q_{\min}$, is eliminated. We repeat the procedure with $N-1$ data points, etc., until no further inconsistencies are encountered.

Let us return to the data of table 2.3. We choose $Q_{0,\text{cut}} = 0.05$ and apply the above procedure. In the first step with $N = 15$ data points we have $Q_{\text{cut}} = 0.0034$ and eliminate data point #7 for which we find $Q_{\min} = 0.47 \times 10^{-5}$. With the resulting 14 data points we have $Q_{\text{cut}} = 0.0037$ and eliminate data point #8 for which we find $Q_{\min} = 0.0012$. The remaining 13 data points are consistent with one another ($Q_{\text{cut}} = 0.0039$ and $Q_{\min} = 0.027$ for data point #14) and our final estimate is

$$\theta_0(\text{sec}) = 1.714 \pm 0.047 . \qquad (2.37)$$

Comparing this with θ_0 of equation (2.35), the Gaussian difference test gives $Q = 0.43$, *i.e.*, perfect agreement in the statistical sense. In comparison, by simply combining all 15 data one gets $\theta_0 = 1.824 \pm 0.042$ and $Q = 0.08$. This is also consistent on the 5% exclusion level. Nevertheless, the complete data set contains measurements which should be eliminated, because they are inconsistent with the (vast) majority of the other measurements (unless, of course, there are convincing non-statistical reasons which explain what went wrong with the majority of measurements). To verify the analysis is left to the reader as assignment 9.

2.1.4 *Assignments for section 2.1*

(1) Use the subroutine `error_f.f` of ForLib to reproduce the results of table 2.1.

(2) Take Marsaglia's initial seeds and generate subsequently (a) $1,024 = 2^{10}$ uniformly distributed random numbers, (b) $2\,048 = 2^{11}$ normally distributed random numbers and (c) $1\,024$ Cauchy distributed random numbers. Use `steb0.f` to verify the following values for the sample means and their error bars:

$$0.4971 \pm 0.0091 \text{ (uniform)}, \tag{2.38}$$

0.0129 ± 0.0219 (normal) and -14.3 ± 16.0 (Cauchy). As expected, for the Cauchy distribution we find no convergence of the mean, whereas there is convergence of its median as we have seen in table 1.9.

(3) Use the same random numbers as in the previous assignment to generate $k = 1, \ldots, 10$ data, which are arithmetic means of N_k uniformly distributed random numbers with $N_k = 2^k$ for $k = 1, \ldots, 9$, and $N_{10} = 2$ (these values are chosen to add up to $\sum_k N_k = 2^{10}$). Use the subroutine `steb1.f` to calculate the mean and its error bar. **Result:**

$$0.4971 \pm 0.0068 . \tag{2.39}$$

Note that the arithmetic mean is identical with the one of the uniform random numbers of the previous assignment, whereas the error bar deviates. However, the $s_{\bar{x}}^2$ values calculated by applying either equation (2.21) or (2.15) agree in the average. This is demonstrated by the next assignment.

(4) Generate $k = 1, \ldots, 6$ data, which are arithmetic means of N_k uniformly distributed random numbers with $N_k = 2^k$ for $k = 1, \ldots, 5$, and $N_6 =$

2. Use `steb1.f` to calculate the mean and its error bar, and then `steb0.f` to calculate the same from the combined statistics of

$$N = \sum_{k=1}^{6} N_k = 64$$

data. Repeat everything 10 000 times and collect the variances and error bars of the mean values in arrays. Use `steb0.f` to calculate the mean variance, the mean error bar, and their corresponding error bars with respect to the 10 000 data. **Result:** With `steb1.f`, corresponding to equation (2.21), we get

$$\overline{s_{\overline{x}}^2} = (1.313 \pm 0.008) \times 10^{-3} \ \text{ and } \ \overline{s_{\overline{x}}} = (3.466 \pm 0.011) \times 10^{-2} \ , \quad (2.40)$$

whereas with `steb0.f`, corresponding to equation (2.15), we obtain

$$\overline{s_{\overline{x}}^2} = (1.302 \pm 0.002) \times 10^{-3} \ \text{ and } \ \overline{s_{\overline{x}}} = (3.603 \pm 0.002) \times 10^{-2} \ . \quad (2.41)$$

Whereas the $\overline{s_{\overline{x}}^2}$ values are consistent within statistical errors, this is not the case for the $\overline{s_{\overline{x}}}$ results. The reason is that $s_{\overline{x}}^r$ is a biased estimator of $\sigma_{\overline{x}}$ for which the small statistics data spoil the $\overline{s_{\overline{x}}}$ estimate (2.40) considerably.

What we really care about are confidence probabilities of intervals as given in table 2.1, and we have to work out finite N corrections. This problem is addressed in section 2.3.

(5) Generate the same random variables as in assignment 3, but use now `steb0.f` for each value of k to calculate the error bar for the corresponding N_k data. Use the data together with their thus calculated error bars as input for `steb2.f`, `W(1)<0`, to obtain a weighted average. **Result:**

$$0.4983 \pm 0.0090 \ . \quad (2.42)$$

Now, not only the error bar, but also the mean value of the estimate deviates from the result of equation (2.38). The reason is that estimators are used for the weights. That the actually estimated mean value is closer to the true one, 0.5, than in case of (2.38) is a pure accident, as is the fact that the error bar is almost identical with the one in (2.38). The results of table 2.10 (assignment 3 in section 2.4) show that combining estimates with `steb2.f`, `W(1)<0`, is problematic when the quality of each estimate is low.

(6) Use the procedure `gau_dif.f` to reproduce table 2.2.

(7) Apply the Gaussian difference test to the numbers of equations (2.40) and (2.41). This is, of course, not legitimate as we calculated (2.40) and (2.41) from the same random numbers. But for this exercise we dot not care, because we expect[3] the effect (if any) to be minor. **Result:** The Gaussian difference test gives 18% likelihood that the discrepancy between the two $s_{\bar{x}}^2$ variance values is due to chance. This is not significant. However, the likelihood that the discrepancy between the two $\overline{s_{\bar{x}}}$ standard deviations is due to chance comes out to be 0%, and all available subsequent digits are also zero, *i.e.*, the two standard deviations disagree really.

(8) (A) Generate 10 000 pairs $(x_{1,i}, x_{2,i})$, $i = 1, \ldots, 10\,000$ of independent Gaussian random numbers with mean $\hat{x} = \hat{x}_1 = \hat{x}_2 = 0$ and variance $\sigma = \sigma_1 = \sigma_2 = 1$. Perform the Gaussian difference test for each pair and histogram the likelihoods Q_i, $i = 1, \ldots, 10\,000$. How does the histogram look? **Result:** As we compare mean values from identical distributions, Q is uniformly distributed in the range $(0, 1]$. We obtain a histogram which is very similar to that of figure 1.2.

(B) Repeat the previous assignment for $|\hat{x}_1 - \hat{x}_2| = n\sigma$, $n = 1, 2, \ldots$. How do the histograms change? **Result:** Each histogram takes now its maximum at $Q = 0$ and falls off monotonically towards $Q = 1$. With increasing n the maximum become more and more pronounced.

(9) Calculate the estimate (2.37) for the deflection of light by the sun along the lines explained in the text.

2.2   The χ^2 Distribution

The distribution of the random variable

$$(\chi^r)^2 = \sum_{i=1}^{N} (y_i^r)^2 , \qquad (2.43)$$

where each y_i^r is normally distributed, defines the χ^2 **distribution** with N degrees of freedom. Its cumulative distribution function is

$$F(\chi^2) = P\left[(\chi^r)^2 \le \chi^2\right] =$$

$$(2\pi)^{-N/2} \int_{y_1^2 + \cdots + y_N^2 \le \chi^2} dy_1 \ldots dy_N \, e^{-\frac{1}{2}(y_1^2 + \cdots + y_N^2)} . \qquad (2.44)$$

---

[3] Warning: Be careful with such expectations.

The importance of the χ^2 distribution stems from its relationship with the sample variance $(s_x^r)^2$. Provided the fluctuations of the underlying data are Gaussian, the study of the distribution of $(s_x^r)^2$ can be reduced to the χ^2-distribution with $f = N - 1$ degrees of freedom

$$(\chi_f^r)^2 = \sum_{i=1}^{N-1} (y_i^r)^2 = \frac{(N-1)\,(s_x^r)^2}{\sigma_x^2} = \sum_{i=1}^{N} \frac{(x_i^r - \overline{x}^r)^2}{\sigma_x^2}\,. \tag{2.45}$$

We first proof that $(\chi_f^r)^2$ of this equation is indeed χ^2 distributed. Next, we calculate the χ^2 distribution function, *i.e.* the integral (2.44), explicitly and find its associated probability density by differentiation. Applications, like confidence levels when σ_x^2 is estimated from N Gaussian data and the error of the error bar problem, are addressed in further sections. Assignment 3 gives a numerical illustration of equation (2.45).

2.2.1 Sample variance distribution

The fluctuations of $(\chi_f^r)^2$ defined by (2.45) do not depend on either the mean $\widehat{x}$ or the variance σ_x^2. Therefore, without restricting the generality of our arguments we chose

$$\widehat{x} = 0 \quad \text{and} \quad \sigma_x^2 = 1 \tag{2.46}$$

and get

$$(\chi_f^r)^2 = \sum_{i=1}^{N} (x_i^r - \overline{x}^r)^2 = \sum_{i=1}^{N} (x_i^r)^2 - N\,(\overline{x}^r)^2 = \sum_{i=1}^{N} (x_i^r)^2 - \frac{1}{N} \left(\sum_{i=1}^{N} x_i^r \right)^2. \tag{2.47}$$

As the variables x_i^r are normally distributed, the distribution function of $(\chi_f^r)^2$ becomes

$$F_f(\chi^2) = P\left[(\chi_f^r)^2 \leq \chi^2 \right] = (2\pi)^{-N/2} \int_{u \leq \chi^2} dx_1 \ldots dx_N \, e^{-\frac{1}{2}(x_1^2 + \ldots x_N^2)} \tag{2.48}$$

where

$$u = \sum_{i=1}^{N} x_i^2 - \frac{1}{N} \left(\sum_{i=1}^{N} x_i \right)^2. \tag{2.49}$$

The surface $u = \chi^2$ of this integration domain looks complicated. The difficulty of this integrations is overcome by an orthonormal transformation to new coordinates $y_1, \ldots, y_N$:

$$y_i = \sum_{j=1}^{N} a_{ij} x_j \quad \text{with} \quad \sum_{i=1}^{N} y_i^2 = \sum_{i=1}^{N} x_i^2 . \tag{2.50}$$

The orthonormality relation

$$\sum_{k=1}^{N} a_{ki} a_{kj} = \delta_{ij} \tag{2.51}$$

implies $\det|a_{ij}| = \pm 1$, because $\det\left(A^{\dagger}A\right) = 1$ holds for the matrix $A = (a_{ij})$. As the transformation is linear, this determinant agrees with the Jacobian

$$\frac{\partial(y_1, \ldots, y_N)}{\partial(x_1, \ldots x_N)} = \det|a_{ij}| = \pm 1 . \tag{2.52}$$

We choose the transformation so that the plus sign holds and have

$$dx_1 \ldots dx_N = dy_1 \ldots dy_N . \tag{2.53}$$

Equation (2.49) suggests to take

$$y_N = \frac{1}{\sqrt{N}} (x_1 + \cdots + x_N) \tag{2.54}$$

as one of the new variables. Provided we can supplement this choice of y_N to an orthogonal transformation, we have

$$u = \sum_{i=1}^{N} y_i^2 - y_N^2 = \sum_{i=1}^{N-1} y_i^2 , \tag{2.55}$$

i.e., the variable u (2.49) does not depend on y_N. The distribution function (2.48) becomes

$$F_f(\chi^2) = (2\pi)^{-(N-1)/2} \int_{u = y_1^2 + \cdots + y_{N-1}^2 \leq \chi^2} dy_1 \ldots dy_{N-1} \, e^{-\frac{1}{2}(y_1^2 + \cdots y_{N-1}^2)} \tag{2.56}$$

where the dy_N integration has already been carried out. The constant resulting from this integration is determined by the normalization $F_f(\chi^2) \to 1$ for $\chi^2 \to \infty$. By comparison with equation (2.44) we see that $F_f(\chi^2)$ is the χ^2-**distribution function with** $f = N - 1$ **degrees of freedom.** Equation (2.56) exhibits that all the y_i^r random variables are independent, *i.e.*, their probability densities factor into a product (1.62). Because y_N^r differs from the mean only by a constant factor $1/\sqrt{N}$, this implies: **The**

mean $\bar{x}^r$ and the estimator of its variance $(s_{\bar{x}}^r)^2 = (s_x^r)^2/N$ are independent random variables.

It remains to show that coordinates $y_2, \ldots, y_{N-1}$ exist which supplement y_N to an orthonormal transformation. This is a consequence of a well-known **theorem of linear algebra:**

Every last row $\sum_{i=1}^{N} a_{Ni}^2$ of a transformation matrix $A = (a_{ij})$ can be supplemented to an orthonormal linear transformation.

Proof: Let us first note that the orthonormality relation (2.51) implies that the inverse transformation (2.50) is performed by the transpose matrix $A^\dagger$:

$$x_i = \sum_{j=1}^{N} a_{ji}\, y_j = \sum_{k=1}^{N}\sum_{j=1}^{N} a_{ji}\, a_{jk}\, x_k = \sum_{k=1}^{N} \delta_{ik}\, x_k = x_i \;. \qquad (2.57)$$

Hence, the relation (2.51) is equivalent to

$$\sum_{k=1}^{N} a_{ik}\, a_{jk} = \delta_{ij} \;. \qquad (2.58)$$

We construct now a matrix $A = (a_{ij})$ which fulfills the orthonormality relation in this form. The equation

$$a_{(N-1)1}\, a_{N1} + a_{(N-1)2}\, a_{N2} + \cdots + a_{(N-1)(N-1)}\, a_{N(N-1)} + a_{(N-1)N}\, a_{NN} = 0$$

has non-trivial solutions. For instance, when $a_{Nk} \neq 0$ holds, take all $a_{(N-1)j}$ with $j \neq k$ arbitrary, e.g. zero, and calculate $a_{(N-1)k}$. Next

$$\sum_{i=1}^{N} a_{(N-1)i}^2 = 1$$

is achieved through multiplication of our initial $a_{(N-1)i}$ values with a suitable scale factor $a_{(N-1)i} \to \lambda a_{(N-1)i}$. In the same way we find a non-trivial solution for the matrix elements $a_{(N-l)i}$, $l = 2, \ldots, N-1$. Namely,

$$\sum_{i=1}^{N} a_{(N-l)i}\, a_{(N-k)i} = 0 \;\; \text{with} \;\; k = 0, \ldots, l-1$$

leads to l equations which have non-trivial solutions as we have (a) $N > l$ free parameters to our disposal and (b) each previous row has non-zero elements due to the normalization $\sum_i a_{(N-k)i}^2 = 1$. For $a_{(N-l)i}$ this normalization is then ensured by multiplying with the appropriate scale factor. This concludes the proof.

2.2.2   *The χ^2 distribution function and probability density*

The χ^2 distribution function (2.44) with N degrees of freedom is related to the incomplete Gamma function (A.5) of appendix A.1. To derive the relation we introduce N-dimensional spherical coordinates

$$dx_1...dx_N \; = \; d\Omega \; r^{N-1} dr \;\; \text{with} \;\; \int d\Omega \; = \; \text{finite} \qquad (2.59)$$

and equation (2.44) becomes

$$F(\chi^2) \; = \; \text{const} \int_0^{\chi^2} e^{-\frac{1}{2}r^2} r^{N-1} dr \; .$$

The substitution $t = r^2/2$ yields

$$F(\chi^2) \; = \; \text{const}' \int_0^{\frac{1}{2}\chi^2} e^{-t} \, t^{a-1} \, dt \;\; \text{where} \;\; a = \frac{N}{2}$$

and the normalization condition $F(\infty) = 1$ determines: const$' = 1/\Gamma(a)$. Consequently, we have

$$F(\chi^2) \; = \; \frac{1}{\Gamma(a)} \int_0^{\frac{1}{2}\chi^2} e^{-t} \, t^{a-1} \, dt = P\left(\frac{N}{2}, \frac{\chi^2}{2}\right) \qquad (2.60)$$

and `gamma_p.f` from `ForLib` enables us to calculate the χ^2 distribution function. The χ^2 **probability density** follows by differentiation

$$f(\chi^2) \; = \; \frac{dF(\chi^2)}{d\chi^2} \; = \; \frac{e^{-\frac{1}{2}\chi^2} \left(\frac{1}{2}\chi^2\right)^{a-1}}{2\,\Gamma(a)} \;\; \text{where} \;\; a = \frac{N}{2} \; . \qquad (2.61)$$

Integration by parts determines the mean value

$$< \chi^2 >= \frac{\int_0^\infty du\, e^{-\frac{1}{2}u} u^a}{\int_0^\infty du\, e^{-\frac{1}{2}u} u^{a-1}} = \frac{2a \int_0^\infty du\, e^{-\frac{1}{2}u} u^{a-1}}{\int_0^\infty du\, e^{-\frac{1}{2}u} u^{a-1}} = N \; , \qquad (2.62)$$

where N is the number of degrees of freedom. Of major interest is the probability density of χ^2 **per degree of freedom (pdf)**, $(\chi_{\text{pdf}}^r)^2 = (\chi^r)^2/N$. It follows from equation (1.100) to be:

$$f_N(\chi_{\text{pdf}}^2) \; = \; Nf(N\chi_{\text{pdf}}^2) \; = \; \frac{a\, e^{-a\chi_{\text{pdf}}^2} \left(a\chi_{\text{pdf}}^2\right)^{a-1}}{\Gamma(a)} \;\; \text{where} \;\; a = \frac{N}{2} \; . \tag{2.63}$$

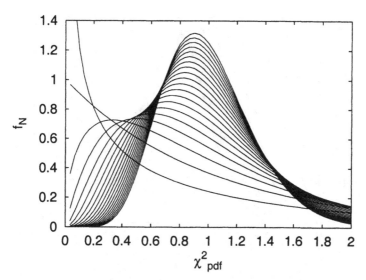

Fig. 2.1 Probability densities (2.63) for the χ^2 per degree of freedom distribution with $N = 1, \ldots, 20$ degrees of freedom, see assignment 2. Picking the curves at $\chi^2_{\text{pdf}} = \chi^2/N = 1$: Increasing $f_N(\chi^2_{\text{pdf}})$ values correspond to increasing $N = 1, 2, \ldots, 20$.

For $N = 1, 2, \ldots, 20$ this probability density is plotted in figure 2.1, and we can see the central limit theorem at work. The distribution function for χ^2 pdf is given by

$$F_N(\chi^2_{\text{pdf}}) = F(N\chi^2_{\text{pdf}}) \tag{2.64}$$

and figure 2.2 depicts the corresponding peaked distribution functions (1.41).

All these functions and the q-tiles x_q (1.34) for the χ^2 distribution are provided as Fortran functions in ForLib. The χ^2 probability density, distribution function, peaked distribution function and q-tiles are, respectively, calculated by the routines chi2_pd.f, chi2_df.f, chi2_qdf.f and chi2_xq.f. For the χ^2 pdf distribution the corresponding Fortran functions are chi2_pdf_pd.f, chi2_pdf_df.f, chi2_pdf_qdf.f and chi2_pdf_xq.f.

As a reference for the correct implementation of the Fortran code, we

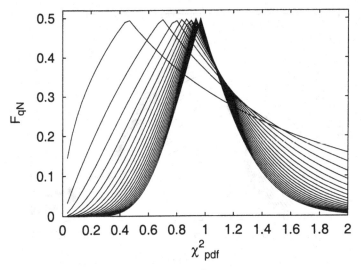

Fig. 2.2 Peaked distribution functions for the χ^2 per degree of freedom distribution with $N = 1, \ldots, 20$ degrees of freedom, see assignment 2. Picking the curves at $\chi^2_{\text{pdf}} = \chi^2/N = 0.4$: Increasing $F_{qN}(\chi^2_{\text{pdf}})$ values correspond to decreasing $N = 20, 19, \ldots, 1$.

collect in table 2.4 values for the χ^2 distribution function (2.60) and associated fractiles (1.34). In the same way table 2.5 presents values for the χ^2 pdf distribution function (2.64) and corresponding fractiles. It is left to assignments 1 and 2 to verify these values and to reproduce figures 2.1 and 2.2.

2.2.3 *Assignments for section 2.2*

(1) Use the program chi2_tabs.f of ForProg/student to verify tables 2.4 and 2.5. Which subroutine calls belong to which table section?

(2) Use chi2pdf_pd.f and chi2pdf_qdf.f to reproduce figures 2.1 and 2.2.

(3) Equation (2.45) relates the χ^2 distribution to the estimator of the sample variance $(s_{\bar{x}}^r)^2$. To illustrate this relationship, use the default seed of Marsaglia's random number generator and generate 10 000 pairs of

Table 2.4 Some values and fractiles (q-tiles) for the χ^2 distribution function (2.60), see assignment 1.

```
CHI2 distribution function
```

CHI2 \ N	1	2	3	4	5	6	7	8
1.00	.683	.393	.199	.090	.037	.014	.005	.002
2.00	.843	.632	.428	.264	.151	.080	.040	.019
3.00	.917	.777	.608	.442	.300	.191	.115	.066
4.00	.954	.865	.739	.594	.451	.323	.220	.143
5.00	.975	.918	.828	.713	.584	.456	.340	.242
6.00	.986	.950	.888	.801	.694	.577	.460	.353
7.00	.992	.970	.928	.864	.779	.679	.571	.463
8.00	.995	.982	.954	.908	.844	.762	.667	.567

```
Fractiles for the CHI2 distribution
```

N \ q	.025	.150	.500	.850	.975
1	.001	.036	.455	2.072	5.024
2	.051	.325	1.386	3.794	7.378
3	.216	.798	2.366	5.317	9.348
4	.484	1.366	3.357	6.745	11.143
5	.831	1.994	4.351	8.115	12.833
6	1.237	2.661	5.348	9.446	14.449
7	1.690	3.358	6.346	10.748	16.013
8	2.180	4.078	7.344	12.027	17.535

random numbers with rmagau.f. For each pair calculate the error bar with steb0.f. Square each of these 10 000 error bars and calculate the $q = 0.15$ fractiles x_q and x_{1-q} of their empirical distribution (rounded the results are $x_{.15} = 0.0347$ and $x_{.85} = 2.0458$). Compare with the fractiles expected from the χ^2 pdf distribution (table 2.5).

2.3 Gosset's Student Distribution

We already encountered the question: What happens with the Gaussian confidence limits when we replace the variance $\sigma_{\bar{x}}^2$ by its estimator (2.14)

Table 2.5 Some values and fractiles (q-tiles) for the χ^2 per degree of freedom distribution function (2.64), see assignment 1.

```
CHI2 per degree of freedom distribution function
```

(CHI2/N) \ N	1	2	3	4	5	6	7	8
.25	.383	.221	.139	.090	.060	.041	.028	.019
.50	.520	.393	.318	.264	.224	.191	.165	.143
.75	.614	.528	.478	.442	.414	.391	.371	.353
1.00	.683	.632	.608	.594	.584	.577	.571	.567
1.25	.736	.713	.710	.713	.717	.723	.729	.735
1.50	.779	.777	.788	.801	.814	.826	.838	.849
1.75	.814	.826	.846	.864	.881	.895	.907	.918
2.00	.843	.865	.888	.908	.925	.938	.949	.958

```
Fractiles for the CHI2 per degree of freedom distribution
```

N \ Q	.025	.150	.500	.850	.975
1	.001	.036	.455	2.072	5.024
2	.025	.163	.693	1.897	3.689
3	.072	.266	.789	1.772	3.116
4	.121	.342	.839	1.686	2.786
5	.166	.399	.870	1.623	2.567
6	.206	.444	.891	1.574	2.408
7	.241	.480	.907	1.535	2.288
8	.272	.510	.918	1.503	2.192

$s_{\overline{x}}^2$ in statements like

$$\frac{|\overline{x} - \widehat{x}|}{\sigma_{\overline{x}}} < 1.96 \text{ with } 95\% \text{ probability.}$$

For sampling from a Gaussian distribution the answer was given by Gosset [Student (1908)], who published his article under the pseudonym *Student*. Following his approach, we study the distribution of the random variable

$$t^r = \frac{\overline{x}^r - \widehat{x}}{s_{\overline{x}}^r} \tag{2.65}$$

where the distribution of $\overline{x}^r$ is Gaussian. The distribution of t^r does not depend on the values of $\widehat{x}$ and $\sigma_{\overline{x}}$. As before (2.46), without restriction of generality we choose $\hat{x} = 0$ and $\sigma_{\bar{x}} = 1$. Further, we introduce $y_N^r = \bar{x}^r \sqrt{N}$ to replace $s_{\bar{x}}^r$ by s_x^r and obtain

$$t^r = \frac{\bar{x}^r}{s_{\bar{x}}^r} = \frac{y_N^r \sqrt{N}}{s_x^r \sqrt{N}} = \frac{y_N^r}{s_x^r} = \frac{y_N^r \sqrt{f}}{\chi_f^r} \tag{2.66}$$

where we use that $(s_x^r)^2$ is given by the χ^2 distribution with $f = N - 1$ degrees of freedom. The **Student distribution function** becomes

$$F(t) = P\left[\frac{y_N^r \sqrt{f}}{\chi_f^r} < t\right] = P\left[y_N^r < c\chi_f^r\right] \quad \text{with} \quad c = \frac{t}{\sqrt{f}}. \tag{2.67}$$

The crucial observation is now that we know from equation (2.56) that y_N^r and $(s_x^r)^2$ are independent random variables. Hence, the probability density of their product $y_N^r(\chi_f^r)^2$ is just the product of the probability densities $g(y) = \exp(-y^2/2)$ of y_N^r (which follows from the definition of y_N^r and equation (1.101) etc.) and $f(\chi^2)$ (2.61) of $(\chi_f^r)^2$. Using for shortness of notation the variable z instead of χ^2, we have

$$g(y) f(z) \sim e^{-\frac{1}{2}y^2} z^{\frac{1}{2}f-1} e^{-\frac{1}{2}z}$$

and the Student distribution function (2.67) has the $y < cz$ integral representation

$$F(t) = \text{const} \int_0^\infty dz\, z^{\frac{1}{2}f-1} e^{-\frac{1}{2}z} \int_{-\infty}^{cz} dy\, e^{-\frac{1}{2}y^2}, \quad c = \frac{t}{\sqrt{f}}.$$

The substitution $y = x\sqrt{z}$ removes the z-dependence of the integration range and

$$F(t) = \text{const} \int_0^\infty dz\, z^{\frac{1}{2}f-1} e^{-\frac{1}{2}z} \int_{-\infty}^c dx\, e^{-\frac{1}{2}x^2 z}$$

$$= \text{const} \int_{-\infty}^c dx \int_0^\infty dz\, z^{\frac{1}{2}f-1} e^{-\frac{1}{2}(1+x^2)z}.$$

Introducing $z' = (1 + x^2)z/2$ as new integration variable, the z-integration leads to the gamma function (A.1) and

$$F(t) = \text{const}\, \Gamma\left(\frac{f+1}{2}\right) \int_{-\infty}^c dx \left(\frac{1+x^2}{2}\right)^{-\frac{f+1}{2}}$$

$$= \frac{\text{const}'}{\sqrt{f}} \int_{-\infty}^t dt' \left(1 + \frac{t'^2}{f}\right)^{-\frac{f+1}{2}}. \tag{2.68}$$

The **probability density of the Student distribution** is therefore

$$f(t) = \frac{\text{const}'}{\sqrt{f}} \left(1 + \frac{t^2}{f}\right)^{-\frac{f+1}{2}} . \qquad (2.69)$$

The fall-off is a power law $|t|^{-f}$ for $|t| \to \infty$, instead of the exponential fall-off of the normal distribution (1.30). The constant follows from the normalization condition $F(\infty) = 1$ to be

$$\text{const}' = \frac{\sqrt{f}}{\int_{-\infty}^{+\infty} dt' \left(1 + \frac{t'^2}{f}\right)^{-\frac{f+1}{2}}} = \frac{1}{\sqrt{f}\, B(1/2, f/2)} , \qquad (2.70)$$

where the substitution $t'^2 = (1-x)f/x$ has been used to relate the integral in the denominator to the beta function (A.9) $B(x, y)$ of appendix A.1. By the same substitution the integral (2.68) becomes related to the incomplete beta function (A.11) $B_I(x; a, b)$ of appendix A.1

$$F(t) = \begin{cases} B_I(x; a, b) \text{ for } t \le 0, \\ 1 - B_I(x; a, b) \text{ for } t > 0, \end{cases} \quad \text{with } x = \frac{f}{f + t^2}, \ a = \frac{f}{2} \text{ and } b = \frac{1}{2}$$

$$(2.71)$$

so that the numerical evaluation of $F(t)$ becomes based on B_I. The Fortran functions provided for the Student probability density, distribution function, peaked distribution function and q-tiles are, respectively, `stud_pd.f`, `stud_df.f`, `stud_qdf.f` and `stud_xq.f`.

Table 2.6 exhibits confidence limits of the Student distribution (2.71)

$$p = p(n) = F(ns) - F(-ns) = 1 - 2F(-ns) \qquad (2.72)$$

where s is the measured standard deviation. The last row of the table gives the corresponding Gaussian results (2.2). For very small samples, say $N \le 4$ we find substantial deviations from the Gaussian confidence levels, in particular on the level of two or more standard deviations. Up to two standard deviations reasonable approximations of Gaussian confidence limits are obtained for $N \ge 16$ data. If desired, the Student distribution function can always be used to calculate the exact confidence limits. Therefore, when the central limit theorem applies, we may use the subroutine `bining.f` (1.113) to bin a larger set of non-Gaussian data into `NBINS` ≥ 16 approximately Gaussian data and, in this way, **reduce the error analysis to Gaussian methods.**

For given confidence limits, chosen to be the Gaussian 1σ to 5σ results, Student confidence intervals are collected in table 2.7. For instance, for

Table 2.6 Student distribution: Confidence probabilities for given confidence intervals, see assignment 1.

N \ S	1.0000	2.0000	3.0000	4.0000	5.0000
2	.50000	.70483	.79517	.84404	.87433
3	.57735	.81650	.90453	.94281	.96225
4	.60900	.86067	.94233	.97199	.98461
5	.62610	.88388	.96006	.98387	.99251
6	.63678	.89806	.96990	.98968	.99590
7	.64408	.90757	.97599	.99288	.99755
8	.64938	.91438	.98006	.99481	.99843
12	.66120	.92920	.98792	.99791	.99960
16	.66683	.93605	.99103	.99884	.99984
24	.67228	.94256	.99361	.99944	.99995
32	.67495	.94567	.99471	.99963	.99998
40	.67653	.94750	.99531	.99973	.99999
48	.67757	.94870	.99569	.99978	.99999
56	.67831	.94955	.99595	.99981	.99999
64	.67886	.95018	.99614	.99983	1.0000
INFINITY:	.68269	.95450	.99730	.99994	1.0000

16 events the $P = 0.954$ Student confidence interval is seen to be about 10% larger ($2.1812\,\sigma$ instead of $2\,\sigma$) than the Gaussian confidence interval. The Fortran function stud_df.f has been used to calculate table 2.6 and stud_xq.f to calculate table 2.7, see assignment 1.

2.3.1 *Student difference test*

This test is a generalization of the Gaussian difference test of section 2.1.3. The Student difference test takes into account that only a finite number of events are sampled. As before it is assumed that the events are drawn from a normal distribution. Let the following data be given

$$\overline{x} \text{ calculated from } M \text{ events, } i.e., \quad \sigma_{\overline{x}}^2 = \sigma_x^2/M \quad (2.73)$$

$$\text{and } \overline{y} \text{ calculated from } N \text{ events, } i.e., \quad \sigma_{\overline{y}}^2 = \sigma_y^2/N \ . \quad (2.74)$$

Unbiased (2.7) estimators of the variances are

$$s_{\overline{x}}^2 = s_x^2/M = \frac{\sum_{i=1}^{M}(x_i - \overline{x})^2}{M\,(M-1)} \text{ and } s_{\overline{y}}^2 = s_y^2/N = \frac{\sum_{j=1}^{N}(y_j - \overline{y})^2}{N\,(N-1)} \ . \quad (2.75)$$

Table 2.7 Student distribution: Confidence intervals for given confidence probabilities, see assignment 1.

N \ P	.682689	.954500	.997300	.999937	.999999
2	1.8373	13.968	235.80	10050.	
3	1.3213	4.5265	19.207	125.64	1320.7
4	1.1969	3.3068	9.2189	32.616	156.68
5	1.1416	2.8693	6.6202	17.448	56.848
6	1.1105	2.6486	5.5071	12.281	31.847
7	1.0906	2.5165	4.9041	9.8442	22.020
8	1.0767	2.4288	4.5300	8.4669	17.103
12	1.0476	2.2549	3.8499	6.2416	10.257
16	1.0345	2.1812	3.5864	5.4804	8.2671
24	1.0222	2.1147	3.3613	4.8767	6.8332
32	1.0164	2.0839	3.2609	4.6216	6.2682
40	1.0130	2.0662	3.2042	4.4813	5.9682
48	1.0108	2.0546	3.1677	4.3926	5.7826
56	1.0092	2.0465	3.1423	4.3314	5.6566
64	1.0080	2.0405	3.1235	4.2867	5.5654
INFINITY:	1.0000	2.0000	3.0000	4.0000	5.0000

As for the Gaussian case (2.28) the random variable of the difference distribution is

$$D = \bar{x} - \bar{y} \quad \text{with} \quad \sigma_D^2 = \sigma_{\bar{x}}^2 + \sigma_{\bar{y}}^2 . \tag{2.76}$$

The expectation value of D is

$$\widehat{D} = 0 \quad \text{because of} \quad \widehat{x} = \widehat{y} . \tag{2.77}$$

Under the **additional assumption** that the true variances are equal

$$\sigma_x^2 = \sigma_y^2 = \sigma^2 \tag{2.78}$$

an estimator s_D^2 of σ_D^2 can be defined, so that the distribution of D/s_D becomes reduced to the Student distribution (2.69).

When equation (2.78) holds, it makes no longer sense to use the estimators (2.75) from partial data. Instead, an unbiased estimator of the variance from all data is (take the expectation value to verify the statement)

$$s^2 = \frac{\sum_{i=1}^{M}(x_i - \bar{x})^2 + \sum_{j=1}^{N}(y_j - \bar{y})^2}{M + N - 2} = \frac{(M-1)\,s_x^2 + (N-1)\,s_y^2}{M + N - 2} . \tag{2.79}$$

Estimators of $\sigma_{\bar{x}}^2$ and $\sigma_{\bar{y}}^2$ from all data are then

$$s_{a,\bar{x}}^2 = \frac{s^2}{M} \quad \text{and} \quad s_{a,\bar{y}}^2 = \frac{s^2}{N} . \tag{2.80}$$

This allows us to define an unbiased estimator of σ_D^2 as

$$s_D^2 = s_{a,\bar{x}}^2 + s_{a,\bar{y}}^2 = \left(\frac{1}{M} + \frac{1}{N} \right) s^2 . \tag{2.81}$$

Following [Student (1908)] and [Fisher (1926)] we consider the distribution of the random variable

$$t = \frac{D}{s_D} = \frac{\bar{x} - \bar{y}}{s_D} . \tag{2.82}$$

Note that $(M + N - 2)s^2$ (2.79) is a χ^2 distribution with $f = M + N - 2$ degrees of freedom, because

$$\chi_x^2 = (M - 1)\, s_x^2 \quad \text{and} \quad \chi_y^2 = (N - 1)\, s_y^2$$

are χ^2 distributions with $M - 1$ and $N - 1$ degrees of freedom. From the definition of the χ^2 distribution (2.43) it follows that the sum

$$\chi^2 = \chi_x^2 + \chi_y^2$$

is a χ^2 distribution with $f = M + N - 2$ degrees of freedom and the relation

$$s_D^2 = \sigma_D^2 \, \frac{\chi^2}{f}$$

holds in the sense of random variables. The mean of s^2 is σ_D^2, the mean of the χ^2 distribution is f (2.62), and the fluctuations match too.

For t defined by equation (2.82) we find

$$t = y \sqrt{\frac{f}{\chi^2}} \quad \text{with} \quad y = \frac{D}{\sigma_D}, \; f = M + N - 2 . \tag{2.83}$$

This is identical to the situation of equation (2.66) and the discussion which leads to the Student distribution applies again. Thus, the quotient (2.82) is Student distributed with $f = M + N - 2$ degrees of freedom and the probability

$$P(|\bar{x} - \bar{y}| > d) \tag{2.84}$$

is determined by the Student distribution function (2.71) in the same way as the probability (2.31) is determined by the normal distribution. The likelihood that the observed difference is due to chance, compare its definition (2.32), becomes

$$Q = 2\,F(t) \,. \tag{2.85}$$

The Student distribution function $F(t)$ is implemented by the routine stud_df.f. To perform the Student difference test interactively the program

$$\text{stud\_dif.f} \tag{2.86}$$

is provided in ForProc/stud_dif. It returns the probability that the difference between $\bar{x}$ and $\bar{y}$ is due to chance. Besides the mean values and their error bars now also the numbers of measurements on which the estimates rely are required on input. The program stud_dif.f relies on a call to

$$\text{STUDDIF(XM1, EB1, NDAT1, XM2, EB2, NDAT2, Q)} \tag{2.87}$$

which is provided as studdif.f in ForLib.

Table 2.8 Numerical examples for the Student difference test: $\bar{x}_1 = 1.00 \pm 0.05$ from M data and $\bar{x}_2 = 1.20 \pm 0.05$ from N data. Compare table 2.2 for the corresponding Gaussian difference test.

M	512	64	16	16	8	16	4	3	2
N	512	64	16	8	8	4	4	3	2
Q	0.0048	0.0054	0.0083	0.020	0.023	0.072	0.030	0.047	0.11

Example applications of stud_dif.f are given in table 2.8. They use the mean values and error bars for which the Gaussian difference test reported in table 2.2 gives $Q = 0.0047$. To the extent that the Gaussian difference test is rightfully applied, there is almost no likelihood that their difference is due to chance. The dependence of this statement on some numbers of generated data is displayed in table 2.8. For $M = N = 512$ the Q value is practically identical with the Gaussian result, for $M = N = 16$ it has almost doubled and for $M = N = 8$ it is more than four times larger. Acceptable likelihoods, above a 5% cut-off, are only obtained for $M = N = 2$ (11%) and for $M = 16$, $N = 4$ (7%). At the first glance the latter result is a surprise, because its Q value is smaller than for $M = N = 4$. One might have expected that the result from $16 + 4 = 20$ data is closer to the Gaussian limit than the result from only $4 + 4 = 8$ data. The explanation is that for $M = 16$, $N = 4$ data one would expect the $N = 4$ error bar to be two times larger than the $M = 16$ error bar, whereas the error bars used are identical. The averaging procedure (2.79), based on the crucial assumption (2.78), leads to a $s_{a,\bar{y}}^2(N = 4)$ estimate, which is larger than the measured error bar $s_{\bar{y}}^2$, thus explaining the 7% value.

The last issue leads to the question: Under the assumption that data are sampled from the same normal distribution, when are two measured error bars consistent? The answer is given by the F-test (also known as variance ratio test) discussed in section 2.5.

2.3.2   *Assignments for section 2.3*

(1) Use stud_tabs.f of ForProg to reproduce tables 2.6 and 2.7. Name the subroutines on which the results are based.
(2) Use stud_pd.f to plot the Student probability densities (2.69) for $f = 1, \ldots, 7$ degrees of freedom and compare them with the Gaussian probability density (1.30). Use stud_qdf.f and gau_qdf.f to create the same plot for the peaked distribution functions.
(3) Use the procedure stud_dif.f to reproduce table 2.8.

2.4   The Error of the Error Bar

We already emphasized in section 2.2 that the error of the error bar is determined by the χ^2 distribution. One may work with error bars or with variances. **For normally distributed data** the interesting point is that the number of data alone determines the errors of these quantities exactly, *i.e.*, **one does not have to rely on estimators.** This is a consequence of equation (2.45). Only the number of data N enters in the relation

$$\frac{(s_x^r)^2}{\sigma_x^2} = \frac{(\chi_{N-1}^r)^2}{N - 1} \tag{2.88}$$

which determines the distribution of $(s_x^r)^2$ relative to its true value σ_x^2. The χ^2 pdf distribution function (2.64) yields q-tiles (1.34) so that

$$\frac{(s_x^r)^2}{\sigma_x^2} \leq \chi_{\text{pdf},q}^2 \quad \text{with probability} \quad q \,. \tag{2.89}$$

Here $\chi_{\text{pdf},q}^2$ depends on the number of degrees of freedom, $N - 1$. Some $\chi_{\text{pdf},q}^2$ q-tiles are already displayed in table 2.5. The situation in practice is that an estimate s_x^2 of σ_x^2 exists. For a given probability p, say $p = 0.95$, we would like to know a confidence interval $[s_{min}^2 < s_x^2 < s_{max}^2]$ so that

$$\sigma_x^2 \in [s_{min}^2, s_{max}^2] \text{ with probability } p. \tag{2.90}$$

Equation (2.89) implies

$$\frac{s_x^2}{\chi_{pdf,q}^2} \le \sigma_x^2 \text{ with probability } q. \tag{2.91}$$

Choosing $q = (1 - p)/2$ the desired confidence interval becomes[4]

$$s_{min}^2 = \frac{s_x^2}{\chi_{pdf,(1-q)}^2} \le \sigma_x^2 \le \frac{s_x^2}{\chi_{pdf,q}^2} = s_{max}^2. \tag{2.92}$$

Namely, the probability $\sigma_x^2 \le s_x^2/\chi_{pdf,(1-q)}^2$ is $1 - (1 - q) = q$ and the probability $\sigma_x^2 \ge s_x^2/\chi_{pdf,q}^2$ is q, *i.e.*, the likelihood for σ_x^2 to be inside the confidence interval is $1 - 2q = p$.

Table 2.9 The confidence intervals (2.92), s_{min}^2 and s_{max}^2, for variance estimates $s_x^2 = 1$ from NDAT data, see assignment 1.

CONFIDENCE INTERVALS FOR VARIANCE ESTIMATES FROM NDAT DATA.

		q	q	q	1-q	1-q
NDAT=2**K		.025	.150	.500	.850	.975
2	1	.199	.483	2.198	27.960	1018.255
4	2	.321	.564	1.268	3.760	13.902
8	3	.437	.651	1.103	2.084	4.142
16	4	.546	.728	1.046	1.579	2.395
32	5	.643	.792	1.022	1.349	1.768
64	6	.726	.844	1.011	1.224	1.467
128	7	.793	.885	1.005	1.149	1.300
256	8	.847	.916	1.003	1.101	1.199
512	9	.888	.939	1.001	1.069	1.135
1024	10	.919	.956	1.001	1.048	1.093
2048	11	.941	.969	1.000	1.033	1.064
4096	12	.958	.978	1.000	1.023	1.045
8192	13	.970	.984	1.000	1.016	1.031
16384	14	.979	.989	1.000	1.012	1.022

---

[4] Alternatively, one may consider to determine s_{min}^2 and s_{max}^2 values so that, for a given p, $s_{max}^2 - s_{min}^2 = $ Minimum. Normally it seems not worthwhile to go through the extra effort and one sticks with the symmetric choice (2.92).

Table 2.10 Difficulties encountered when combining binned data with `steb2.f` and using estimated variances as weights, see assignment 3.

Binsize:		2	3	4	8	16	32
Data:	19200	9600	6400	4800	2400	1200	600
Means:	.4988	.6915	.5555	.5039	.4948	.4983	.4983
Errors:	.0021	.0000	.0007	.0013	.0019	.0020	.0021

For $p = 0.7$, $p = 0.95$ and the median $s_x^2/\chi^2_{\text{pdf},0.5}$ table 2.9 displays some σ_x^2 confidence bounds as defined by equation (2.91) for $s_x^2 = 1$. The number of data points is $N = 2^K$ with $K = 1, 2, \ldots, 14$. The subroutine `sebar_e.f` of ForLib returns the upper and lower confidence limits of error bars. Its use is a follows: On input the number of data N and the desired confidence level PC have to be provided. A call to

$$\text{SEBAR\_E(N, PC, EBUP, EBDO)} \qquad (2.93)$$

returns EBUP and EBDO, the upper and the lower bounds which contains the true error bar with likelihood PC, assuming the estimated error bar is one. If the estimated error bar is EB, the probability that the true error bar is larger than EB*EBUP is (1-PC)/2 as is the probability that the true error bar is smaller than EB*EBDO, *i.e.*,

$$\text{EB} * \text{EBDO} \leq \text{EB} \leq \text{EB} * \text{EBUP} \quad \text{with probability } PC. \qquad (2.94)$$

For large N the χ^2 pdf distribution approaches the Gaussian distribution (see figure 2.1) and we can use `gau_xq.f` (1.37) instead of `chi2pdf_xq.f`. This is implemented in the subroutine `sebar_e_as.f`, which has the same arguments as `sebar_e.f`. For $N \geq 17\,000$ the asymptotic subroutine is used inside `sebar_e.f`, because the Gamma function part of the exact calculation runs for large N into numerical problems. Convergence to the asymptotic relation is slow and its use is only recommended for $N > 16\,000$, see assignment 2.

We notice from table 2.9 that for small data sets the fluctuations of the error bar are large. At the 95% confidence level one order of magnitude is exceeded for $N = 4$. For $N = 16$ we are still dealing with a factor larger than two. Interestingly, these large fluctuation of the variance allow for the confidence levels of the Student distribution, which are already almost Gaussian for $N = 16$, see table 2.7. The fluctuations average kind of out as far as confidence intervals are concerned.

However, when weighting data with the inverse of their estimated variance (2.23), to combine them with the subroutine `steb2.f` (2.25), disaster can happen due to large fluctuations of the estimators. An example is given in table 2.10. There, 19 200 uniformly distributed random numbers are generated and the first column contains mean values and error bars as obtained by applying the subroutine `steb0.f` (2.17) to all 19 200 data. The result 0.4988 ± 0.0021 is consistent with the true value $\hat{x} = 0.5$. For the subsequent columns the same 19 200 data are binned (1.113) and, relying on the estimated variances (2.23), the binned data are combined with `steb2.f`. Up to binsize ≤ 8 the results are quite disastrous. For binsize ≤ 3 the error is severely underestimated and up to binsize 8 an artificial two σ discrepancy with the true mean value is produced.

On the 95% confidence level one needs about 1 000 statistically independent Gaussian data to determine the true variance σ_x^2 with an accuracy of approximately $\pm10\%$. Compare the values of table 2.9 for $q = 0.025$, $q = 0.975$ and NDAT=1024. When we deal with non-Gaussian data for which the central limit theorem applies, we need even more values: First, we have to bin them to get an approximately normal distribution. Then, we still need 1 000 bins to estimate the variance with an accuracy of 10% at the 95% confidence level.

To estimate the error of the variance becomes particularly important and difficult for data from a Markov chain time series, with which we deal in the subsequent chapters. To compare the efficiency of different Markov chain Monte Carlo algorithms, binning is may be used to reduce the autocorrelations between the data points to a tolerable level. Depending on the autocorrelation time, the number of (original) data needed can be very large, see chapter 4. The algorithm with the smaller error bars wins. To find out, whether two algorithms differ in a statistically significant way, may then be decided by the variance ratio test discussed next.

2.4.1   *Assignments for section 2.4*

(1) Use the definitions (2.92) and `chi2pdf_xq.f` from ForLib to reproduce the $s_{\min}^2$ and $s_{\max}^2$ values of table 2.9.
(2) Use 95% confidence limits and the NDAT values of table 2.9 to compare results of the subroutine `sebar_e.f` with its asymptotic approximation implemented by the subroutine `sebar_e_as.f`.
(3) Combination of data from small samples: Use the Marsaglia random number generator with the standard seed and generate 19 200 random

numbers. Group the time series data into bins of 2 to 32 subsequent events and estimate the variance of each bin with `steb0.f`. Use the estimated variances as weights, see equation (2.23), and combine the bins with `steb2.f` to find the results of table 2.10. Try to understand what happens when (far) more than 19 200 random numbers are used, while the binsizes are kept as before.

2.5 Variance Ratio Test (F-test)

We assume that two sets of normal data are given together with estimates of their variances $\sigma_{x_1}^2$ and $\sigma_{x_2}^2$:

$$\left(s_{x_1}^2, N_1\right) \quad \text{and} \quad \left(s_{x_2}^2, N_2\right) .\tag{2.95}$$

We would like to test whether the ratio

$$F = \frac{s_{x_1}^2}{s_{x_2}^2}\tag{2.96}$$

differs from $F = 1$ in a statistically significant way. The notation F for the ratio (2.96) is a bit unfortunate, because $F(x)$ is already used for distribution functions. However, it seems not to be a good idea to change the established convention. The F-test investigates the distribution of (2.96) under the assumption $\sigma_{x_1}^2 = \sigma_{x_2}^2$. The assumption is rejected when it leads to a too small likelihood. Let $f_1 = N_1 - 1$ and $f_2 = N_2 - 1$, equation (2.88) relates the χ^2 distribution to our estimators:

$$\chi_{f_1}^2 = \frac{f_1 \, s_{x_1}^2}{\sigma_{x_1}^2} \quad \text{and} \quad \chi_{f_2}^2 = \frac{f_2 \, s_{x_2}^2}{\sigma_{x_2}^2} .$$

As $\sigma_{x_1}^2 = \sigma_{x_2}^2$ holds, we find

$$\frac{\chi_{f_1}^2}{\chi_{f_2}^2} = \frac{f_1}{f_2} F .\tag{2.97}$$

Up to a factor, the F distribution is given by the ratio of two χ^2 distributions. The probability densities for the distributions $\chi_{f_i}^2$, $i = 1, 2$ are (2.61)

$$g_i(t) = c_i \, t^{\frac{1}{2}f_i - 1} e^{-\frac{1}{2}t} \quad \text{with} \quad c_i = \Gamma\left(\frac{1}{2}f_i\right)^{-1} 2^{-\frac{1}{2}f_i} .\tag{2.98}$$

Therefore, the probability

$$\frac{\chi^2_{f_1}}{\chi^2_{f_2}} < w \tag{2.99}$$

is given by

$$H(w) = c_1 c_2 \int \int t^{\frac{1}{2}f_1 - 1} e^{-\frac{1}{2}t} u^{\frac{1}{2}f_2 - 1} e^{-\frac{1}{2}u} \, dt \, du \tag{2.100}$$

$$\text{with } t > 0, \ u > 0 \text{ and } t < uw,$$

where the conditions on t and u enforce the integration range corresponding to (2.99). This may be re-written as

$$H(w) = c_1 c_2 \int_0^\infty du \int_0^{uw} dt \, t^{\frac{1}{2}f_1 - 1} u^{\frac{1}{2}f_2 - 1} e^{-\frac{1}{2}t - \frac{1}{2}u} \tag{2.101}$$

$$= c_1 c_2 \int_0^\infty du \int_0^w dy \, u^{\frac{1}{2}f - 1} y^{\frac{1}{2}f_1 - 1} e^{-\frac{1}{2}uy - \frac{1}{2}u} \quad \text{with } f = f_1 + f_2 \,,$$

where the last equal sign relies on the substitution $t = uy$. The u-integration can now be carried out and gives

$$\int_0^\infty du \, u^{\frac{1}{2}f - 1} e^{-\frac{y+1}{2}u} = \left(\frac{2}{y+1}\right)^{\frac{1}{2}f} \Gamma\left(\frac{1}{2}f\right) \,.$$

Finally, we obtain

$$H(w) = c \int_0^w dy \, y^{\frac{1}{2}f_1 - 1} (y + 1)^{-\frac{1}{2}f} \tag{2.102}$$

$$\text{with } c = \Gamma\left(\frac{1}{2}f_1\right)^{-1} \Gamma\left(\frac{1}{2}f_2\right)^{-1} \Gamma\left(\frac{1}{2}f\right)$$

from which with $y = f_1 F/f_2$ the probability density of the F-ratio (2.96) is read off. The integral is solved by the incomplete beta function (A.11) of appendix A.1, which we already encountered in connection with the Student distribution function (2.71)

$$H(w) = 1 - B_I\left(\frac{1}{w+1}, \frac{1}{2}f_2, \frac{1}{2}f_1\right) \,. \tag{2.103}$$

Note that the median of this distribution is at $F = 1$. Based on equations (2.102) and (2.103) we can now program the F-ratio (2.96) probability density, distribution function, peaked distribution function and q-tiles.

Following our previously established notation, the corresponding Fortran routines in ForLib are F_pd.f, F_df.f, F_qdf.f and F_xq.f. For instance, the F-ratio distribution function is

$$F\text{\_df}(F) = H\left(\frac{f_1}{f_2} F\right) .$$

(2.104)

The likelihood that the observed difference is due to chance becomes

$$Q = \begin{cases} 2\,F\text{\_df}(F) \text{ for } F \leq 1, \\ 2 - 2\,F\text{\_df}(F) \text{ for } F > 1. \end{cases}$$

(2.105)

But for the asymmetry of F_df(F) this is as for the Gaussian (2.32) and the Student (2.85) difference test. For probability q the fractile of F is $F_q = $ F_xq(q) and a confidence interval of probability p, $p > 0.5$ is given by

$$F_q \leq F \leq F_{1-q} \text{ with } q = \frac{1}{2}(1-p) .$$

(2.106)

To perform the F-ratio test interactively, the program

$$F\text{\_test.f}$$

(2.107)

is provided in ForProc/F_test. It relies on a subroutine call to

$$FTEST(EB1, NDAT1, EB2, NDAT2, Q)$$

(2.108)

of ForLib. **Error bars** and not variances **are used as arguments.** Examples are collected in table 2.11. The first three columns of this table should be compared with the $M = 16$ and $N = 16, 8, 4$ results of the Student difference tests of table 2.8: The low statistics of 16 versus 4 data does not indicate conclusively that the error bars are inconsistent, but with 64 versus 16 data of the fifth column the inconsistency has become obvious. High accuracy determinations of the error bars are tedious: Columns six and seven show that more than 1 000 data are needed to detect a discrepancy of 5%. The remaining columns deal with an error bar factor of two, which can be caught with only 16 data.

It makes sense to combine the Student difference test and the F-test into one interactive procedure. This is done by the program

$$F\text{\_stud.f}$$

(2.109)

of ForProc/F_stud, which relies on the subroutines (2.87) and (2.108).

Table 2.11 Numerical examples for the F-test, see assignment 2.

$\triangle\bar{x}_1$	1.0	1.0	1.0	1.0	1.0	1.0	1.0	1.0	1.0	1.0
N_1	16	16	16	32	64	1024	2048	32	1024	16
$\triangle\bar{x}_2$	1.0	1.0	1.0	1.0	1.0	1.05	1.05	2.0	2.0	2.0
N_2	16	8	4	8	16	1024	2048	8	256	16
Q	1.0	0.36	0.28	0.06	0.005	0.12	0.027	0.90	0.98	0.01

2.5.1   *F ratio confidence limits*

A situation of practical importance is that an estimator $\overline{F}$ for the F-ratio (2.96) exits, which relies on NDA1 data for the numerator and NDA2 data for the denominator. Similarly as discussed after equation (2.90) for the error of the error bar, we would like to determine a confidence interval so that $F \in [F_{\min}, F_{\max}]$ with probability p. We define $q = (1 - p)/2$ and find

$$F_{\min} = \frac{\overline{F}}{\text{F\_XQ}(1 - q)} \leq F \leq \frac{\overline{F}}{\text{F\_XQ}(q)} = F_{\max} \qquad (2.110)$$

where we use our Fortran function notation for the quantiles of the F distribution.

Table 2.12 Some confidence intervals (2.110), $F_{\min}(q)$ and $F_{\max}(1 - q)$, for F ratio estimates from NDA1=1024 data for the numerator, NDA2 data for the denominator, and $\overline{F} = 1$, see assignment 3.

SOME CONFIDENCE INTERVALS FOR F RATIO ESTIMATES.

NDA2	q .025	q .150	1-q .850	1-q .975
64	0.676	0.813	1.192	1.396
128	0.759	0.865	1.140	1.282
256	0.819	0.900	1.105	1.208
512	0.859	0.923	1.082	1.160
1024	0.885	0.937	1.067	1.130
2048	0.900	0.946	1.058	1.113
4096	0.909	0.951	1.053	1.103
8192	0.914	0.953	1.051	1.098
16384	0.916	0.955	1.049	1.095

To give an example, we show results for $\overline{F} = 1$, NDA1=1024 and various values of NDA2 in table 2.12. Note that for NDA2 $\gg$ NDA1 the NDAT=1024 confidence intervals of the variance estimates of table 2.9 are approached, because in this limit the fluctuations of the denominator become negligible compared to the fluctuations of the numerator. An application, discussed in chapter 4.1.2, are the errors of integrated autocorrelation time estimates from a binning procedure.

2.5.2 Assignments for section 2.5

(1) Plot the F-ratio probability densities and peaked distribution functions for the following degrees of freedom: $(f_1 = 16, f_2 = 16)$, $(f_1 = 32, f_2 = 32)$ and $(f_1 = 32, f_2 = 16)$. This is similar to figures 2.1 and 2.2 for the χ^2 pdf distribution, see assignment 2 of subsection 2.2.3. Plot also the F-ratio distribution function for the same degrees of freedoms and test the function F_xq.f.
(2) F-test: Reproduce the numbers of table 2.11.
(3) Use equation (2.110) and F_xq.f from ForLib to reproduce the $F_{\min}$ and $F_{\max}$ values of table 2.12.

2.6 When are Distributions Consistent?

There are many applications in which one would like to compare data either to an exactly known distribution or two data sets with one another. We review two tests for these situations. The χ^2 test is well suited for discrete distributions and can, to some extent, also be used for histogrammed data of continuous distributions. The Kolmogorov test applies directly to continuous distributions when all data are sorted. For the one-sided Kolmogorov test one considers the positive (or the negative) deviation of the empirical distribution function from the exact one. The two-sided test uses the absolute value of the maximum difference between the exact and the sampled distribution functions. We extend the two-sided test also to the comparison of empirical data sets.

2.6.1 χ^2 Test

Suppose we have events with n distinct outcomes, which are collected in a histogram of n entries. Let n_i be the number of events recorded in the ith

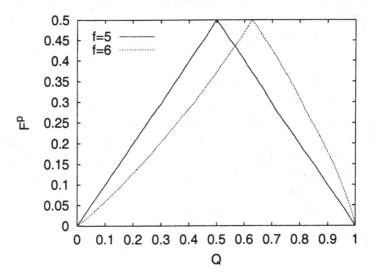

Fig. 2.3 χ^2 test for a dice: Relying on 10 000 experiments, each performing 1 000 throws of a dice, the peaked Q-distribution from the χ^2 distribution with $f = 5$ degrees of freedom is compared with the one from $f = 6$ degrees of freedom. It is seen that $f = 6$ overestimates the Q probability (2.112).

histogram entry and x_i be the number expected according to some known distribution (note that the n_i are integers, while the x_i are in general not). The χ^2 **statistic** is

$$\chi^2 = \sum_{i=1}^{n} \frac{(n_i - x_i)^2}{x_i} \tag{2.111}$$

where the sum is over the histogram entries. The terms in the sum are not individually normal. However, when either the number of histogram entries is large ($n \gg 1$), or the number of events for each entry is large ($n_i \gg 1$), then the χ^2 distribution (2.43) provides a good approximation to (2.111).

Assume we know the probability p_i for each event, then the x_i are given by $x_i = N p_i$, where n is the total number of data. The χ^2 distribution with $f = n - 1$ **degrees of freedom** has to be used, because we have the relation

$$N = \sum_{i=1}^{n} n_i = \sum_{i=1}^{n} x_i$$

between the sampled events and their predictions. The probability that the difference between the data and the prediction is due to chance is then

$$Q = 1 - \text{CHI2\_DF(CHI2)} , \qquad (2.112)$$

see subsection 2.2.2. In assignment 1 this is illustrated for a dice ($p_i = 1/6$, $i = 1, \ldots, n = 6$). Figure 2.3 shows the peaked Q-distribution function from 10 000 experiments, where in each experiment the dice is thrown 1 000 times. This distribution is compared with a false one, obtained by using $f = n$ instead of $f = n - 1$ as the number of degrees of freedom. We see that $f = n$ leads to too high Q values.

If the event probabilities p_i are not exactly known, but come from a model with adjustable parameters, the number of degrees of freedom has to be reduced by one for each adjustable parameter. Let us further point out, that there are situations where the total number of events fluctuates. For instance, when the probability to generate an event per time unit is known and the experiment runs over a certain time. In such a situation $f = n$ has to be used.

The χ^2 test can also be applied to compare two histogrammed data sets, both having n entries. Let $n_{1,i}$ be the number of events in entry i for the first data set and $n_{2,i}$ be the number of events in entry i for the second data set. When the total numbers of events agree for both data sets, the χ^2 statistic becomes

$$\chi^2 = \sum_{i=1}^{n} \frac{(n_i^1 - n_i^2)^2}{n_i^1 + n_i^2} . \qquad (2.113)$$

Again, $f = n - 1$ has to be used for the number of degrees of freedom. Note that the denominator of (2.113) is not the average of $n_{1,i}$ and $n_{2,i}$, but their sum. The reason is that the variance of the difference of two normal quantities is the sum of their individual variances. Entries i With $n_{1,i} + n_{2,i} = 0$ have to be omitted.

If the total numbers of events do not agree, equation (2.113) has to be rescaled to

$$\chi^2 = \sum_{i=1}^{n} \frac{(n_i^1 - n_i^2 N_1/N_2)^2}{n_i^1 + n_i^2 (N_1/N_2)^2} \qquad (2.114)$$

where N_1 is the total number of events in data set one and N_2 the total number of events in data set two. For the dice tests are performed in assignment 2.

2.6.2   The one-sided Kolmogorov test

An exact theory for the question how much the empirical cumulative distribution function (1.45) can differ from the true one was developed by [Kolmogorov (1933)]. The **one-sided Kolmogorov test** considers either the positive or the negative deviation of a sample from the exact distribution and, due to [Birnbaum and Tingey (1951-1953)], rigorous analytical results exist for all sample sizes. In the presentation we follow, in part, [Van der Waerden (1969)]. Let us assume that the data are sorted so that

$$x_1 \leq x_2 \leq \cdots \leq x_n \qquad (2.115)$$

is the order statistic (1.42). First, we refine the definition (1.45) of the empirical cumulative distribution function $\overline{F}(x)$ of chapter 1. Let $x_0 = -\infty$ and $x_{n+1} = +\infty$, we introduce

$$\overline{F}_1(x) \leq \overline{F}_2(x) = \overline{F}(x)$$

where

$$\overline{F}_1(x) = \frac{i}{n} \text{ for } x_i < x \leq x_{i+1}, \ i = 0, 1, \ldots, n, n+1 \qquad (2.116)$$

and

$$\overline{F}_2(x) = \overline{F}(x) = \frac{i}{n} \text{ for } x_i \leq x < x_{i+1}, \ i = 0, 1, \ldots, n, n+1 \qquad (2.117)$$

is the previous definition (1.45). The functions $\overline{F}_1(x)$ and $\overline{F}_2(x)$ differ only at the points x_i, but for small values of n the one-sided Kolmogorov test is sensitive to this. In the following we examine the positive deviations $F(x) - \overline{F}_1(x)$. Afterwards, it is easy to see that all arguments apply as well to the positive deviations $\overline{F}_2(x) - \overline{F}(x)$. Let us define

$$\triangle \ = \ \triangle_1 \ = \max_x \left\{ F(x) - \overline{F}_1(x) \right\} . \qquad (2.118)$$

A continuous monotone transformation of the x-axis leaves the difference $F(x) - \overline{F}_1(x)$ unchanged. Under the (not very restrictive) assumption that the distribution function $F(x)$ is itself continuous

$$x' \ = \ F(x) \qquad (2.119)$$

constitutes such a transformation. Let F' be the cumulative distribution function of the new variable x' and $\overline{F'}_i$, $(i = 1, 2)$ be the corresponding empirical distribution functions, we have then

$$F(x) = F'(x') = F'[F(x)] \quad \text{and} \quad \overline{F}_i(x) = \overline{F'}_i(x') = \overline{F'}_i[\overline{F}_i(x)], \ (i = 1, 2) \ .$$
$$(2.120)$$

This means, the mapping (2.119) is simply to the uniform distribution function.

For the one-sided Kolmogorov test one computes the probability Q that Δ of equation (2.118) exceeds a bound, say $\epsilon \geq 0$. We have $Q = 1$ for $\epsilon = 0$. All cumulative distribution functions agree for the smallest allowable value of x (where they are zero) and the largest allowable value of x (where they are one). Of interest are then the $Q < 1$ values for $\epsilon > 0$. The random variables underlying the $x'_1, \ldots, x'_n$ data are all assumed to be independent and each of them is distributed according to the uniform probability density (1.8). We have

$$Q = \int_0^1 \cdots \int_0^1 dx'_1 \, dx'_2 \ldots dx'_n \quad \text{with the constraint} \ \Delta > \epsilon \qquad (2.121)$$

or

$$Q = n! \, q \quad \text{where} \qquad (2.122)$$

$$q = \int_0^{x'_2} dx'_1 \int_0^{x'_3} dx'_2 \cdots \int_0^{x'_{n-1}} dx'_{n-1} \int_0^1 dx'_n \quad \text{with the constraint} \ \Delta > \epsilon \ .$$
$$(2.123)$$

$F(x)$ is monotonically increasing and at each point x'_i the function $\overline{F'}_1(x')$ jumps from $(i-1)/n$ to i/n. It is clear that the maximum Δ will be attained at one of these points, say x'_{i_m}, and is

$$\Delta = x'_{i_m} - (i_m - 1)/n \ . \qquad (2.124)$$

Therefore, the event $[\Delta > \epsilon]$ occurs whenever one of the differences $x'_i - (i-1)/n$ is larger than ϵ. Let the probability that this event occurs for an index j, but not for $i < j$, be q_j. For distinct j these events are mutually exclusive and the probability q is

$$q = q_1 + q_2 + \cdots + q_n \qquad (2.125)$$

so that we only have to calculate the probabilities q_j. By its definition q_j

is the probability of the event

$$[\ 0 < x_1 < \cdots < x_n < 1;\ x_j - (j-1)/n > \epsilon;\ x_i - (i-1)/n \le \epsilon \text{ for } i < j\ ].$$
(2.126)

For $j = 1$ it becomes

$$[\ 0 < x_1 < x_2 < \cdots < x_n < 1;\ x_1 > \epsilon;\ ] \tag{2.127}$$

so that all the x_i, $(i = 1, \ldots, n)$ lie between ϵ and one. The probability that this happens is $(1 - \epsilon)^n$ and, therefore,

$$q_1 = \frac{1}{n!}(1 - \epsilon)^n . \tag{2.128}$$

Now, let $j = k+1 > 1$. The inequalities in (2.126) split into those containing $x_1, \ldots, x_k$

$$[\ 0 < x_1 < x_2 < \cdots < x_k < 1;\ x_i \le \epsilon + (i-1)/n \text{ for } i = 1, \ldots, k\ ] \tag{2.129}$$

and those containing $x_j, \ldots, x_n$

$$[\ 0 < x_j < x_{j+1} < \cdots < x_n < 1;\ x_j > \epsilon + (j-1)/n\ ] . \tag{2.130}$$

The $x_j, \ldots, x_n$ are independent of the $x_1, \ldots, x_k$ and the probability of the event (2.126) is the product of the probabilities of the events (2.129) and (2.130), denoted by p_k and r_k, respectively

$$q_j = q_{k+1} = p_k\, r_k . \tag{2.131}$$

The probability r_k follows along the same lines which led to (2.128) for q_1 and

$$r_k = \frac{1}{(n-k)!}\left(1 - \epsilon - \frac{k}{n}\right)^{n-k} . \tag{2.132}$$

Here it is assumed that $1 - \epsilon - k/n$ is positive. If this is not the case, the event (2.130) is impossible and r_k becomes zero. The probability p_k of the event (2.129) is

$$p_k = \int_0^{\epsilon} dx_1 \int_{x_1}^{\epsilon+1/n} dx_2 \int_{x_2}^{\epsilon+2/n} dx_3 \ldots \int_{x_{k-1}}^{\epsilon+(k-1)/n} dx_k . \tag{2.133}$$

For $k = 1$ we find easily $p_1 = \epsilon$. Calculating p_2 and p_3 explicitly, we are led to the conjecture

$$p_k = \frac{\epsilon}{k!} \left(\epsilon + \frac{k}{n} \right)^{k-1} . \tag{2.134}$$

This equation will now be proven by induction on k. For $k + 1$ equation (2.133) reads

$$p_{k+1} = \int_0^\epsilon dx_1 \int_{x_1}^{\epsilon+1/n} dx_2 \int_{x_2}^{\epsilon+2/n} dx_3 \ldots \int_{x_k}^{\epsilon+k/n} dx_{k+1} . \tag{2.135}$$

Using the transformation $y_{i-1} = -x_1 + x_i$ we substitute new variables $y_1, y_2, \ldots, y_k$ for $x_2, x_3, \ldots, y_j$ and find

$$p_{k+1} = \int_0^\epsilon dx_1 \, I_k \quad \text{with} \tag{2.136}$$

$$I_k = \int_0^{-x_1+\epsilon+1/n} dy_1 \int_{y_1}^{-x_1+\epsilon+2/n} dy_2 \ldots \int_{y_{k-1}}^{-x_1+\epsilon+k/n} dy_k .$$

We define now $\epsilon' = -x_1 + \epsilon + 1/n$ and see that the integral I_k has the same form as p_k in (2.133) with ϵ replaced by ϵ'. Therefore, according to the induction hypothesis

$$I_k = \frac{\epsilon'}{k!} \left(\epsilon' + \frac{k}{n} \right)^{k-1} = \frac{-x_1 + \epsilon + 1/n}{k!} \left(-x_1 + \epsilon + \frac{(k+1)}{n} \right)^{k-1} . \tag{2.137}$$

We substitute this into equation (2.136) and perform the integration. The result

$$p_{k+1} = \frac{\epsilon}{(k+1)!} \left(\epsilon + \frac{k+1}{n} \right)^k$$

proofs the induction hypothesis for (2.134). We find q_j by substituting the result (2.132) for r_k and (2.134) for p_k into equation (2.131):

$$q_j = \frac{\epsilon}{k! \, (n-k)!} \left(\epsilon + \frac{k}{n} \right)^{k-1} \left(1 - \epsilon - \frac{k}{n} \right)^{n-k} , \quad j = 1, \ldots, n . \tag{2.138}$$

From (2.125) we obtain q and from (2.122) the Kolmogorov probability Q

as first derived by [Birnbaum and Tingey (1951-1953)]

$$Q = \sum_{k=0}^{K} \binom{n}{k} \epsilon \left(\epsilon + \frac{k}{n}\right)^{k-1} \left(1 - \epsilon - \frac{k}{n}\right)^{n-k} . \qquad (2.139)$$

Here the upper summation limit K is determined by the condition that $1 - \epsilon - k/n$ may not be negative

$$K = \mathtt{INT}\,[n(1 - \epsilon)] \qquad (2.140)$$

where INT is the Fortran INT function. For large n it is simpler to use instead of (2.139) the asymptotic expansion

$$Q \approx e^{-2n\epsilon^2} \qquad (2.141)$$

which was derived by [Smirnov (1939)].

Numerical demonstrations

The subroutine `kolm1.f` of `ForLib` gives the Fortran implementation of the upper one-sided Kolmogorov test, equation (2.139), and the corresponding lower one-sided test. A call to

$$\mathtt{KOLM1(N, Fxct, DEL1, DEL2, Q1, Q2)} \qquad (2.142)$$

needs on input the array Fxct of dimension N, which contains the exact (analytical) cumulative distribution function at the values of the sorted data (2.115). The subroutine returns with DEL1 the maximum positive deviation (2.118) of the exact distribution function from the lower empirical distribution function (2.116) and with DEL2 the maximum deviation in the negative direction with respect to the empirical distribution function $\overline{F}_2$, *i.e.,*

$$\Delta_2 = \max_x \{\overline{F}_2(x) - F(x)\} . \qquad (2.143)$$

The probabilities (2.139) resulting from the upper and lower one-sided Kolmogorov tests are returned as Q1 and Q2, respectively. A call to

$$\mathtt{KOLM1\_AS(N, Fxct, DEL1, DEL2, Q1, Q2)} \qquad (2.144)$$

has the same form as a call to KOLM1, but uses Smirnov's asymptotic approximation (2.141). It should only be used when a data set is so large that the exact calculation with `kolm1.f` becomes impractical, say for $N > 1\,000$.

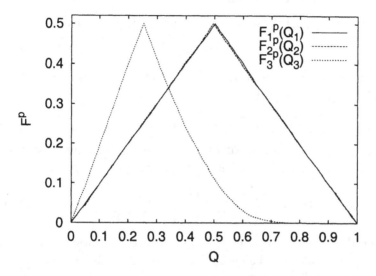

Fig. 2.4 Peaked distribution functions $F_1^p(Q_1)$, $F_2^p(Q_2)$ and $F_3^p(Q_3)$ for 10 000 repetitions of Kolmogorov tests, each on a sample of $N = 5$ Cauchy random numbers, see assignment 4. The variables Q_1, Q_2 are obtained from KOLM1 of equation (2.142) and Q_3 is defined by equation (2.145).

In assignment 3 the probabilities Q_1 and Q_2 are calculated for examples with eight and one hundred random numbers. The purpose of assignment 4 is to demonstrate that, independently of the number of data used, the probabilities Q_1 and Q_2 are uniformly distributed. We create $N = 5$ Cauchy-distributed random numbers and perform the two one-sided Kolmogorov tests. This is repeated ten thousand times and the resulting arrays $Q_1(i)$ and $Q_2(i)$, $i = 1, \ldots, 10^4$ are sorted. The right-hand side of figure 2.4 depicts the thus obtained peaked distribution functions (1.41), $F_1^p(Q_1)$ and $F_2^p(Q_2)$. Within the accuracy of the plot the functions F_1^p and F_2^p are almost indistinguishable from one another and both are well consistent with the uniform distribution, compare figure 1.7 (non-believers can run Kolmogorov tests). It is notable that this result holds universally, *i.e.*, independently of the analytical form of the exact distribution function and independently of the number of available data, down to a single data point (see the assignment). Using the large number of $N = 1\,000$ points per data set, it is seen that Smirnov's approximation (2.141) gives then also the desired distribution.

2.6.3   The two-sided Kolmogorov test

Also depicted in figure 2.4 is the peaked distribution function of

$$Q_3 = \min(Q_1, Q_2) \,. \tag{2.145}$$

This leads us to the two-sided Kolmogorov test, where one is interested in the probability $Q(\triangle_3; N)$ that

$$\triangle_3 = \max\{\triangle_1, \triangle_2\} \tag{2.146}$$

($\triangle_1$ and $\triangle_2$ as calculated by KOLM1) exceeds a certain value. For $N = 5$ data we can read off this probability from figure 2.4 by mapping

$$Q_3 \;\rightarrow\; F(Q_3) = \begin{cases} F_3^{\mathrm{P}}(Q_3) & \text{for } Q_3 \le 0.5; \\ 1 - F_3^{\mathrm{P}}(Q_3) & \text{for } Q_3 \ge 0.5. \end{cases} \tag{2.147}$$

Although the analytical solution is not known, straightforward MC calculations allow to tabulate the $Q(\triangle_3; N)$ functions and the results are universally valid for all distributions.

For $N \ge 4$ a reasonably accurate asymptotic approximation exists

$$Q_{as} = \sum_{j=1}^{\infty} (-1)^{j-1} e^{-2\,(j\triangle_3)^2\, S_N} \tag{2.148}$$

where[5], due to [Stephens (1970)],

$$S_N = \left(\sqrt{N} + 0.155 + \frac{0.24}{\sqrt{N}} \right)^2 \,.$$

The subroutine kolm2_as.f of ForLib gives the Fortran implementation. A call to

$$\text{KOLM2\_AS(N, Fxct, DEL, Q)} \tag{2.149}$$

needs the same input as kolm1.f (2.142), and returns with DEL the maximum deviation from the exact distribution function and with Q the approximation (2.148) to the probability that this deviation is due to chance.

---

[5]In the original form of Kolmogorov $S_N = N$ is used, which needs values of N larger than about forty.

Assignment 5 compares results of the asymptotic equation (2.148) with the (up to statistical fluctuations) exact estimates of assignment 4. In this illustration the worst discrepancy found is only a few percent of the value. Namely, for $N = 5$ and $Q_3 = 0.03775$ the values $Q_{as} = 0.0757$ and $F(Q_3) = 0.0726$ (rounded in the last digit) differ by 0.0031. As expected, for a large N value the agreement is even better. It is also notable that for $Q \leq 0.5$ the formula $Q = 2Q_3$ provides a very good approximation.

Typical for many applications is assignment 6: The asymptotic two-sided Kolmogorov test (2.148) is used to compare an exactly known distribution function with a (sufficiently large) empirical data set. Our data points are sums of two uniformly distributed random numbers. First, we use two subsequent Marsaglia random numbers for this purpose and the test gives the results which are expected for independent random numbers. Next we add two random numbers from sequences of Marsaglia random numbers, which differ by a slight change of the initial seed. For the test these random numbers behave also as if they were independent. This is of practical importance for parallel computing, where each process may initialize the generator with a different seed.

The sensitivity of the Kolmogorov test to deviations from a cumulative distribution is best around the median. It is less sensitive at the ends of the distribution, where F is near zero or one. By mapping the distribution on a circle, [Kuiper (1962)] developed a test which is equally sensitive at all values, see *Numerical Recipes* [Press *et al.* (1992)] for a brief discussion.

Comparison of two empirical distributions

The Kolmogorov test can also be used to test the assumption that two empirical distributions are the same. For sufficiently large data sets this is achieved by replacing N in equation (2.149) by the effective number of data

$$N_{\text{effective}} = \frac{N_1 N_2}{N_1 + N_2} \qquad (2.150)$$

where N_1 and N_2 are the numbers of data in the first and second data set respectively. This is implemented by the subroutine kolm2_as2.f of ForLib. A call to

$$\text{KOLM2\_AS2}(\text{N1}, \text{N2}, \text{DAT1}, \text{DAT2}, \text{DEL}, \text{Q}) \qquad (2.151)$$

needs on input the data arrays DAT1 and DAT2 together with their dimensions N1 and N2. It returns with DEL the maximum deviation between their

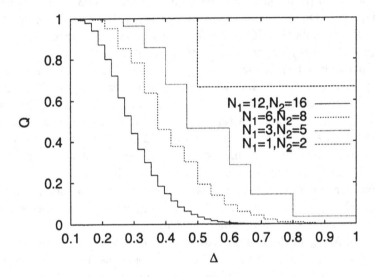

Fig. 2.5 MC results for the probability $Q = 1 - F(\Delta)$ that the difference between two identically distributed samples of N_1 and N_2 data is due to chance, see assignment 7. The (N_1, N_2) values, as depicted from up to down, label the curves from left to right.

empirical distribution functions and with Q the approximation (2.148) to the probability that this deviation is due to chance, where N is replaced by $N_{\text{effective}}$ (2.150).

For small values of N_1 and or N_2, the relevant probabilities can be estimated by a MC simulation. One simply repeats the generation of N_1 and N_2 equally distributed data many times and deduces Q from the cumulative distribution function $F(\Delta)$ of Δ. The subroutine kolm2_del2.f should then be called according to the syntax

$$\text{KOLM2\_DEL2}(N1, N2, DAT1, DAT2, DEL) . \qquad (2.152)$$

The arguments are the same as in equation (2.151), only that Q is missing. Figure 2.5 shows MC estimates of $Q = 1 - F(\Delta)$ for various values of N_1 and N_2 (as low as 1 and 2), see assignment 7. For two choices of somewhat larger (N_1, N_2) values the asymptotic approach (2.151) is compared with a

MC simulation in assignment 8.

2.6.4   *Assignments for section 2.6*

(1) A throw of a dice creates an integer in the range $i = 1, \ldots, 6$ with probability $p_i = 1/6$. Generate 10 000 data sets, each consisting of 1 000 throws of the dice, calculate for each data set the likelihood (2.112) that the difference with the predicted values is due to chance and plot the peaked distribution function for Q. Repeat the same, but calculate the Q values from the χ^2 distribution with $f = n$, instead of (correctly) $f = n - 1$, degrees of freedoms. Figure 2.3 plots both peaked distribution functions together. Repeat the above using only 100 throws of the dice per experiment and explain whatever happens to the figure.

(2) Use equation (2.113) to perform the comparison of the previous assignment between two data sets, each generated 10 000 times with a statistics of 1 000 throws of the dice. Afterwards, reduce the statistics of the second data set to 10 000 repetitions of 500 throws and rely on equation (2.114) for the analysis.

(3) Use `kolm1.f` of `ForLib` to calculate the upper (2.139) and lower probabilities, Q_1 and Q_2, of the one-sided Kolmogorov test for the following data: (a) The first eight Marsaglia random numbers as given in table 1.2. (b) The first one hundred Marsaglia random numbers as used in figure 1.7. (c) The first eight Cauchy random numbers as generated by `rmacau.f` of `ForLib`. (d) The first one hundred Cauchy random numbers of `rmacau.f`. The (Q_1, Q_2) values obtained are: (a) (0.29,0.93), (b) (0.14,0.80), (c) (0.27,0.66) and (d) (0.22,0.23).

(4) Use `rmacau.f` of `ForLib` with the standard initialization to generate 10 000 data sets of $N = 5$ Cauchy-distributed random numbers. Perform the two one-sided Kolmogorov tests on each data set. Sort the thus obtained $Q_1(i)$ and $Q_2(i)$, $i = 1, \ldots, 10^4$ arrays and plot their peaked distribution functions (F_1^p and F_2^p of figure 2.4 should result). Repeat this exercise with uniform and with Gaussian random numbers, as well as with just $N = 1$ data point per data set. Finally, use `kolm1_as.f` (2.144) with the large number of $N = 1 000$ points per data set, to see that Smirnov's asymptotic expansion (2.141) approximates then the desired Q distribution.

(5) Perform the asymptotic two-sided Kolmogorov test (2.148) for the first $N = 5$ uniformly distributed Marsaglia random numbers and compare Q_{as} with the result from the Monte Carlo simulation of the two-sided

Kolmogorov test in the previous assignment. Repeat this for the next three sets of $N = 5$ uniformly distributed random numbers. Afterwards repeat this analysis (including the Monte Carlo simulation of the previous problem) for $N = 80$. Finally, compare the Q_{as} and $F(Q_3)$ values with $2Q_3$.

(6) Recall the distribution function $F(y) = \int_{-\infty}^{y} dy'\, g(y')$ for the sum of two uniformly distributed random numbers (1.66). (a) Use the first 2 000 Marsaglia random number with the default seeds to generate 1 000 events, where each event is the sum of two successive random numbers. Use the two-sided asymptotic Kolmogorov test (2.148) to test this data set against the exact distribution. The result is the somewhat high value of $Q = 0.9899$. To test whether this high value is an accident or of statistical significance, repeat the investigation 1 000 times (using altogether 2 000 000 random numbers). Sort the obtained Q-values and plot their peaked distribution functions. (b) Repeat the investigation of (a) for pairs of Marsaglia random numbers which are generated with different seeds,

$$(\texttt{iseed1} = 1,\ \texttt{iseed2} = \texttt{irpt} - 1)\ \text{and}\ (\texttt{iseed1} = 1,\ \texttt{iseed2} = \texttt{irpt})\,. \tag{2.153}$$

Here $\texttt{irpt} = 1, \ldots, 1\,000$ refers to the repetitions of the test and the first 1 000 random numbers of each sequence are used. (c) Apply the asymptotic two-sided Kolmogorov test to compare the empirical Q distributions obtained in (a) and (b) with the uniform distribution function (1.8). The results are: $Q = 0.93$ (a) and $Q = 0.78$ (b).

(7) Take the N_1 and N_2 values of figure 2.5 and use uniform random numbers to generate one million data sets for each case. Use the subroutine $\texttt{kolm2\_del2.f}$ (2.152) to calculate Δ for each pair and sort the values to obtain the empirical cumulative distributions functions. Plot the probabilities that the discrepancy is due to chance as depicted in figure 2.5. Repeat this exercise with Cauchy or Gaussian random numbers. Of what nature are the differences between the obtained figures?

(8) Consider the data sets $(N_1 = 12, N_2 = 16)$ and $(N_1 = 72, N_2 = 96)$. Repeat the analysis of the previous assignment for uniform random numbers and use $\texttt{kolm2\_as.f}$ (2.151) to calculate the asymptotic likelihood (2.148). Compare the MC estimates and the asymptotic results around $Q = 0.05$. You should find that in the first case the accuracy of the asymptotic formula is about 10% and it improves to about 3% for the second case.

2.7 The Jackknife Approach

Jackknife estimators allow to correct for the bias and the error of the bias. The method was introduced in the 1950s in papers by [Quenouille (1956)] and [Tukey (1958)]. A detailed review has been given by [Efron (1982)] and a more comprehensive one by [Miller (1974)]. We focus on a simple Fortran implementation [Berg (1992a)].

The jackknife method is **recommended as the standard** for error bar calculations. In unbiased situations the jackknife and the usual error bars agree. Otherwise the jackknife estimates are improvements, so that one cannot loose by relying on the jackknife approach. Assume we have data

$$x_i, \quad (i = 1, ..., N) \tag{2.154}$$

which are uncorrelated ($\langle x_i x_j \rangle = 0$ for $i \neq j$) and correspond to a random variable x. Then

$$\bar{x} = \frac{1}{N} \sum_{i=1}^{N} x_i$$

is an unbiased (2.7) estimator of the expectation value $\hat{x} = \langle \bar{x} \rangle = \langle x \rangle$. Typically, bias problems show up when one likes to estimate a non-linear function of $\hat{x}$:

$$\hat{f} = f(\hat{x}) . \tag{2.155}$$

A terribly bad estimator of $\hat{f}$ is

$$\bar{f}_{\text{bad}} = \frac{1}{N} \sum_{i=1}^{N} f_i \text{ where } f_i = f(x_i) . \tag{2.156}$$

Namely, we never converge towards the correct answer:

$$\text{bias}(\bar{f}_{\text{bad}}) = \hat{f} - \langle \bar{f}_{\text{bad}} \rangle = O(1) .$$

A correct estimator of $\hat{f}$ is

$$\bar{f} = f(\bar{x}) . \tag{2.157}$$

Typically, the bias is now of order $1/N$:

$$\text{bias}(\bar{f}) = \hat{f} - \langle \bar{f} \rangle = \frac{a_1}{N} + \frac{a_2}{N^2} + O(\frac{1}{N^3}) . \tag{2.158}$$

Unfortunately, we lost the ability to estimate the variance $\sigma^2(\overline{f})$ via the standard equation

$$s^2(\overline{f}) = \frac{1}{N}s^2(f) = \frac{1}{N(N-1)}\sum_{i=1}^{N}(f_i - \overline{f})^2 , \qquad (2.159)$$

because (2.156) $f_i = f(x_i)$ is not a valid estimator of $\widehat{f}$.

At this point it has to be mentioned that it is in non-trivial applications almost always a bad idea to to use the error propagation formulas (1.86) with the aim to deduce $\Delta \overline{f}$ from $\Delta \overline{x}$. When the function $f(\widehat{x})$ is nonlinear, statistical fluctuations are often a problem to the application of the error propagation formulas, which rely on the linear approximation. Jackknife methods are not only easier to implement, but also more precise and far more **robust**, *i.e.*, stable under statistical fluctuations.

The error bar problem for the estimator (2.157) is overcome by using **jackknife estimators** $\overline{f}^J$, f_i^J, defined by

$$\overline{f}^J = \frac{1}{N}\sum_{i=1}^{N} f_i^J \text{ with } f_i^J = f(x_i^J) \text{ and } x_i^J = \frac{1}{N-1}\sum_{k\neq i} x_k . \quad (2.160)$$

The estimator for the variance $\sigma^2(\overline{f}^J)$ is

$$s_J^2(\overline{f}^J) = \frac{N-1}{N}\sum_{i=1}^{N}(f_i^J - \overline{f}^J)^2 . \qquad (2.161)$$

Straightforward algebra shows that in the unbiased case the jackknife estimator of the variance (2.161) reduces to the normal variance of the mean (2.159). For instance, let $f_i = x_i$, it follows

$$\overline{x}^J - x_k^J = \sum_{i\neq k}\left(\frac{1}{N} - \frac{1}{N-1}\right)x_i + \frac{1}{N}x_k = -\frac{1}{N(N-1)}\sum_{i\neq k}x_i + \frac{1}{N}x_k$$

$$= -\frac{1}{(N-1)}\overline{x} + \frac{1}{N(N-1)}x_k + \frac{1}{N}x_k = -\frac{1}{N-1}(\overline{x} - \overline{x}_k)$$

and therefore

$$s_J^2(\overline{x}^J) = \frac{(N-1)}{N}\sum_{k=1}^{N}(\overline{x} - x_k^J)^2 = \frac{1}{N(N-1)}\sum_{k=1}^{N}(\overline{x} - x_k)^2 = s^2(\overline{x}) .$$

Assignment 1 gives a numerical illustration of this equation.

In case of a bias equation (2.158) implies that the bias of each jackknife estimator $\overline{f}^J$ is of order $1/N$:

$$\text{bias}\,(\overline{f}^J) = \hat{f} - \langle \overline{f}^J \rangle = \frac{a_1}{N-1} + \frac{a_2}{(N-1)^2} + O\left(\frac{1}{N^3}\right) . \qquad (2.162)$$

In our Fortran code the calculation of jackknife error bars is implemented by the subroutines datjack.f and stebj0.f of ForLib. The jackknife data x_i^J, $(i = 1, \ldots, N)$ of equation (2.160) are generated by a call to

$$\text{DATJACK(N, X, XJ)} \qquad (2.163)$$

where X is the array of the original data and XJ the array of the corresponding jackknife data (estimators). An array FJ, corresponding to f_i^J of equation (2.160), has then to be defined by the user. Relying on the jackknife variance definition (2.161) a subsequent call to

$$\text{STEBJ0(N, FJ, FM, FV, FE)} \qquad (2.164)$$

uses the array FJ as input to calculate the mean FM, the variance FV and the error bar FE of the mean.

On the basis of equation (2.162) the bias converges faster than the statistical error and is normally ignored. In exceptional situations the constant a_1 of equation (2.162) may be large. Bias corrections should be done as soon as the bias reaches the order of magnitude of the statistical error bar. Let us give an example. Assume that our average relies on sixteen statistically independent data x_i, $(i = 1, \ldots, 16)$ and that the jackknife estimators are normally distributed around $\langle \overline{f}^J \rangle$, whereas the correct mean value is $\hat{f} = \langle \overline{f}^J \rangle + \text{bias}(\overline{f}^J)$. In the unbiased case the probability to find the unknown exact result within two standard deviations is determined by the Student distribution (2.68) to be

$$P\left[\overline{f}^J - 2s_J(\overline{f}^J) \leq \hat{f} \leq \overline{f}^J + 2s_J(\overline{f}^J)\right] = 0.936 .$$

Suppose bias $(\overline{f}^J) = \sigma(\overline{f}_J)$. This reduces this confidence level to 0.829, meaning the chance of an error ($\hat{f}$ outside two standard deviations) is almost three times larger.

An estimator for the bias is easily constructed. Combining equations (2.158) and (2.162) we get

$$a_1 = N(N-1)\left(\langle \overline{f} \rangle - \langle \overline{f}^J \rangle\right) + O\left(\frac{1}{N}\right)$$

and

$$\bar{b} = (N-1)(\bar{f} - \bar{f}^J) \tag{2.165}$$

defines an estimator for the bias up to $O\left(1/N^2\right)$ corrections. The subroutine bias.f of ForLib implements this equation. Relying on bias.f our jackknife error bar subroutine stebj1.f generalizes stebj0.f. A call to

$$\text{STEBJ1(N, FJ, FMEAN, FJM, FMM, FV, FE)} \tag{2.166}$$

returns the results of (2.164), where the more accurate notation FJM replaces now FM, FMEAN is the mean $\bar{f}$ from all data and FMM the bias corrected estimator $\bar{f}^c$ of equation (2.167) below. Variance and error bar are calculated by stebj1.f with respect to FJM and may differ from those of the bias corrected estimator FMM. The error analysis of FMM requires second level jackknife estimators defined in the following subsection. If one needs only some re-assurance that the bias is small, like in assignments 1 and 2, it is recommended to rely on stebj1.f instead of using the more sophisticated second level estimators.

2.7.1   Bias corrected estimators

It follows from equation (2.165) that

$$\bar{f}^c = \bar{f} + \bar{b} \tag{2.167}$$

is a bias-corrected estimator of $\widehat{f}$ in the sense that

$$\text{bias}\,(\bar{f}^c) = \widehat{f} - \bar{f}^c = O\left(\frac{1}{N^2}\right) \tag{2.168}$$

holds. The estimator $\bar{b}$ is affiliated with a statistical error which may have the same order of magnitude as $\bar{b}$ itself and is correlated with the statistical error $s_J(\bar{f}^J)$. To estimate the statistical error of $\bar{f}^c$ we define **second level jackknife estimators**

$$f_{i,j}^J = f(x_{i,j}^J) \quad \text{with} \quad x_{i,j}^J = \frac{1}{N-2}\sum_{k\neq i,j} x_k \quad \text{for } i \neq j . \tag{2.169}$$

A jackknife sample of bias estimators and bias corrected estimators is then given by

$$b_i^J = \frac{1}{N-1}\sum_{k\neq i}(f_i^J - f_{i,k}^J) \quad \text{and} \quad f_i^{c,J} = f_i^J + b_i^J . \tag{2.170}$$

Equation (2.161) may be applied to the $f_i^{c,J}$ so that $s_J^2(\overline{f}^{c,J})$ estimates the variance of the bias corrected estimator

$$\overline{f}^{c,J} = \frac{1}{N} \sum_{i=1}^{N} f_i^{c,J} . \qquad (2.171)$$

Similarly, one may estimate the error of the estimated bias. In the usual jackknife way $s_J^2(\overline{f}^{c,J})$ takes correlations between $s_J^2(\overline{f}^J)$ and $s_J^2(\overline{b}^J)$ already properly into account.

The subroutines datjack2.f and stebjj1.f of ForLib implement these equations. A call to

$$\text{DATJACK2}(N, X, XJJ) \qquad (2.172)$$

returns with XJJ the second level jackknife estimators $x_{i,j}^J$ of equation (2.169) for $i < j$ (the entries $i \geq j$ of the $(N-1) \times N$ dimensional array XJJ are not filled). The array XJJ has then to be converted into FJJ, corresponding to $f_{i,j}^J$ of (2.169), and serves as input for

$$\text{STEBJJ1}(N, XJJ, XJ, XJMN, XMM, XV, XE) \qquad (2.173)$$

which returns with XJMM the array of bias corrected estimators $f_i^{c,J}$ of equation (2.170). Their mean, variance and error bar are then XMM, XV and XE, respectively. Further, stebjj1.f returns another N dimensional array, XJJ, which collects the mean values of the second level estimators.

A simple example for a biased estimator is $\overline{x}^2$ as estimator for the mean squared $f(\widehat{x}) = \widehat{x}^2$. In this case (2.12):

$$f^{ub}(\widehat{x}) = \overline{x}^2 - \frac{1}{N-1}\left(\overline{x^2} - \overline{x}^2\right) \qquad (2.174)$$

is an unbiased estimator, i.e., $\langle f^{ub} \rangle = \widehat{x}^2$. In assignment 2 we generate some random numbers and calculate for the mean squared the bad estimator (2.156), the unbiased estimator (2.174), the standard jackknife estimator (2.160) and the double jackknife biased improved estimator (2.170). The results are listed after the assignment and we make the observation that in this example the second level jackknife estimator is **identical** with the known exact, unbiased estimator. This result may also be verified analytically. Of course, in realistic applications an exact unbiased estimator is normally unknown. For our simple illustration the exact result is $\widehat{x}^2 = 0.25$

and the bias correction is seen to be negligible when compared with the statistical error. To show that this is not always the case, we give an example for which the bias is much larger than the statistical error.

In assignment 3 we are dealing with estimating

$$f(\widehat{x}) = e^{a_0 |\widehat{x}|} / \widehat{x^4}, \quad a_0 > 0 \tag{2.175}$$

for normally distributed random numbers. The exact result is $f(\widehat{x}) = 1/3$ (independent of a_0), because $\widehat{x} = 0$ and, according to equation (1.57), $\widehat{x^4} = 3\sigma^2$ with $\sigma^2 = 1$ for the normal distribution. For the function (2.175) the bias of the estimator (2.157) has two sources: (a) Due to a fluctuation the denominator can get close to zero and the result may blow up. If the denominator is estimated from a sufficiently large sample, this is quite unlikely as the value $\widehat{x^4} = 3$ is sufficiently far away from zero. (b) The absolute sign of $\widehat{x}$ in the exponent means that for any finite sample the estimate will be larger than the true value, $\exp(a_0|\widehat{x}|) = 1$, and this bias gets worse with increasing a_0. Assignment 3 demonstrates for $a_0 = 2$ and 320 data a considerable bias: The standard jackknife estimator exhibits an average bias of about 13% which is reduced to about 4% by the second level bias improved jackknife estimators. In both cases the estimated statistical error is much smaller, about 0.5%. With an increasing number of data the bias can be made to disappear. On the other side, when there are very few data, the second level jackknife estimates may encounter stability problems.

The illustrations given in this section are for pedagogical purposes. More realistic examples with data from Monte Carlo simulations are considered in chapters 4, 5 and 6.

2.7.2   *Assignments for section 2.7*

(1) Generate the first 320 random numbers of Marsaglia's generator with the default seed. Calculate the mean and its error bar with `steb0.f`. Afterwards calculate the corresponding 320 jackknife data with `datjack.f` and use `stebj0.f` to calculate the mean and its error bar from the jackknife data. In both cases the (rounded) result is 0.5037 ± 0.0163. The square of this mean is 0.25371. Use `stebj1.f` to estimate the bias of this quantity to be 0.00026.

(2) Use the first 16 random numbers of Marsaglia's generator with the default seed to calculate the following estimators for the squared mean $\widehat{x}^2 = 0.25$: The bad estimator (2.156), the unbiased estimator (2.174),

the standard jackknife estimator (2.160) and the double jackknife biased improved estimator (2.170). The results are:

```
BIASED ESTIMATE FROM ALL DATA:              0.300213
BAD, BIASED ESTIMATOR (FALSE):              0.358675   +/- 0.068700
UNBIASED ESTIMATOR:                         0.296316
STANDARD, BIASED JACKKNIFE ESTIMATOR:       0.300473   +/- 0.068520
SECOND LEVEL
BIAS-CORRECTED JACKKNIFE ESTIMATOR:         0.296316   +/- 0.068662
```

(3) Generate the first 320 random numbers of our Gaussian random number generator with its standard seed. Use bining.f to bin them down to 32 numbers and repeat the binning procedure for the fourth moments of the same 320 numbers. Calculate from the binned data the following estimators for the function of equation (2.175): The estimate (without error bar) from all data (2.157), the bad estimator (2.156), the unbiased estimator (2.174), the standard jackknife estimator (2.160) and the double jackknife biased improved estimator (2.170). Your results should be:

```
BIASED ESTIMATE FROM ALL DATA:              0.382451
BAD, BIASED ESTIMATOR:                      1.503532   +/- 0.075568
STANDARD, BIASED JACKKNIFE ESTIMATOR: 0.383602   +/- 0.075568
BIAS-CORRECTED JACKKNIFE ESTIMATOR:   0.333867   +/- 0.240994
```

Repeat the previous calculation two thousand times (*i.e.*, use the first 2000×320 Gaussian random numbers). With respect to the 2000 repetitions, use steb0.f to calculate (conventional) averages and error bars of the four quantities considered. Those should read (remember, the exact result is $f(\widehat{x}) = 1/3 = 0.33\overline{3}$):

```
BIASED ESTIMATE FROM ALL DATA:              0.376332   +/- 0.001648
BAD, BIASED ESTIMATOR:                      1.384823   +/- 0.009037
STANDARD, BIASED JACKKNIFE ESTIMATOR: 0.377288   +/- 0.001646
BIAS-CORRECTED JACKKNIFE ESTIMATOR:   0.347185   +/-. 0.001788
```

2.8 Determination of Parameters (Fitting)

Given are N data points

$$(x_i, y_i), \ i = 1, \ldots, N \tag{2.176}$$

together with their standard deviations

$$\sigma_i = \sigma(y_i) \tag{2.177}$$

and the x_i are error free. In most applications the standard deviations are approximated (replaced) by the statistical error bars of the data points. We like to **model the data** with a function, called **fit**, of M free parameters

$$y = y(x; a_1, \ldots, a_M) .$$
(2.178)

The problem is to decide whether the model is consistent with the data and to estimate the parameters and their error bars. This is often called **fitting**. There exists plenty of literature on the subject and a large number of fitting routines for all kind of purposes are available. For readers who want to dig beyond the material covered in this section, the following sources are recommended.

(1) *Numerical Recipes* [Press *et al.* (1992)] explains a number of approaches and gives references for further reading.
(2) MINUIT is a robust and efficient general purpose fitting package which is part of the CERNLIB. It is freely available on the Web, although some registration has to be done. Go to www.cern.ch and find your way.
(3) For certain advanced topics the book by [Miller (2002)] on subset selection in regression may be consulted. Miller entertains Fortran code for this book on the website users/bigpond/net.au/miller (google finds it when searching for "Alan Miller Fortran").

Here we confine ourselves to two methods, simple linear fitting and the Levenberg-Marquardt method for nonlinear fitting. These turn out to be sufficient to cover the application in our later chapters. Both methods are based on the **maximum likelihood** approach.

Assume the parameters $a_1, \ldots, a_M$ are known and the function $y(x; a_1, \ldots, a_M)$ is the exact law for the data. For Gaussian data y_i the probability density becomes

$$\rho(y_1, \ldots, y_N) = \prod_{i=1}^{N} \frac{\exp\{-[y_i - y(x_i; a_1, \ldots, a_M)]^2 / (2\sigma_i^2)\}}{2\sigma_i}$$
(2.179)

and its maximum is at

$$\chi^2 = \sum_{i=1}^{N} \left(\frac{y_i - y(x_i; a_1, \ldots, a_M)}{\sigma_i^2} \right)^2 = \text{minimum.}$$
(2.180)

For the situation where the exact parameters are unknown, but the model function (2.178) is conjectured, χ^2 fitting turns this around and determines the parameters $a_1, \ldots, a_M$ by minimizing (2.180). This defines the maximum likelihood estimate of the parameters. It assumes that the measurement errors are Gaussian and exactly known. In practice this will seldom be the case. Even when the data are Gaussian distributed, the σ_i will normally not be exact, but just estimators s_i, which are Student instead of normally distributed. In addition, the data themselves will rarely be exactly Gaussian, although binning (1.112) can often be employed to have at least approximately Gaussian data. The strength of the least square fitting method is that it leads still to satisfactory results in many situations where the Gaussian assumptions are only approximately valid.

Of course, the confidence probabilities will be effected when the exact standard deviations are replaced by estimators. An analytical calculation of the corrections, like for the Student distribution, is out of reach. Fortunately, the corrections are for most applications small enough that some conservative rounding is all one needs. For the rare situations where on needs precise estimates of the confidence intervals, it is reassuring to know that MC simulations of the fitting situations allow to calculate them [Berg (1992b)].

2.8.1 *Linear regression*

We consider the problem of fitting N data points y_i, $(i = 1, \ldots, N)$ to a straight line

$$y = y(x) = a_1 + a_2\, x \, . \tag{2.181}$$

This problem is often called **linear regression**, a terminology that originated in social sciences. With (2.181) the χ^2 function becomes

$$\chi^2 \;=\; \chi^2(a_1, a_2) \;=\; \sum_{i=1}^{N} \left(\frac{y_i - a_1 - a_2\, x_i}{\sigma_i} \right)^2 . \tag{2.182}$$

Once the fit parameters a_1 and a_2 are determined, we assume that $y = a_1 + a_2\, x$ is the exact law and ask, similarly as before for the Gaussian (2.32) and the Student (2.85) difference tests, what it the likelihood that the discrepancy between the fit and the data is due to chance? The answer is given by the **goodness of fit** Q, which is defined by

$$Q \;=\; F(\chi^2) \;\; \text{with} \;\; f = N - 2 \,, \tag{2.183}$$

where $F(\chi^2)$ is the χ^2 distribution (2.44). The number of degrees of freedom is $f = N - 2$, because we have to subtract the number of fit parameters from the number N of independent data.

We would like to attach error bars to the estimated fit parameters. At the minimum of the $\chi^2(a_1, a_2)$ function (2.182) its derivatives with respect to a_1 and a_2 vanish,

$$0 = \frac{\partial \chi^2}{\partial a_1} = -2 \sum_{i=1}^{N} \frac{y_i - a_1 - a_2 x_i}{\sigma_i^2} = -2 (S_y - a_1 S - a_2 S_x) \qquad (2.184)$$

and

$$0 = \frac{\partial \chi^2}{\partial a_2} = -2 \sum_{i=1}^{N} x_i \frac{y_i - a_1 - a_2 x_i}{\sigma_i^2} = -2 (S_{xy} - a_1 S_x - a_2 S_{xx}) \;,$$
$$(2.185)$$

where we have introduced the notation

$$S = \sum_{i=1}^{N} \frac{1}{\sigma_i^2} \;, \quad S_x = \sum_{i=1}^{N} \frac{x_i}{\sigma_i^2} \;, \quad S_y = \sum_{i=1}^{N} \frac{y_i}{\sigma_i^2} \;,$$

$$S_{xy} = \sum_{i=1}^{N} \frac{x_i \, y_i}{\sigma_i^2} \quad \text{and} \quad S_{xx} = \sum_{i=1}^{N} \frac{x_i^2}{\sigma_i^2} \;. \qquad (2.186)$$

The equations (2.184) and (2.185) are easily solved

$$a_1 = \frac{S_{xx} S_y - S_x S_{xy}}{D} \quad \text{and} \quad a_2 = \frac{S S_{xy} - S_x S_y}{D} \quad \text{with} \quad D = S S_{xx} - (S_x)^2 \;.$$
$$(2.187)$$

On the right-hand side of these equations the y_i are random variables and the x_i are constants. The relationship between the parameters a_i and the random variables is linear (1.81) (entering only through S_y and S_{xy}). Therefore, linear error propagation (1.86) determines the variances of the parameter estimates a_1 and a_2

$$\sigma^2(a_i) = \sum_{i=1}^{N} \sigma_i^2 \left(\frac{\partial a_i}{\partial y_i} \right)^2 \;, \quad (i = 1, 2) \qquad (2.188)$$

and their covariance (1.89)

$$\text{cov}(a_1, a_2) = \sum_{i=1}^{N} \sum_{j=1}^{N} \sigma_i \, \sigma_j \left(\frac{\partial a_1}{\partial y_i} \right) \left(\frac{\partial a_2}{\partial y_j} \right) \;. \qquad (2.189)$$

Equations (2.188) and (2.189) are exact to the extent that the σ_i are the exact variances of normally distributed random variables y_i^r. The derivatives are found from the solutions (2.187)

$$\frac{\partial a_1}{\partial y_i} = \frac{S_{xx} - S_x \, x_i}{\sigma_i^2 \, D} \quad \text{and} \quad \frac{\partial a_2}{\partial y_i} = \frac{S \, x_i - S_x}{\sigma_i^2 \, D} \ .$$

Squaring, summing over i, using (2.186) and dividing out $1/(\sigma_i^2 D)$ and $1/(\sigma_i \sigma_j D)$ factors yields

$$\sigma^2(a_1) = \frac{S_{xx}}{D}, \ \ \sigma^2(a_2) = \frac{S}{D} \quad \text{and} \quad \text{cov}(a_1, a_2) = -\frac{S_x}{D} \ . \tag{2.190}$$

To avoid round-off errors in the numerical implementation, one has to be careful when calculating the determinant D, because it occurs in the denominators of the relevant equations. Introducing

$$S_{tt} = \sum_{i=1}^{N} t_i^2 \quad \text{with} \ \ t_i = \frac{1}{\sigma_i} \left(x_i - \frac{S_x}{S} \right) \tag{2.191}$$

equation (2.187) becomes

$$a_2 = \frac{1}{S_{tt}} \sum_{i=1}^{N} \frac{t_i \, y_i}{\sigma_i} \quad \text{and} \quad a_1 = \frac{S_y - S_x \, a_2}{S} \ . \tag{2.192}$$

Further, we get

$$\sigma^2(a_1) = \frac{1}{S} \left(1 + \frac{S_x^2}{S \, S_{tt}} \right), \ \ \sigma^2(a_2) = \frac{1}{S_{tt}} \quad \text{and} \quad \text{cov}(a_1, a_2) = -\frac{S_x}{S \, S_{tt}} \ . \tag{2.193}$$

Linear regression is implemented by our subroutine fit_1.f of ForLib. A call to

$$\text{FIT\_L(NDAT, X, Y, SGY, A, SGA, CHI2, Q, COV)} \tag{2.194}$$

returns the fit parameters a_i and their estimated standard deviations in the arrays A(2) and SGA(2). Further, the χ^2 of the fit (2.182), the corresponding goodness of the fit (2.183) and the covariance (2.189) are returned as CHI2, Q and COV, respectively. On input the data arrays X, Y and SGY, each of length NDAT, have to be provided. The array Y contains the data points and SGY their standard deviations, which are normally approximated by the error bars. Besides, the user can assign to SGY whatever values deem appropriate to achieve a particular task (*e.g.*, give all data points equal weights by assigning the same constant value to all standard deviations).

2.8.1.1 *Confidence limits of the regression line*

To find the confidence limits of the regression line (2.181), we have to calculate the eigenvalues and eigenvectors of the covariance matrix (1.87). For the situation at hand the covariance matrix is the symmetric 2×2 matrix

$$(c_{ij}) = \begin{pmatrix} \sigma^2(a_1) & \text{cov}(a_1, a_2) \\ \text{cov}(a_1, a_2) & \sigma^2(a_2) \end{pmatrix} . \tag{2.195}$$

The subroutine `eigen_2x2.f` computes the eigenvalues and eigenvectors of a 2×2 matrix. A call to

$$\text{EIGEN\_2X2(AMAT, EVAL, EVCT)} \tag{2.196}$$

needs on input the matrix array `AMAT(2,2)`. If real eigenvalues are encountered, it returns the eigenvalues λ_1 and λ_2 in the array `EVAL(2)` and the corresponding eigenvectors as columns of the matrix array `EVCT(2,2)`. In accordance with the notation of equation (1.93) the elements of the column eigenvector $\vec{b}_i$, $(i = 1, 2)$ are given by b_{ji}, $(j = 1, 2)$. The two eigenvector directions $\vec{b}_1$ and $\vec{b}_2$ contribute as follows to the standard deviation of the regression line:

$$s(y) = \sqrt{s_{b_1}^2(y) + s_{b_2}^2(y)} \tag{2.197}$$

with

$$s_{b_1}^2(y) = \lambda_1 (b_{11} + b_{21}\, x)^2 \quad \text{and} \quad s_{b_2}^2(y) = \lambda_2 (b_{12} + b_{22}\, x)^2 . \tag{2.198}$$

Graphically the eigenvectors may be depicted as the directions of the major and the minor axes of an ellipse in the $a_1 - a_2$ plane, called **confidence ellipse**. When we regard a_1 and a_2 as random variables, the directions of the confidence ellipse axes define the diagonal random variables (1.98) for which the covariance matrix vanishes. We denote them by b_i^r, $(i = 1, 2)$

$$b_1^r = b_{11}\, a_1^r + b_{21}\, a_2^3 \quad \text{and} \quad b_2^r = b_{21}\, a_1^r + b_{21}\, a_2^3 . \tag{2.199}$$

The variances $\sigma^2(b_i)$ of the diagonal random variables b_1^r and b_2^r are the eigenvalues of the covariance matrix (2.195). The standard deviations $\sigma(b_i)$ conventionally define the lengths of the semi-minor and semi-major axes of the confidence ellipse. If the distribution is normal, the likelihood to find the correct parameter pair (a_1, a_2) within the confidence ellipse is

$$P^0_{c-\text{ellipse}} = \frac{1}{\sqrt{2\pi\,\lambda_1}}\,\frac{1}{\sqrt{2\pi\,\lambda_2}} \int_{\text{ellipse}} dx_1\,dx_2\,\exp\left(-\frac{x_1^2}{2\,\lambda_1}\right)\exp\left(-\frac{x_2^2}{2\,\lambda_2}\right)$$

$$= 2\pi \int_0^1 r\,dr\,\exp\left(-\frac{r^2}{2}\right) = 1 - \exp\left(-\frac{1}{2}\right) = 0.393\,. \qquad (2.200)$$

The substitutions done for calculating this integral are from x_1, x_2 to $x_1' = x_1/\sqrt{\lambda_1}$, $x_2' = x_2/\sqrt{\lambda_1}$ and from there to polar coordinates. Manipulations with ellipses become easy by realizing that they are distorted circles.

Let us turn to the graphical presentation of the results. A call to the subroutine

$$\text{FIT\_LGNU(IUD, NDAT, X, Y, SGY, A, SGA, COV, PROB)} \qquad (2.201)$$

prepares data sets for gnuplots of the regression line $y(x)$, its one standard deviation confidence limits $y(x) \pm s(y)$, and the confidence ellipse. With two exceptions the arguments of FIT_LGNU are the same as for FIT_L (2.194). New are IUD, the unit number used to write the data sets, and PROB, the probability content of the confidence ellipse to be drawn. For PROB ≤ 0 the probability content 0.393 of equation (2.200) is chosen. Other choices of $P_{c-\text{ellipse}} = $ PROB are possible and the variances (eigenvalues) in equation (2.200) are then re-scaled by a common factor, so that the desired probability $P_{c-\text{ellipse}}$ is matched. The data sets created for the regression line are lfit.d1 and lfit.d2. The data sets for the confidence ellipse are ellipse.d1 and ellipse.d2. To prepare the plots of the confidence ellipse, the routine fit_lgnu.f calls two other subroutines, first EIGEN_2x2 (2.196) and then

$$\text{ELLIPSE(IUD, A, SGA, SGB, EIGVS, PROB)}\,. \qquad (2.202)$$

With the exception of the array SGB(2) the arguments of ellipse.f are a subset of those of FIT_LGNU (2.201). The array SGB(2) contains the standard deviations of the diagonal parameters (2.199), *i.e.*, the square roots of the eigenvalues of the covariance matrix.

2.8.1.2 *Related functional forms*

The fitting routine fit_l.f covers all two parameter fits which can be reduced to the form (2.181). This includes, for instance,

$$y(x) = c_1 + c_2\,x^\alpha \quad \text{for given}\ \ \alpha \qquad (2.203)$$

and

$$y(x) = c_1 \, x^{-c_2} = c_1 \, \exp[-c_2 \, \ln(x)] \qquad (2.204)$$

as is illustrated in the following .

User package

The folder fit_1 of ForProg contains our routines for linear regression. The main program is lfit.f. Drivers for gnuplots are the fit.plt, lfit.plt and ellipse.plt files.

In the program lfit.f a subroutine

$$\text{SUBL}(IUO, N, X, Y, EY) \qquad (2.205)$$

is called before performing the fit. Its purpose is to allow for transformations of related functional forms to the linear form (2.181). The arguments of the SUBL routine (2.205) are the data unit IUO on which informative messages are written, the number of data N and the arrays X(N), Y(N) and EY(N), which contain the data points (2.176) with their error bars (2.177).

After fitting the parameters, the main program makes a call to

$$\text{SUBPLOT}(IUO, IUD, N, X, A) \qquad (2.206)$$

which creates the data files data.d and plot.d used by the gnuplot driver fit.plt. The arguments IUO, N and X are as for SUBL (2.205). The other arguments are IUD, the unit number used to write the data files, and the fit parameter array A(2) as calculated by FIT_L (2.194). To calculate the plot curve, the SUBPLOT subroutine converts the fit parameters to those used in the initial functional form.

Table 2.13 Subroutines for inclusion in lfit.f, which are provided in ForProg/fit_1.

Subroutine	Fit form	Transformation
subl_1ox.f	$y = c_1 + c_2/x \to yx = c_1 x + c_2$	$a_1 = c_2,\ a_2 = c_1$
subl_exp.f	$y = c_1 \, e^{-c_2 x} \to \ln(y) = \ln(c_1) - c_2 x$	$a_1 = \ln(c_1),\ a_2 = -c_2$
subl_linear.f	$y = a_1 + a_2 \, x$	none
subl_power.f	$y = c_1 \, x^{-c_2} \to \ln(y) = \ln(c_1) - c_2 \ln(x)$	$a_1 = \ln(c_1),\ a_2 = -c_2$

The subroutines SUBL and SUBPLOT are transferred into the main program lfit.f via one

$$\text{include file\_name.f} \qquad (2.207)$$

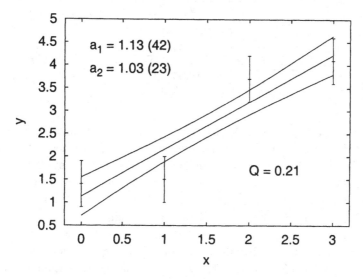

Fig. 2.6 Linear fit of the data of [Brandt (1983)] from table 2.14, see assignment 1. The curves are the regression line and its upper and lower standard deviations (2.197). Fit parameters and goodness of fit are also shown.

statement. **In table 2.13 we list the fit forms for which routines are provided** in ForProg/fit_1. By including the subroutine file, the transformation subroutine SUBL and the plot routine SUBPLOT are both transferred into lfit.f. For customized applications the user has to write own SUBL and SUBPLOT routines. To allow for the same structure of lfit.f in all applications, the file subl_linear.f, whose subroutines do not change the data at all, has to be included for the original linear fit (2.181).

The following examples illustrate the use of some of the routines of table 2.13. The data files for these examples are the *.D files in ForProg/fit_1.

2.8.1.3 *Examples*

(1) Figure 2.6 shows the regression lines, the fit parameters and the goodness of fit (2.183) obtained for illustrative data from the book by

Table 2.14 Data from [Brandt (1983)], [Berg and Neuhaus (1992)], and [Bhanot *et. al.* (1987)]. The one standard deviation estimators s_i of the y_i data points are given in parenthesis and refer to the last digits of the y_i.

		Brandt	Berg and Neuhaus		Bhanot et al.	
i	x_i	y_i	$x_i = L$	$y_i = F_L^s$	$x_i = L$	$y_i = \mathrm{Im}(u)$
1	0.0	1.4 (5)	16	0.1086 (07)	4	0.087739 (5)
2	1.0	1.5 (5)	24	0.1058 (08)	5	0.060978 (5)
3	2.0	3.7 (5)	34	0.1039 (12)	6	0.045411 (5)
4	3.0	4.1 (5)	50	0.1016 (08)	8	0.028596 (5)
5			70	0.0995 (12)	10	0.019996 (5)
6			100	0.0990 (12)		

[Brandt (1983)]. The data are reproduced in our table 2.14. The estimated errors of the fit parameters are given in parenthesis. Our fit is performed using the subroutine fit_1.f (2.194) with subl_linear.f of table 2.13 included. The regression plots are prepared by a call to the subroutine fit_1gnu.f (2.201). The standard confidence ellipse (2.200) for this fit is shown in figure 2.7 (assignment 1).

(2) The upper part of figure 2.8 gives an example for the two parameter fit to equation (2.203). Monte Carlo data of [Berg and Neuhaus (1992)] (see table 2.14) for the interface tension of the two dimensional 10-state Potts model are fitted to the behavior

$$y = c_1 + c_2 x^{-1} \qquad (2.208)$$

where $x = L$ and L^2 is the lattice size. The behavior (2.208) is converted to the linear fit (2.181) by multiplying both sides of (2.208) with x. The appropriate subroutine is subl_1ox.f of table 2.13. The upper part of figure 2.8 is prepared by the plotting routine of the subl_1ox.f package. The fit parameter c_1 of equation (2.208) is the desired infinite volume estimate of the interface tension and the result of our fit is (assignment 2)

$$F^s = c_1 = a_1 = 0.0978 \pm 0.0008 . \qquad (2.209)$$

(3) The lower part of figure 2.8 gives an example for the two parameter fit to equation (2.204). Monte Carlo data of [Bhanot *et al.* (1987)] for the imaginary part $\mathrm{Im}(u)$ of the first partition function zero of the three dimensional Ising model are reproduced in table 2.14. This quantity follows the behavior $y = c_1 x^{c_2}$ with $y = \mathrm{Im}(u)$ and $x = L$ where L^3 is the lattice size. The main interest is in the fit parameter $c_2 = -1/\nu$,

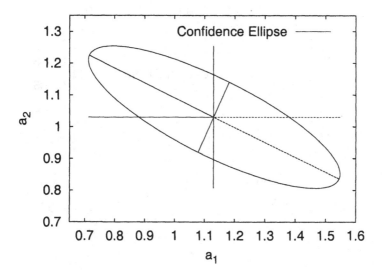

Fig. 2.7 The $P_{c-\text{ellipse}} = P^0_{c-\text{ellipse}} = 0.393$ confidence ellipse (2.200) for the fit of figure 2.6, see assignment 1.

where ν is the critical exponent of the correlation length. Taking the log on both sides of equation (2.204) gives the linear relation

$$\ln(y) = \ln(c_1) - c_2 \ln(x) . \qquad (2.210)$$

The package `subl_exp.f` in `ForProb/fit_l` contains the subroutines `SUBL` and `SUBPLOT` for this case. Conversion of the Bhanot *et al.* data of table 2.14 to the form (2.210) is a bit tricky, because the errors of the data propagate non-linearly. For our fit purposes the transformed data points are defined by

$$\ln(y_i) = \frac{1}{2} \left[\ln(y_i + \sigma_i) + \ln(y_i - \sigma_i) \right] \quad \text{and}$$
$$\sigma[\ln(y_i)] = \ln(y_i + \sigma_i) - \ln(y_i) . \qquad (2.211)$$

The interested reader may follow up on this issue by doing assignment 3. It is notable that the fit depicted in the lower part of figure 2.8 has a

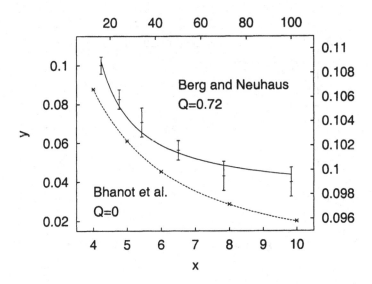

Fig. 2.8 Two parameter fits $y = a_1 + a_2\,x^{-1}$ (2.208) for data of [Berg and Neuhaus (1992)] and $y = c_1\,x^{c_2}$ (2.204) for data of [Bhanot *et al.* (1987)]. The scale on top and on the right-hand side of the figure belongs to the Berg–Neuhaus data (assignment 2) and the scale on the bottom and on the left-hand side to the Bhanot *et al.* data (assignment 3).

very high $\chi^2 \approx 470$ per degree of freedom and, therefore, a goodness of fit $Q = 0$. The reason are the very small error bars of the data, which make the fit sensitive to higher order finite size corrections to equation (2.204). The optical impression of the fit is only good, because the accuracy of the plot does not resolve the discrepancy between the fit and the data. The critical exponent corresponding to the fit is

$$\frac{1}{\nu} = -c_2 = -a_1 = 1.6185 \pm 0.0002 \quad \text{or} \quad \nu = 0.61786 \pm 0.00008 \;.$$
$$(2.212)$$

Due to the $Q = 0$ problem, we cannot trust the small error bar. One may now attempt to include higher order finite size corrections. This is done at the end of the next subsection.

2.8.2 *Levenberg-Marquardt fitting*

We now consider the general case (2.178) of a non-linear model $y = y(x; \vec{a})$, where $\vec{a}$ is the column vector of the fit parameters $a_1, \ldots, a_M$. We want to minimize the $\chi^2 = \chi^2(\vec{a})$ merit function (2.180) with respect to the parameters. Let us assume that the parameters of the nth iteration are given by the vector $\vec{a}_n$. For $\vec{a}$ in the neighborhood of $\vec{a}_n$ we can approximate $\chi^2(\vec{a})$ by the quadratic form[6]

$$\chi^2(\vec{a}) = \chi^2(\vec{a}_n) + (\vec{a}^\dagger - \vec{a}_n^\dagger) \cdot \vec{b} + \frac{1}{2}(\vec{a}^\dagger - \vec{a}_n^\dagger) A (\vec{a} - \vec{a}_n) \qquad (2.213)$$

where the coefficients of the vector $\vec{b} = (b_j)$ and of the $M \times M$ *Hessian matrix*[7] $A = (a_{jk})$ are given by the first and second derivatives of $\chi^2(\vec{a})$ at $\vec{a} = \vec{a}_n$:

$$b_j = \frac{\partial \chi^2(\vec{a})}{\partial a_j}\bigg|_{\vec{a}=\vec{a}_n} = -2 \sum_{i=1}^{N} \frac{[y_i - y(x_i; \vec{a})]}{\sigma_i^2} \frac{\partial y(x_i; \vec{a})}{\partial a_j}\bigg|_{\vec{a}=\vec{a}_n}, \, j = 1, \ldots, M$$

$$(2.214)$$

and

$$a_{jk} = \frac{\partial^2 \chi^2(\vec{a})}{\partial a_j \partial a_k}\bigg|_{\vec{a}=\vec{a}_n} \quad (j = 1, \ldots, M, \, k = 1, \ldots, M) \qquad (2.215)$$

$$= 2 \sum_{i=1}^{N} \frac{1}{\sigma_i^2} \left[\frac{\partial y(x_i, \vec{a})}{\partial a_j} \frac{\partial y(x_i, \vec{a})}{\partial a_k} - [y_i - y(x_i; \vec{a})] \frac{\partial^2 y(x_i; \vec{a})}{\partial a_j \partial a_k} \right]\bigg|_{\vec{a}=\vec{a}_n}.$$

In the **Newton–Raphson method**[8] one determines the next iteration point, $\vec{a}_{n+1}$, from the condition $\nabla \chi^2(\vec{a})\big|_{\vec{a}=\vec{a}_{n+1}} = 0$, which yields the linear equation

$$0 = \vec{b} + A(\vec{a}_{n+1} - \vec{a}_n). \qquad (2.216)$$

Solving for $\vec{a}_{n+1}$ gives

$$\vec{a}_{n+1} = \vec{a}_n - A^{-1} \vec{b}. \qquad (2.217)$$

---

[6] As in section 1.7.4, the dagger distinguishes row vectors from the column vectors.

[7] The Hessian matrix is the matrix of the second derivatives of a function of, say, M variables.

[8] The basic idea (in one dimension) was already known to Arabic mathematicians. Newton re-invented it and Raphson gave a systematic description in his analysis book [Verbeke and Cools (1995)].

Therefore, we have to calculate the inverse Hessian matrix A^{-1}. This can be done with the subroutine mat_gau.f, which implements Gauss-Jordan elimination as described in appendix A.2. However, it may happen that equation (2.213) provides a poor approximation of the local shape of the function $\chi^2(\vec{a})$. In that case, all one can do is to move a suitable step down in the direction of the gradient, *i.e.*,

$$\vec{a}_{n+1} \;=\; \vec{a}_n - c\,\vec{b} \qquad\qquad (2.218)$$

where the constant c is small enough not to overshoot the downhill direction. This is called a **steepest descent method**.

A method which varies smoothly between the inverse Hessian (2.217) and the steepest decent (2.218) definitions of $\vec{a}_{n+1}$ was proposed by [Levenberg (1944)]. Steepest decent is used far from the minimum and it is switched to the Hessian definition when the minimum is approached. This was then modified and perfected by [Marquardt (1963)].

Let us begin with some technical details. In the following we shall use the matrix

$$A_{\text{mat}} \;=\; (a_{jk}) \ \ \text{with} \ \ a_{jk} = \sum_{i=1}^{N} \frac{1}{\sigma_i^2} \left[\frac{\partial\, y(x_i, \vec{a})}{\partial\, a_j} \frac{\partial\, y(x_i, \vec{a})}{\partial\, a_k} \right]\Bigg|_{\vec{a}=\vec{a}_n} \qquad (2.219)$$

instead of (2.215). In the range where the Newton-Raphson iteration (2.217) is useful, the second derivatives of $y(x_i; \vec{a})$ in equation (2.215) can be neglected, because they are multiplied with the fluctuations of the data around the fit. Sufficiently close to a meaningful fit this averages out and further away the Newton-Raphson step makes little sense in the first place. We also removed the factor of two in front of the sum in (2.215) from the definition of the matrix elements. The iteration equations (2.217) and (2.218) are unaffected when we redefine the vector $\vec{b}$ and the matrix A by a factor of common magnitude. For $\vec{b}$ we absorb the factor -2 into the definition. Instead of (2.214) we use

$$b_j \;=\; \sum_{i=1}^{N} \frac{[y_i - y(x_i; \vec{a})]}{\sigma_i^2} \frac{\partial\, y(x_i; \vec{a})}{\partial\, a_j}\Bigg|_{\vec{a}=\vec{a}_n} \qquad (2.220)$$

for the vector components of $\vec{b}$. The remnant (2.219) of the Hessian matrix is often called the *curvature matrix*.

Levenberg suggested to combine the Newton-Raphson (2.217) and the steepest descent (2.218) steps on the level of the linear equation (2.216)

$$(A_{\text{mat}} + \lambda \mathbb{1}) \, \Delta \vec{a} \; = \; \vec{b} \; \text{ with } \; \Delta \vec{a} \; = \; \vec{a}_{n+1} - \vec{a}_n$$

where (2.220) defines $\vec{b}$. The parameter λ multiplies the unit matrix and can be adjusted. Up to the modification of the matrix A, small λ leads back to equation (2.216) and, therefore, the Newton-Raphson step (2.217), whereas large λ approaches the steepest descent equation (2.218) with $c = 1/\lambda$. Marquardt made an important practical improvement by replacing the unit matrix with the diagonal elements of A_{mat}. The equation becomes

$$\sum_{j=1}^{N} \left(a_{ij} + \lambda \, \delta_{ij} \, a_{ij} \right) \Delta a_j \; = \; b_i \tag{2.221}$$

where δ_{ij} is the Kronecker delta (1.92). Marquardt's insight is that, due to the last term of the quadratic approximation (2.213), the diagonal of the matrix A_{mat} contains more information than the gradient alone. For given λ, the solution $\Delta \vec{a}$ of equation (2.221) is calculated with the subroutine mat_gau.f of appendix A.2. To adjust λ, the method uses a trust-region approach: Initially a small value of λ (10^{-3} is our default) is used. Then, each time when equation (2.221) is solved $\chi^2(\vec{a}_{n+1})$ is calculated. If $\chi^2(\vec{a}_{n+1}) \geq \chi^2(\vec{a}_n)$ holds, λ is multiplied by a substantial factor (10 is our default), otherwise λ is divided by this factor. The iteration is stopped when the relative improvement in χ^2 becomes marginal, *i.e.*,

$$0 < \frac{\chi^2 - \chi^2_{\text{try}}}{\chi^2} < \epsilon \,, \tag{2.222}$$

where our default value for ϵ is 10^{-4}. The left-hand side of this equation is needed to avoid a termination of the iteration when the direction is simply not yet well adjusted. Further, care is taken to ensure that the routine still works when the number of data agrees with the number of fit parameters, NDAT=NFIT. In that case equation (2.222) cannot be applied, because χ^2 itself approaches zero. Instead we use $\chi^2 < \epsilon^2$.

User package

Marquardt's method is implemented in our general fit subroutine fit_g.f of ForLib. A call to

FIT_G(NDAT, X, Y, SIGY, NFIT, A, COVAR, AMAT, CHI2, Q, SUBG, LPRI) (2.223)

needs on input the data arrays X(NDAT), Y(NDAT), SIGY(NDAT), an initial guess of the fit parameter array A(NFIT), the dimensions NDAT and NFIT, and the print variable LPRI. For LPRI=.TRUE. the actual value of χ^2 will be printed after each iteration. Most important, the user has to provide the fit subroutine SUBG in the form

$$\text{SUBG(XX, A, YFIT, DYDA, NFIT)} . \qquad (2.224)$$

On input SUBG needs the argument XX of the model function (2.178) together with the fit parameter array A(NFIT) and its dimension NFIT. On output SUBG returns with YFIT the value of the fit function and in the array DYDA(NFIT) its derivatives with respect to the parameters.

The fit routine fit_g.f overwrites on output the array A(NFIT) with the best estimate of the fits parameters and returns in COVAR(NFIT,NFIT) their covariance matrix (1.87). Its diagonal elements are the estimated standard deviations (error bars) of the fit parameters. The array AMAT(NDAT,NDAT) contains the curvature matrix (2.219) at the estimated minimum of χ^2. CHI2 returns χ^2 and Q is the corresponding goodness of fit (2.183). Note that the number of degrees of freedom is now NDAT-NFIT, *i.e.*, CHI2/(NDAT-NFIT) corresponds (approximately) to χ^2 per degree of freedom (compare table 2.5 for selected fractiles).

The subroutine FIT_G (2.223) delegates the calculation of the curvature matrix A_{mat} (2.219) and the vector $\vec{b}$ (2.220) to a further subroutine,

$$\text{FIT\_GC(NDAT, X, Y, SIGY, NFIT, A, AMAT, DYDA, CHI2, Q, SUBG)} . \qquad (2.225)$$

As the only place where FIT_GC is ever used is within fit_g.f, it is simply included in this file of ForLib. Note, that the parameter λ of equation (2.221) is kept inside the subroutine fit_g.f. If desired, the default schedule can be modified there.

After the fit a call to the subroutine

$$\text{FIT\_GGNU(IUO, IUD, NDAT, NFIT, X, Y, SIGY, A)} \qquad (2.226)$$

prepares the data files data.d and gfit.d for plotting with the gnuplot driver file gfit.plt.

The routines are similarly organized as those for linear fitting. **The folder fit_g of ForProg contains the main program gfit.f, the gnuplot driver program gfit.plt and the SUBG routines (2.224) which are listed in table 2.15.** The chosen SUBG routine has to be transferred into the main program via an include statement of the form (2.207). Some

Table 2.15 Subroutines for inclusion in gfit.f, which are provided in ForProg/fit_g. NFIT denotes the number of parameters which has to be set in gfit.f.

Subroutine	Fit form	NFIT
subg_1ox.f	$y = a_1 + a_2/x$	2
subg_exp.f	$y = a_1 e^{a_2 x}$	2
subg_exp2.f	$y = a_1 e^{a_2 x} + a_3 e^{a_4 x}$	4
subg_linear.f	$y = a_1 + a_2 x$	2
subg_power.f	$y = a_1 x^{a_2}$	2
subg_power2.f	$y = a_1 x^{a_2} + a_3 x^{a_4}$	4
subg_power2f.f	$y = a_1 x^{a_2} (1 + a_3 x^{a_4})$	4
subg_power2p.f	$y = e^{a_1} x^{a_2} + e^{a_3} x^{a_4}$	4

*.D data files are also contained in ForProg/fit_g. Details and the use of some of the SUBG routines are explained in the following examples.

2.8.2.1 *Examples*

It is assignment 4 to repeat the fits of figures 2.6 and 2.8 using the subroutine (2.223). For figure 2.8 the nonlinear functions should now be used, and not their linear equivalents.

To give one more elaborate illustration, we use the [Bhanot *et al.* (1987)] data of table 2.14 to extend the estimate (2.212) of ν from the two parameter fit (2.204) to the four parameter fit

$$y(x) = a_1 x^{a_2} (1 + a_3 x^{a_4}) . \qquad (2.227)$$

The critical exponent from this fit is

$$\frac{1}{\nu} = -a_2 = -1.598 \pm 0.003 \quad \text{or} \quad \nu = 0.6257 \pm 0.0012 \qquad (2.228)$$

and the goodness of fit is $Q = 0.74$. This is now consistent in contrast to the fit depicted in figure 2.8. To see that the nice looking fit of figure 2.8 is indeed not particularly good, we subtract the fitted values from the data and plot the results, for which we give the error bars of the data, in figure 2.9. This is done for the two parameter fit as well as for our four parameter fit. We see that the results from the four parameter fit are nicely consistent with zero, whereas this is not the case for the two parameter fit. It comes as no surprise that the ν estimate (2.228) is closer than the ν estimate (2.212) to the value $\nu = 0.6295\,(10)$, which Bhanot *et al.* give as their best value. It is assignment 5 to reproduce figure 2.9 and the results of equation (2.228).

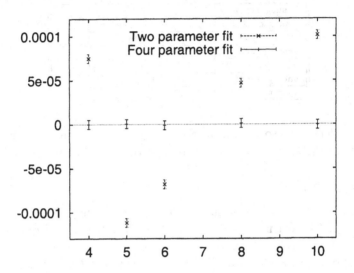

Fig. 2.9 Data of Bhanot *et al.*, see table 2.14, minus the fitted values: 1. For the two parameter fit (2.204) and 2. for the four parameter fit (2.227).

A few final remarks:

(1) When the subroutine fit_g.f does not work properly, the most likely reason is an error in the derivatives of the trial function. Use an algebraic program like Maple or Mathematica to check them again.

(2) Another reason why the subroutine fit_g.f returns occasionally strange results is that the initial guess values of the parameters are too far away from the relevant minimum. Convergence may then be into a local minimum, which can be so bad that the goodness of fit Q comes out as an underflow. The only remedy is to find better start values. If they exist, that can be a subtle task, worse otherwise. A method which searches for a global minimum, like simulated annealing [Kirkpatrick *et al.* (1983); Press *et al.* (1992)], can provide help in some cases.

(3) The Levenberg-Marquardt method can, of course, also be used for gen-

eral linear fits to a polynomial

$$y(x) = a_1 + a_2 x + \cdots + a_{n+1} x^n , \qquad (2.229)$$

although more direct methods are then available. See for instance the section about general linear least square fits in *Numerical Recipes* [Press *et al.* (1992)].

(4) Sometimes one wants to restrict the range of some of the fit parameters. From the program files listed in table 2.15 subg_power2p.f is an example. It gives an implementation of a power law fit with two exponents and positive linear coefficients. This can be far more stable than the fit of subg_power2.f. Of course, one has to be sure that the coefficients are positive.

2.8.3 *Assignments for section 2.8*

(1) Use lfit.f of ForProg/fit_1 to reproduce the results of figures 2.6 and 2.7. Note that lfit.f reads the input data using the default file name of the unit IUD which is preset to IUD=10. Under Linux this means, you have to copy the data of brandt.D to fort.10 before you run lfit.f. Gnuplot users may generate plots with the drivers lfit.plt and ellipse.plt provided in ForProg/fit_1. .

(2) Copy the data file berg.D to fort.10 and use lfit.f to find the estimate (2.209). Do not forget to change the include statement for subl_linear.f in lfit.f to subl_1ox.f. Finally, have a look at the plot generated by fit.plt.

(3) Copy the data file bhanot.D to fort.10 and use lfit.f to find the estimate (2.212). Do not forget to include subl_power.f in lfit.f and have a look at the plot generated by fit.plt.

(4) Use the general Levenberg-Marquardt fitting routine fit_g.f (2.223) of ForProg/fit_g to repeat the fits of the previous three assignments. Use the functions in their nonlinear form whenever this is appropriate. The SUBG (2.224) routines needed are subg_linear.f, subg_1ox and subg_power.f of ForProg/fit_g.

(5) Use the general Levenberg-Marquardt fitting routine fit_g.f of ForProg/fit_g together with subg_power2f.f to reproduce figure 2.9 and the ν estimate of equation (2.228).

Chapter 3

Markov Chain Monte Carlo

So far we sampled statistically independent events. This requires a method to generate the desired probability distribution directly. But for many interesting problems such a method does not exist. It may still be possible to generate statistically independent event, but only in such a way that the statistically important contributions are missed, *i.e.*, they occur only with tiny probabilities. The way out are in many cases Markov chain Monte Carlo (MC) simulations[1]. A **Markov process** generates event $n + 1$ stochastically from event n, without information about event $n - 1$ needed at that point. A **MC algorithm** arranges the statistical weights of the Markov process so that the desired probability distribution is asymptotically (*i.e.*, for large n) approached. This enables the **importance sampling** discussed in this chapter. For statistical physics the problem to generate configurations of the Gibbs canonical ensemble with their correct Boltzmann weights (also called Gibbs weights) was solved in the well-known work by Nicholas Metropolis, Arianna Rosenbluth, Marshall Rosenbluth, Augusta Teller and Edward Teller [Metropolis *et al.* (1953)]. A price paid in MC simulations is that subsequent configurations become correlated – possibly strongly. This has consequences for the error analysis, which are discussed in chapter 4.

In this chapter we focus on practical aspects of performing MC calculations. Some basic knowledge of *statistical physics* will help the understanding, but is not an absolutely necessary prerequisite. Preliminaries together with the 2d Ising model are summarized in the next section. Good statistical physics textbooks are those by [Landau and Lifshitz (1999)], by [Huang (1987)] and by [Pathria (1972)].

---

[1] We follow the notation of the physics literature, where MC is widely used to denote Markov chain Monte Carlo simulations. In mathematics and computer science the more accurate notation MCMC is preferred.

Convergence of importance sampling (including the Metropolis method) is proven in section 3.2. Additional details and the heat bath algorithm are discussed for the $O(3)$ σ-model, which is also known as the Heisenberg ferromagnet. Section 3.3 introduces generalized Potts models in arbitrary dimensions and presents Fortran code for MC simulations of these models. Potts models, which include the Ising model as a special case, are sufficiently involved to exhibit many relevant features of MC simulations, while they are simple enough to allow us to focus on the essentials. MC simulations of **continuous systems** face a number of additional difficulties. In section 3.4 we illustrate them using $O(n)$ σ-models. The generalization of the concepts of this chapter to more involved systems is rather straightforward, but not pursued in volume I, because this leads away from the enabling techniques.

3.1 Preliminaries and the Two-Dimensional Ising Model

MC simulations of systems described by the Gibbs canonical ensemble aim at calculating estimators of physical observables at a temperature T. In the following we choose units so that the Boltzmann constant becomes one and, consequently,

$$\beta = 1/T \ . \tag{3.1}$$

Let us consider the numerical calculation of the **expectation value** of an **observable** $\mathcal{O}$. Mathematically all systems on a computer are discrete, because a finite word length has to be used. Hence, the expectation value is given by the sum

$$\widehat{O} = \widehat{O}(\beta) = \langle \mathcal{O} \rangle = Z^{-1} \sum_{k=1}^{K} \mathcal{O}^{(k)} \, e^{-\beta \, E^{(k)}} \tag{3.2}$$

where

$$Z = Z(\beta) = \sum_{k=1}^{K} e^{-\beta \, E^{(k)}} \tag{3.3}$$

is called **partition function**. The index $k = 1, \ldots, K$ labels **all configurations** of the system and $E^{(k)}$ is the (internal) energy of configuration k. The configurations are also called **microstates**. To distinguish the configuration index from other indices, we put it in parenthesis.

A simple model of a ferromagnet is the Ising model [Ising (1925)]. Its energy is given by

$$E_0 = - \sum_{<ij>} s_i \, s_j \; . \tag{3.4}$$

The sum is over the sites of a hypercubic[2] d-dimensional lattice and the Ising spins take the values

$$s_i = \pm 1 \;\; \text{for} \;\; i = 1, \dots, N \;\; \text{with} \;\; N = \prod_{i=1}^{d} n_i \tag{3.5}$$

where n_i are the edge lengths of the lattice. For the Ising model the number of states is $K = 2^N$. Already for a moderately sized lattice, say 100×100, this is a tremendously large number, namely $2^{10\,000}$ which by far exceeds the estimated number of 10^{80} protons in the universe. With the exception of very small systems it is therefore impossible to carry out the sums in equations (3.2) and (3.3). Instead, the large number suggests a statistical evaluation through generating a representative subsample.

On an infinite lattice the $2d$ Ising model was solved by [Onsager (1944)]. See, for instance, [Huang (1987)] for a textbook treatment. A second oder phase transition is found and exact solutions are known for many observables. This makes the $2d$ Ising model an ideal testing ground for simulation methods as well as for perturbative expansions. Of course, the availability of an exact solution is related to the simplicity of the model in $2d$. In $3d$ the solution is not known, what is the typical case for realistic models of physical processes. The $1d$ Ising model is still far simpler than the $2d$ model, and to derive its exact solution [Ising (1925)] is often given as an exercise to graduate students. But the $1d$ model does not exhibit a phase transition at a finite temperature, although some interesting physics is found for $\beta \to \infty$.

Among the prominent features of the $2d$ model are the exact equation for its **critical temperature**, also called **Curie temperature**,

$$\beta_c \;\; = \;\; \frac{1}{T_c} \;\; = \;\; \frac{1}{2} \, \ln(1 + \sqrt{2}) = 0.4406867935 \dots \tag{3.6}$$

and exact results for **critical exponents**, which govern the singularities at the phase transition:

$$\alpha = 0, \;\; \beta = \frac{1}{8}, \;\; \gamma = \frac{7}{4}, \;\; \delta = 15, \;\; \nu = 1 \;\; \text{and} \;\; \eta = \frac{1}{4} \, . \tag{3.7}$$

---

[2]Other lattice geometries are occasionally used in the literature.

The (positive) critical exponents are defined by the behavior of physical observables near the critical temperature:

$$\text{the specific heat } C \sim |t|^{-\alpha}, \tag{3.8}$$

$$\text{the magnetization } M \sim |t|^{\beta}, \tag{3.9}$$

$$\text{the susceptibility } \chi \sim |t|^{\gamma}, \tag{3.10}$$

the $t = 0$ magnetization as function of an external magnetic field

$$M \sim H^{-1/\delta}, \ (t = 0), \tag{3.11}$$

$$\text{the correlation length } \xi \sim |t|^{-\nu} \tag{3.12}$$

and $t = 0$ power law correlations

$$\langle s_i s_j \rangle \sim |\vec{x}_i - \vec{x}_j|^{-d+2-\eta} \text{ for } |\vec{x}_i - \vec{x}_j| \to \infty, \ (t = 0), \tag{3.13}$$

where the **reduced temperature** t is given by

$$t = \frac{T - T_c}{T_c} = \frac{\beta_c}{\beta} - 1. \tag{3.14}$$

In MC simulations the evaluation of critical exponents is possible by many methods, see [Pelissetto and Vicari (2002)] for a review of the subject. Particularly useful are finite size scaling (FSS) extrapolations, relying on the FSS theory originally introduced by [Fisher (1971)].

It appears that the partition function (3.3), being a sum of analytic terms, must be an analytic function of the temperature. Indeed, it is in a finite volume. However, one is interested in the infinite volume limit, where the sum contains an infinite number of terms and analyticity may be lost. If analyticity is lost, this must be due to some subtlety in taking the infinite volume limit, expected to be associated with the development of long-range correlations between spins (or whatever the dynamical variables are). A powerful method to identify non-analytic points is to construct **order parameters**. Usually an order parameter characterizes a symmetry of the system, so that the order parameter is zero (or constant) in one phase and not in the other. For the Ising ferromagnet this is the magnetization, which vanishes in a state with reflection symmetry $s_i \to -s_i$. If one has an order parameter of this kind one can be sure that there is a singularity in passing from one phase to the other, because an analytic function cannot vanish in a finite range without vanishing identically. On the other hand, one finds phase transitions without order parameters.

If the transition is temperature driven and second order, the (infinite volume) correlation length goes, as function of the temperature, continuously to infinity at the critical point. Sufficiently close to the transition point, the correlation length will greatly exceed any microphysical length associated with the problem (*e.g.* the lattice spacing). These observations lead to the concepts of **scale invariance** and **universality**, which form the cornerstone of the modern theory of phase transitions. Scale invariance follows from the idea that non-trivial correlations on length scales much larger than the microscopic one can only depend on the ratio of the length in question to the diverging correlation length. There should be just one fundamental diverging length, since there is only one parameter, which must be tuned to reach the critical point, *i.e*, all divergent quantities should be given as functions of the reduced temperature (3.14). Universality follows from the idea that likely only modes with long-wavelength fluctuations in equilibrium are important which are required by the symmetries of the order parameter. Together these principles tell us that for the investigation of singularities of thermodynamic functions near second order phase transitions, it is sufficient to look for the simplest scale-invariant model in the appropriate dimension with the required symmetry. In this way symmetries define universality classes, which share the same critical exponents.

The order parameter concept works also for predicting that a particular phase transition is first order, but in contrast to the second order case one has no powerful predictions for the physical quantities of a first order transition.

For the purposes of testing our first MC algorithms the paper by [Ferdinand and Fisher (1969)], build on work by [Kaufmann (1949)], is most useful. They derive $2d$ Ising model results for $m \times n$ lattices, which allow for direct comparisons with stochastic results. The program

$$\texttt{ferdinand.f} \qquad (3.15)$$

in the folder **ForProg/Ferdinand** is based on their paper. It calculates important physical observables and prepares their graphical presentation. To give a example, figure 3.1 compares for a 20×20 lattice the exact **expectation value of the energy per spin**

$$\widehat{e}_{0s}(\beta) = \widehat{E_0}(\beta)/N \qquad (3.16)$$

from **ferdinand.f** with its stochastic estimates of a multicanonical simulation (compare assignment 3 of chapter 5.1.9).

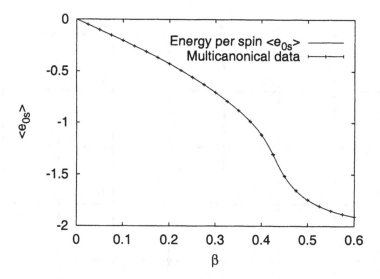

Fig. 3.1 Expectation value of the energy per spin $\widehat{e}_{0s}$ versus β for the 2d Ising model on an 20 × 20 lattice. The full line is the exact result of [Ferdinand and Fisher (1969)]. The data points are from a multicanonical simulation, see assignment 3 in chapter 5.1.9. This figure was first published in [Berg (2003)].

3.1.1 *Lattice labeling*

Let us consider lattices with periodic boundary conditions. It is convenient to label all sites of the lattice by a single integer variable, is, which we call **site number**. The dimension of the lattice is given by another integer variable, nd. The Cartesian coordinates of a site are chosen to be

$$x^i = 0, \ldots, n^i - 1 \quad \text{for} \quad i = 1, \ldots, d. \tag{3.17}$$

The lattice size dimensions n^i are stored in the array nla(nd) (we allow an asymmetric lattice). Let the coordinates of one site be given by an integer array ix(nd). The Fortran function

$$\text{isfun}(\text{ix}, \text{nla}, \text{nd}) \tag{3.18}$$

converts the coordinates of the given site to a site number according to the

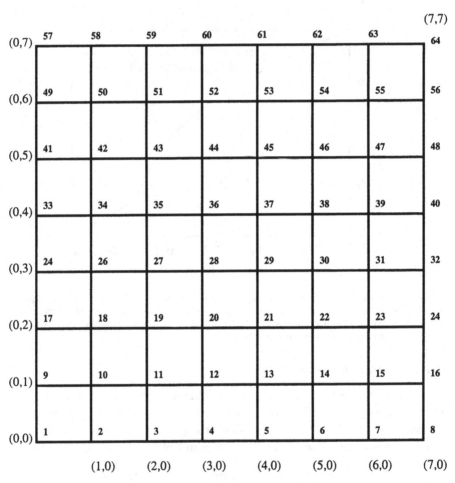

Fig. 3.2 An 8 × 8 lattice in the Cartesian coordinates.

formula

$$\text{is} = 1 + \sum_{id=1}^{nd} \text{ix}(\text{id}) \times \text{nsa}(\text{id}) \text{ with } \text{nsa}(\text{id}) = \begin{cases} 1 & \text{for } \text{id} = 1, \\ \prod_{i=1}^{id-1} \text{nla}(i) & \text{for } \text{id} > 1. \end{cases}$$

$$(3.19)$$

See figure 3.2 for nd=2. Once we have set the values for nla(i) we obtain the number of sites of the lattice, ns, from the function

$$\text{nsfun}(\text{nla}, \text{nd}) = \prod_{i=1}^{\text{nd}} \text{nla}(i) .$$ (3.20)

To get the coordinates for a given site number is is achieved by the subroutine

$$\text{ixcor}(\text{ix}, \text{nla}, \text{is}, \text{nd})$$ (3.21)

which returns in the array ix(nd) the Cartesian coordinates corresponding to the site number is. Used is an iteration procedure relying on Fortran integer division (*i.e.*, $1 = [5/3] = [5/5]$, $0 = [4/5]$, etc.):

$$\text{ix}(\text{nd}) = \frac{(\text{is} - 1)}{[\text{ns} / \text{nla}(\text{nd})]}$$

and for $\text{id} = \text{nd} - 1, \ldots, 1$

$$\text{ix}(\text{id}) = [(\text{js}(\text{id}) - 1) / \text{nspart}(\text{id})] ,$$

where

$$\text{js}(\text{id}) = \text{ns} - \sum_{i=\text{nd}}^{\text{id}+1} \text{ix}(i)\,\text{nspart}(i) \text{ and } \text{nspart}(\text{id}) = \text{ns} \prod_{i=\text{nd}}^{\text{id}+1} (\text{nla}(i))^{-1} .$$

The subroutine

$$\text{ipointer}(\text{ipf}, \text{ipb}, \text{nla}, \text{ix}, \text{ns}, \text{is}, \text{nd})$$ (3.22)

calculates the site numbers of the neighbors of a given site is. Input are nla, ix, ns, is and nd. On output the **pointer arrays**

$$\text{ipf}(\text{id}, \text{is}) \text{ and } \text{ipb}(\text{id}, \text{is})$$

are returned for $\text{id} = 1, \ldots, \text{nd}$, where ipf gives the site numbers of the nd neighbors in forward direction and ipb the site numbers of the nd neighbors in backward direction. To achieve this goal the subroutine ixcor (3.21) is first used to calculate the Cartesian coordinates for the site number is. From it the coordinates of the neighbors are calculated, taking into account the periodic boundary conditions:

(1) For the forward direction (ipf)

$$\text{ix}(\text{id}) \rightarrow \text{mod}(\text{ix}(\text{id}) + 1, \text{nla}(\text{id})), \quad \text{id} = 1, \ldots, \text{nd}.$$

(2) For the backward direction (ipb)

$$ix(id) \rightarrow mod(ix(id) - 1 + nla(id), nla(id)), \quad id = 1, ..., nd.$$

Then, the function isfun (3.18) is used to get the site numbers, which are stored in the arrays ipd and ipb. For a given lattice the pointer arrays ipf and ipb are initialized by a call to the subroutine

$$lat\_init(ns, nd, ipf, ipb, nla, ix, nlink) \qquad (3.23)$$

which initializes also the variable nlink=nd*ns. The routine needs ns, nd and nla assigned on input and ix as workspace and it combines applications of the routines (3.22), (3.18), (3.20) and (3.21).

For $d = 3$ the program

$$lat3d.f \text{ of } ForProg/lat \qquad (3.24)$$

illustrates the use of lat_init.f. Table 3.1 collects the thus obtained site numbers, their coordinates and their pointers for a $2^2 4$ lattice. That the forward and backward pointers agree in some instances is an artifact of the small lattice size, see assignment 1 for more cases. For use in MC simulations the results are stored in the common block lat.com

$$common \text{ /lat/ } ns, nla(nd), ipf(nd, ms), ipb(nd, ms) \qquad (3.25)$$

which is transferred into the programs through

$$include \text{ '../../ForLib/lat.com'}$$

statements, *i.e.*, the file lat.com is kept in the directory ForLib. In this common block ns is the total number of sites, nla stores the lattice size dimensions, ipf and ipb are the forward and backward pointer arrays. The dimensional arguments needed in this common block are assigned in the parameter statement lat.par

$$parameter(nd = 3, ml = 10, ms = ml * *nd, mlink = nd * ms, n2d = 2 * nd)$$
$$(3.26)$$

where nd is the dimension of the system, ml the maximum lattice length encountered, ms the maximum number of sites and mlink the maximum number of links allowed. The parameters nd and ml have to be chosen by the user. The choice in equation (3.26) implies $3d$ lattices with the total number of sites less or equal to 10^3. The statement lat.par is transferred into the programs through include 'lat.par' statements, *i.e.*, the file lat.par is kept in the folder where the program lat3d.f is compiled.

Table 3.1 Bookkeeping for a $2^2 4$ lattice: The site number **is**, its coordinates **ix**, the forward pointers **ipf** and the backward pointers **ipb** are given. These results are obtained with the program **lat3d.f** of the directory **ForProg/lat**, see assignment 1.

is	ix(1)	(2)	(3)	ipf(is,1)	(,2)	(,3)	ipb(is,1)	(,2)	(,3)
1	0	0	0	2	3	5	2	3	13
2	1	0	0	1	4	6	1	4	14
3	0	1	0	4	1	7	4	1	15
4	1	1	0	3	2	8	3	2	16
5	0	0	1	6	7	9	6	7	1
6	1	0	1	5	8	10	5	8	2
7	0	1	1	8	5	11	8	5	3
8	1	1	1	7	6	12	7	6	4
9	0	0	2	10	11	13	10	11	5
10	1	0	2	9	12	14	9	12	6
11	0	1	2	12	9	15	12	9	7
12	1	1	2	11	10	16	11	10	8
13	0	0	3	14	15	1	14	15	9
14	1	0	3	13	16	2	13	16	10
15	0	1	3	16	13	3	16	13	11
16	1	1	3	15	14	4	15	14	12

The rationale for keeping lat.par and lat.com in different directories is: We do not intend to change lat.com, but if such a change is unavoidable it should propagate to all routines which include lat.com. On the other hand, we expect frequent changes of the lattice parameters of lat.par, and the actual values are chosen for a particular run of the program.

Besides the parameters in lat.par **the array nla needs to be initialized**. This is done by including a file lat.dat. The array nla contains the nd lattice lengths so that

$$\prod_{id=1}^{nd} nla(id) \leq ml^{nd}$$

holds. The purpose of the lat.dat complication is to allow for asymmetric lattices. An example of lat.dat, consistent with the choice (3.26) of lat.par, is

$$\text{data nla}/6, 6, 12/ \tag{3.27}$$

which is permissible because of $6^2 12 = 432 < 10^3$. The most common choice in our examples are symmetric lattices, *e.g.*, data nla/3 * 10/ for (3.26).

The statement `lat.dat` is transferred into the program in the same way as `lat.par`, *i.e.*, `lat.dat` is kept in the subdirectory where the program is compiled.

The routines (3.18), 3.20), (3.21), (3.22) and (3.23) of this section are part of our program library ForLib. We are now ready to turn to simulations of statistical physics models.

3.1.2 *Sampling and Re-weighting*

For the Ising model (and most systems of statistical physics) it is straightforward to **sample statistically independent configurations**. We simply have to generate N spins, each either up or down with 50% likelihood. This is called **random sampling**. On its left-hand side figure 3.3 depicts a thus obtained histogram of 100 000 entries for the $2d$ Ising model **energy per spin**

$$e_{0s}(\beta) = E_0(\beta)/N \ . \tag{3.28}$$

Note that is is very important to distinguish the energy measurements on single configurations (3.28) from the expectation value (3.16) of this quantity. The expectation value $\widehat{e}_{0s}$ is a single number, while e_{0s} fluctuates. From the measurement of many e_{0s} values one finds estimators of its moments (1.59). In particular the mean is of importance, for which the estimator is denoted by $\overline{e}_{0s}$. In contrast to the expectation value $\widehat{e}_{0s}$ itself, its estimator $\overline{e}_{0s}$ fluctuates. The histogram from random sampling corresponds to $\beta = 0$ in the partition function (3.3). Also shown in figure 3.3 are histograms of 100 000 entries each at $\beta = 0.2$ and $\beta = 0.4$, obtained with the MC sampling procedure which is the subject of this chapter. The rather small-sized 20×20 system is used to make the results easily reproducible.

The histogram entries at $\beta = 0$ can be re-weighted so that they correspond to other β values. We simply have to multiply the entry corresponding to energy E by $c_\beta \exp(-\beta E)$ as follows from (3.3), where the constant c_β is adjusted to keep the normalization unchanged (*i.e.*, the sum over the re-weighted histogram values has to agree with the sum over the original histogram values). Similarly histograms corresponding to the Gibbs ensemble at some value β_0 can be re-weighted to other β values. This is implemented by our subroutine

$$\texttt{potts\_rwght.f} \quad \text{of} \quad \texttt{ForLib} \tag{3.29}$$

which is reproduced in table 3.2. This routine is written for Potts models

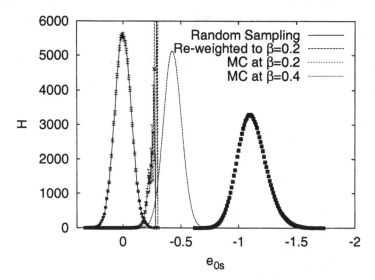

Fig. 3.3 Energy histograms of 100 000 entries each for the 2*d* Ising model on an 20 × 20 lattice are depicted. By Random Sampling we obtain the histogram of statistically independent configurations at $\beta = 0$. Histograms at $\beta = 0.2$ and $\beta = 0.4$ are generated by Markov chain MC. Re-weighting of the $\beta = 0$ random configurations to $\beta = 0.2$ is shown to fail. See assignment 2 of the present section and assignment 2 of section 3.3.5.

(section 3.3), which include the Ising model as a special case. Instead of the configuration energy $E_0^{(k)}$ (3.4) an action variable iact$^{(k)}$ (3.78) is used, which reads for the Ising model

$$\text{iact} = -(E - d\,N)/2 \ . \tag{3.30}$$

Care has to be taken to ensure that the involved arguments of the exponential function do not become too large. This is done in potts_rwght.f by first calculating the mean action, act0m, and re-weighting with respect to the difference from the mean.

For FSS investigations of second order phase transitions the usefulness of the re-weighting has been stressed by [Ferrenberg and Swendsen (1988)]. **Ferrenberg-Swendsen re-weighting** is an important technical tool when

Table 3.2 The subroutine `potts_rwght.f` of ForLib is depicted. It re-weights histograms of the variable iact (3.30), (3.77) from β_0 to β.

```
      SUBROUTINE POTTS_RWGHT(nlink,ha,hae,beta0,beta,hb,hbe)
C Copyright, Bernd Berg, Dec 9 2000.
C Re-weighting of the histogram ha() and its error bars hae() at beta0
C             to the histogram hb() and its error bars hbe() at beta.
      include 'implicit.sta'
      include 'constants.par'
      dimension ha(0:nlink),hae(0:nlink),hb(0:nlink),hbe(0:nlink)
c
      hasum=zero
      act0m=zero
      do iact=0,nlink
        hasum=hasum+ha(iact)
        act0m=act0m+iact*ha(iact)
      end do
      act0m=act0m/hasum
c
      hbsum=zero
      do iact=0,nlink
        hb(iact)=zero
        hbe(iact)=zero
        if(ha(iact).gt.half) then
          a=(beta-beta0)*2*(iact-act0m)
          hb(iact)=exp(log(ha(iact))+a)
          hbe(iact)=exp(log(hae(iact))+a)
          hbsum=hbsum+hb(iact)
        end if ! Else hb(iact) is, as ha(iact), zero.
      end do
      factor=hasum/hbsum
      do iact=0,nlink
        hb(iact)=factor*hb(iact)
        hbe(iact)=factor*hbe(iact)
      end do
c
      RETURN
      END
```

one likes to estimate critical exponents, like those given in equation (3.7), numerically. For the history of re-weighting procedures see [Ferrenberg and Swendsen (1989b)]. A time-series re-weighting program, which prepares jackknife error analysis, can be found in appendix C.

In figure 3.3 re-weighting is done from $\beta_0 = 0$ to $\beta = 0.2$. By comparison to the histogram from a Metropolis MC calculation at $\beta = 0.2$, the result is seen to be disastrous. The reason is easily identified: In the range where the $\beta = 0.2$ histogram takes on its maximum, the $\beta = 0$ histogram has not a single entry, *i.e.*, our naive sampling procedure misses the **important configurations** at $\beta = 0.2$. As stressed in [Ferrenberg and Swendsen (1988)], re-weighting to new β values works only in a range $\beta_0 \pm \Delta\beta$, where $\Delta\beta \to 0$ in the infinite volume limit (details are given below). These are the β values for which the energies measured in the β_0 simulation still cover the expectation values $\widehat{e}_{0s}(\beta)$. For the $\beta_0 = 0$ histogram of 100 000 entries a reasonably large statistic is only available down to $e_{0s} = -0.2$. According to the $\widehat{e}_{0s}(\beta)$ curve depicted in figure 3.1 this would allow to cover the range up to approximately $\beta = 0.1$.

3.1.2.1 *Important configurations and re-weighting range*

Let us determine the important contributions to the partition function (3.3). The partition function can be re-written as a sum over energies

$$Z = Z(\beta) = \sum_E n(E)\, e^{-\beta E} \tag{3.31}$$

where the unnormalized **spectral density** $n(E)$ is defined as the number of microstates k with energy E (remember, on the computer all numbers are discrete). For a fixed value of β, away from first and second order phase transitions, the energy probability density

$$P_\beta(E) = c_\beta\, n(E)\, e^{-\beta E} \tag{3.32}$$

is peaked around the average value $\widehat{E}(\beta)$ with a width proportional to $\sqrt{V}$, where V is the volume of the system. Here c_β is a normalization constant which we choose so that the normalization $\sum_E P_\beta(E) = 1$ holds.

That the width of the energy distribution is $\Delta E \sim \sqrt{V}$ follows from the fact that, away from phase transition points, the fluctuations of the $N \sim V$ lattice spins are only locally correlated, so that the magnitude of a typical fluctuations is $\sim \sqrt{N}$. The proportionality is the same as the one encountered for the sum of uniformly distributed ± 1 random numbers. From this we find also that the re-weighting range is $\Delta\beta \sim 1/\sqrt{V}$, so that $\Delta\beta E \sim \sqrt{V}$ can stay within the fluctuations of the system (the energy E is an extensive quantity $\sim V$).

Interestingly, the re-weighting range increases at second order phase transitions point, because critical fluctuations are larger than non-critical fluctuations. One has $\triangle E \sim V^x$ with $1/2 < x < 1$ and the requirement $\triangle \beta E \sim V^x$ yields $\triangle \beta \sim V^{x-1}$. For first order phase transitions one has a latent heat $\triangle V \sim V$, but this does not mean that the re-weighting range becomes of order one. In essence, the fluctuations collapse, because the two phases become separated by an interfacial tension. One is back to fluctuations within either of the two phases, *i.e.* $\triangle \beta \sim 1/\sqrt{V}$.

The important configurations at temperature $T = 1/\beta$ are at the energy values for which the probability density $P_\beta(E)$ is large. To sample them efficiently, one needs a procedure which generates the configurations with their Boltzmann weights

$$w_B^{(k)} = e^{-\beta E^{(k)}} . \tag{3.33}$$

The number of configurations $n(E)$ and the weights combine then so that the probability to generate a configuration at energy E becomes precisely $P_\beta(E)$ as given by (3.32). In the following we introduce a Markov process which allows to generate configurations with the desired weights.

3.1.3 *Assignments for section 3.1*

(1) Use the program lat3d.f of ForProg/lat to reproduce the results of table 3.1. Next, produce the same table for 2^3 and 3^3 lattices. Compile and run also the program lat4d.f of ForProg/lat and interpret the output.

(2) Reproduce the $\beta = 0$ histogram of figure 3.3 and its associated error bars. Subsequently, re-weight the histogram (including error bars) to $\beta = 0.2$. Note: The histogram relies on 100 000 independent data points, each generated by a call of potts_ran.f (default seeds are used for the random number generator). The error bars are found from the variance of the bimodal distribution (1.58).

3.2 Importance Sampling

For the canonical ensemble **Importance sampling** generates configurations k with probability

$$P_B^{(k)} = c_B \, w_B^{(k)} = c_B \, e^{-\beta E^{(k)}} \tag{3.34}$$

where the constant c_B is determined by the normalization condition $\sum_k P_B^{(k)} = 1$. The **state vector** $(P_B^{(k)})$, for which the configurations are the vector indices, is called **Boltzmann state**. When configurations are **stochastically** generated with probability $P_B^{(k)}$, the **expectation value** (3.2) becomes the **arithmetic average**:

$$\hat{\mathcal{O}} = \hat{\mathcal{O}}(\beta) = \langle \mathcal{O} \rangle = \lim_{N_K \to \infty} \frac{1}{N_K} \sum_{n=1}^{N_K} \mathcal{O}^{(k_n)} . \tag{3.35}$$

When the sum is truncated at some finite value of N_K (as in laboratory measurements), we obtain an **estimator of the expectation value**

$$\overline{\mathcal{O}} = \frac{1}{N_K} \sum_{k=1}^{N_K} \mathcal{O}^{(k_n)} \tag{3.36}$$

whose errors bars have to be determined by statistical analysis.

Normally, we cannot generate configurations k directly with probability (3.34). But they may be found as members of the equilibrium distribution of a dynamic process. In practice Markov chains are used. A **Markov process** is a particularly simple dynamic process, which generates configuration k_{n+1} stochastically from configuration k_n, so that no information about previous configurations $k_{n-1}, k_{n-2}, \ldots$ is needed. The elements of the Markov chain **time series** are the configurations. Assume that the configuration k is given. Let the transition probability to create the configuration l in one step from k be given by $W^{(l)(k)} = W[k \to l]$. The **transition matrix**

$$W = \left(W^{(l)(k)} \right) \tag{3.37}$$

defines the Markov process. Note, that this is a very big matrix, because its labels are the configurations (it is never stored in the computer!). To achieve our goal to generate configurations with the desired probabilities (3.34), the matrix W is required to satisfy the following properties:

(i) **Ergodicity:**

$$e^{-\beta E^{(k)}} > 0 \text{ and } e^{-\beta E^{(l)}} > 0 \quad \text{imply}: \tag{3.38}$$

an integer number $n > 0$ exists so that $(W^n)^{(l)(k)} > 0$ holds.

(ii) **Normalization**:

$$\sum_l W^{(l)(k)} = 1 .$$ (3.39)

(iii) **Balance**:

$$\sum_k W^{(l)(k)} e^{-\beta E^{(k)}} = e^{-\beta E^{(l)}} .$$ (3.40)

Balance means: The Boltzmann state (3.34) is an eigenvector with eigenvalue 1 of the matrix $W = (W^{(l)(k)})$.

To proceed, we have to elaborate a bit on the ensemble notation. By definition, an **ensemble** is a collection of configurations so that to each configuration k a probability $P^{(k)}$ is assigned, $\sum_k P^{(k)} = 1$. The **Gibbs or Boltzmann ensemble** E_B is defined to be the ensemble with probability distribution (3.34). An **equilibrium ensemble** E_{eq} of the Markov process is defined by its probability distribution P_{eq} satisfying

$$W P_{eq} = P_{eq} , \quad \text{in components} \quad P_{eq}^{(l)} = \sum_k W^{(l)(k)} P_{eq}^{(k)} .$$ (3.41)

The following statement and the remark after equation (3.43) are central for the MC method.

Statement: Under the conditions (i), (ii) and (iii) the Boltzmann ensemble is the **only** equilibrium ensemble of the Markov process.

Proof: Let us first define a distance between ensembles. Suppose we have two ensembles E and E', each of which is a collection of many configurations. Denote the probability for configuration k in E by $P^{(k)}$ and in E' by $P'^{(k)}$. We define the **distance** between E and E' to be

$$||E - E'|| = \sum_k |P^{(k)} - P'^{(k)}| ,$$ (3.42)

where the sum goes over all configurations. Suppose that E' resulted from the application of the transition matrix W to the ensemble E. We can compare the distance of E' from the Boltzmann ensemble with the distance of E from the Boltzmann ensemble:

$$\|E' - E_B\| = \sum_l \left| \sum_k W^{(l)(k)} \left(P^{(k)} - P_B^{(k)} \right) \right| \qquad \text{(using balance)}$$

$$\leq \sum_l \sum_k \left| W^{(l)(k)} \left(P^{(k)} - P_B^{(k)} \right) \right| \qquad \text{(using the triangle inequality)}$$

$$= \sum_k \left| P^{(k)} - P_B^{(k)} \right| = \|E - E_B\| . \qquad (3.43)$$

The last line is obtained by making use of the normalization condition $\sum_l W^{(l)(k)} = 1$ and of $W^{(l)(k)} \geq 0$. This shows that the Boltzmann state is a fixed point of the Markov process. We now study the approach of an arbitrary state to the fixed point.

The matrix W by itself is normally not ergodic. Instead, due to the finiteness of the system, a number n exists, so that the matrix $\mathcal{W} = W^n$ is ergodic. This means, all matrix elements of $\mathcal{W}$ are larger than zero. In particular

$$1 > w_{\min} = \min_{k,l} \left(\mathcal{W}^{(k)(l)} \right) > 0 \qquad (3.44)$$

holds. Let us assume that the state vectors P and P' are related by

$$P' = \mathcal{W} P . \qquad (3.45)$$

As shown in the following, this implies the inequality

$$\|E' - E_B\| \leq (1 - w_{\min}) \|E - E_B\| . \qquad (3.46)$$

Let $\epsilon = \|E - E_B\|$. We can decompose the contributions to $\|E - E_B\|$ in $P^{(k)} - P_B^{(k)} > 0$ and in $P^{(k)} - P_B^{(k)} < 0$ terms,

$$\|E - E_B\| = \sum_{k \in K^+} \left(P^{(k)} - P_B^{(k)} \right) + \sum_{k \in K^-} \left(P_B^{(k)} - P^{(k)} \right) .$$

Then

$$\sum_{k \in K^+} \left(P^{(k)} - P_B^{(k)} \right) = \sum_{k \in K^-} \left(P_B^{(k)} - P^{(k)} \right) = \epsilon/2 \qquad (3.47)$$

holds due to the normalization $\sum_k P^{(k)} = \sum P_B^{(k)} = 1$. Using $\mathcal{W} P_B = P_B$, we have

$$||E' - E_B|| =$$

$$\sum_l \left| \sum_{k \in K^+} \mathcal{W}^{(l)(k)} \left(P^{(k)} - P_B^{(k)} \right) - \sum_{k \in K^-} \mathcal{W}^{(l)(k)} \left(P_B^{(k)} - P^{(k)} \right) \right|$$

$$= \sum_{l \in L^+} \left(\sum_{k \in K^+} \mathcal{W}^{(l)(k)} \left(P^{(k)} - P_B^{(k)} \right) - \sum_{k \in K^-} \mathcal{W}^{(l)(k)} \left(P_B^{(k)} - P^{(k)} \right) \right)$$

$$+ \sum_{l \in L^-} \left(\sum_{k \in K^-} \mathcal{W}^{(l)(k)} \left(P_B^{(k)} - P^{(k)} \right) - \sum_{k \in K^+} \mathcal{W}^{(l)(k)} \left(P^{(k)} - P_B^{(k)} \right) \right)$$

$$\leq (1 - w_{\min})\,\epsilon/2 \; + \; (1 - w_{\min})\,\epsilon/2 \; = \; (1 - w_{\min})\,\epsilon \; . \tag{3.48}$$

Here L^+ is defined as the set of configurations l which fulfills the inequality

$$\sum_{k \in K^+} \mathcal{W}^{(l)(k)} \left(P^{(k)} - P_B^{(k)} \right) > \sum_{k \in K^-} \mathcal{W}^{(l)(k)} \left(P_B^{(k)} - P^{(k)} \right) ,$$

while L^- is the set of configurations l with

$$\sum_{k \in K^-} \mathcal{W}^{(l)(k)} \left(P_B^{(k)} - P^{(k)} \right) > \sum_{k \in K^+} \mathcal{W}^{(l)(k)} \left(P^{(k)} - P_B^{(k)} \right) .$$

The inequality (3.48) is then a consequence of equation (3.47), of $\mathcal{W}^{(l)(k)} \leq 1$, and the fact that the smallest matrix element is larger than zero (3.44). Under repeated application of the matrix we obtain a state

$$P^t = \mathcal{W}^t P, \quad \text{with} \quad t = 1, 2, \ldots \tag{3.49}$$

and find for the **approach to the equilibrium ensemble**

$$||E^t - E_B|| \leq \exp(-\lambda t)\, ||E - E_B|| \tag{3.50}$$

where $\lambda = -\ln(1 - w_{\min}) > 0$.

There are many ways to construct a Markov process satisfying (i), (ii) and (iii). In practice most MC algorithms are based on products of steps which satisfy a stronger condition than balance (3.40), namely:

(iii') **Detailed balance:**

$$W^{(l)(k)} e^{-\beta E^{(k)}} = W^{(k)(l)} e^{-\beta E^{(l)}} . \tag{3.51}$$

Using the normalization condition $\sum_k W^{(k)(l)} = 1$ detailed balance implies balance (iii).

At this point we have succeeded to replace the canonical ensemble average by a time average over an artificial dynamics. Calculating then averages over large times, like one does in real experiments, is equivalent to calculating averages of the ensemble. One distinguishes *dynamical universality classes*. The Metropolis and heat bath algorithms discussed in the following fall into the class of so called *Glauber dynamics* [Glauber (1963); Kawasaki (1972)], which imitates the thermal fluctuations of nature to some extent. In a frequently used classification [Chaikin and Lubensky (1997)] this is called *model A*. Cluster algorithms discussed in chapter 5 constitute another universality class [Swendsen and Wang (1987)]. Some recent attention has focused on dynamical universality classes of non-equilibrium systems, see [Ódor (2002)] for a review.

3.2.1 *The Metropolis algorithm*

Detailed balance still does not uniquely fix the transition probabilities $W^{(l)(k)}$. The **Metropolis algorithm** [Metropolis *et al.* (1953)] is a popular choice because of its generality and computational simplicity. It can be used whenever one knows how to calculate the energy of a configuration. The original formulation uses Boltzmann weights (3.33). But, the algorithm generalizes immediately to arbitrary weights, what is of importance for multicanonical simulations, see chapter 5.1. Given a configuration k, the Metropolis algorithm proposes new configurations l with some transition probabilities

$$f(l,k) \quad \text{with normalization} \quad \sum_l f(l,k) = 1 \ . \tag{3.52}$$

We show that the symmetry condition (3.56) below is needed to ensure detailed balance.

The new configuration l is accepted with probability

$$w^{(l)(k)} = \min\left[1, \frac{P_B^{(l)}}{P_B^{(k)}}\right] = \begin{cases} 1 & \text{for } E^{(l)} < E^{(k)} \\ e^{-\beta(E^{(l)}-E^{(k)})} & \text{for } E^{(l)} > E^{(k)}. \end{cases} \tag{3.53}$$

If the new configuration is rejected (*i.e.*, not accepted), the old configuration is kept, and has to be counted again[3]. The **acceptance rate** is defined as the ratio of accepted changes over proposed moves. With this definition we

---

[3]Although this point is already outlined in the original [Metropolis *et al.* (1953)] paper, it has remained a frequent mistake of beginners to count configurations only after accepted changes.

do not count a move as accepted when it proposes the at hand configuration (some authors may count these proposals as accepted).

The Metropolis procedure gives rise to the transition probabilities

$$W^{(l)(k)} = f(l,k)\, w^{(l)(k)} \text{ for } l \neq k \tag{3.54}$$

$$\text{and } W^{(k)(k)} = f(k,k) + \sum_{l \neq k} f(l,k)\,(1 - w^{(l)(k)}) \;. \tag{3.55}$$

Therefore, the ratio $\left(W^{(l)(k)}/W^{(k)(l)}\right)$ satisfies detailed balance (3.51) if

$$f(l,k) = f(k,l) \tag{3.56}$$

holds. Otherwise the probability density $f(l,k)$ is unconstrained and can be chosen conveniently, *i.e.*, there is an amazing flexibility in the choice of the transition probabilities. One can even use acceptance probabilities distinct from those of equation (3.53) and the proposal probabilities are then not necessarily symmetric anymore, see [Hastings (1970)].

At this point, the reader is recommended to go through the simple exercise of writing a Metropolis code which generates the Gaussian distribution (assignment 1). Next we discuss the Metropolis algorithm for a more involved continuous example, which we also use to introduce the heat bath algorithm. The reader interested in getting quickly started with simulations may prefer to jump to section 3.3.

3.2.2   The $O(3)$ σ-model and the heat bath algorithm

Here we give one example of a model with a continuous[4] energy function, the $O(3)$ σ-model. The $2d$ version of this model is of interest to field theorists because of its analogies with the four-dimensional Yang-Mills theory [Polyakov (1975); Belavin and Polykov (1975); Brézin and Zinn-Justin (1976)]. In statistical physics the d-dimensional model is known as the Heisenberg ferromagnet, see [Huang (1987)]. Expectation values are calculated with respect to the partition function

$$Z = \int \prod_i ds_i \; e^{-\beta E(\{s_i\})} \;. \tag{3.57}$$

---

[4]Until digitized by the computer.

The spins

$$\vec{s}_i = \begin{pmatrix} s_{i,1} \\ s_{i,2} \\ s_{i,3} \end{pmatrix} \quad \text{are normalized to} \quad (\vec{s}_i)^2 = 1 \tag{3.58}$$

and are defined on the sites of a lattice of size $n_1 \times n_2 \times \cdots \times n_{d-1} \times n_d$. The measure ds_i is defined by

$$\int ds_i = \frac{1}{4\pi} \int_{-1}^{+1} d\cos(\theta_i) \int_0^{2\pi} d\phi_i \,, \tag{3.59}$$

where θ_i and ϕ_i are polar and azimuth angles defining the spin s_i on the unit sphere. The energy is

$$E = \sum_{<ij>} (1 - \vec{s}_i \vec{s}_j) \,, \tag{3.60}$$

where the sum goes over the nearest neighbor sites of the lattice. We would like to change (update) a single spin $\vec{s}$ (dropping site indices i for the moment). The sum of its $2d$ neighbors is

$$\vec{S} = \vec{s}_1 + \vec{s}_2 + \cdots + \vec{s}_{2d-1} + \vec{s}_{2d} \,.$$

Hence, the contribution of spin $\vec{s}$ to the action is $2d - \vec{s}\vec{S}$. We propose a new spin $\vec{s}'$ with the measure (3.59). This defines the probability function $f(\vec{s}', \vec{s})$, equation (3.52), of the Metropolis process. In practice, we get the new spin $\vec{s}'$ by simply drawing two uniformly distributed random numbers

$$\phi^r \in [0, \pi) \quad \text{for the azimuth angle and}$$

$$\cos(\theta^r) = x^r \in [-1, +1) \quad \text{for the cosine of the polar angle.}$$

The Metropolis algorithm accepts the proposed spin $\vec{s}'$ with probability

$$w(\vec{s} \to \vec{s}') = \begin{cases} 1 & \text{for } \vec{S}\vec{s}' > \vec{S}\vec{s}, \\ e^{-\beta(\vec{S}\vec{s} - \vec{S}\vec{s}')} & \text{for } \vec{S}\vec{s}' < \vec{S}\vec{s}. \end{cases}$$

The algorithm proceeds by choosing another site and repeating the above procedure. If sites are chosen with the uniform probabilities $1/N$ per site, where N is the total number of spins, the procedure fulfills detailed balance: Just combine the spin acceptance probabilities with the probability function $f(\vec{s}', \vec{s})$ and the thus obtained transition probabilities (3.54), (3.55) fulfill (3.51). It is noteworthy that the procedure remains valid when the spins are chosen in the systematic order $1, \ldots, N$, then $1, \ldots, N$ again, and so on. Balance (3.40) still holds, whereas detailed balance (3.51) is violated (see appendix B, exercise 1 to the present section).

The heath bath algorithm

Repeating the Metropolis algorithm again and again for the same spin $\vec{s}$ leads (as has been proven) to the equilibrium distribution of this spin with respect to its neighbors, which reads

$$P(\vec{s}\,';\vec{S}) = \text{const}\, e^{\beta \vec{S}\vec{s}\,'} \quad \text{with} \quad \int P(\vec{s}\,';\vec{S})\, ds' = 1 \, .$$

One would prefer to choose $\vec{s}\,'$ directly with the probability

$$W(\vec{s} \to \vec{s}\,') \ = \ P(\vec{s}\,';\vec{S}) \ = \ \text{const}\, e^{\beta \vec{s}\,' \vec{S}} \, .$$

The algorithm, which creates this distribution, is called the **heat bath** algorithm, because it has the physical interpretation that the spin is brought in contact with a heat bath at the particular temperature $\beta = 1/T$. It was already known to [Glauber (1963)] that heat bath transition probabilities bring the system to equilibrium. The first computer implementation may be due to [Creutz (1980)] in lattice gauge theory, and applied mathematicians [Geman and Geman (1984)] named the algorithm the *Gibbs Sampler*.

Implementation of the heat bath algorithm becomes feasible when the energy function is sufficiently simple to allow for an explicit calculation of the probability $P(\vec{s}\,';\vec{S})$. Setting it up is an easy task for the $O(3)$ σ-model. Let

$$\alpha = \text{angle}(\vec{s}\,',\vec{S}), \quad x = \cos(\alpha) \quad \text{and} \quad S = \beta|\vec{S}| \, .$$

For $S = 0$ a new spin $\vec{s}\,'$ is simply obtained by random sampling. We assume in the following $S > 0$. The Boltzmann weight becomes $\exp(xS)$ and the normalization constant follows from

$$\int_{-1}^{+1} dx\, e^{xS} \ = \ \frac{2}{S} \sinh(S) \, .$$

Therefore, the desired probability is

$$P(\vec{s}\,';\vec{S}) \ = \ \frac{S}{2\sinh(S)}\, e^{xS} \ =: f(x)$$

and the method of chapter 1, equation (1.10), can be used to generate events with the probability density $f(x)$. With

$$y = F(x) = \int_{-1}^{x} dx'\, f(x') = \int_{-1}^{x} dx'\, \frac{S}{2\sinh(S)}\, e^{x'S} = \frac{\exp(+xS) - \exp(-S)}{2\sinh(S)}$$

a uniformly distributed random number $y^r \in [0,1)$ translates into

$$x^r = \cos\alpha^r = \frac{1}{S}\ln\left[\exp(+S) - y^r\exp(+S) + y^r\exp(-S)\right].\qquad(3.61)$$

Finally, one has to give $\vec{s}'$ a direction in the plane orthogonal to S. This is done by choosing a random angle β^r uniformly distributed in the range $0 \le \beta^r < 2\pi$. Then, $x^r = \cos\alpha^r$ and β^r completely determine $\vec{s}'$ with respect to $\vec{S}$. Before storing $\vec{s}'$ in the computer memory, we have to calculate coordinates of $\vec{s}'$ with respect to a Cartesian coordinate system, which is globally used for all spins of the lattice. This is achieved by a linear transformation. We define

$$\cos\theta = \frac{S_3}{S},\ \sin\theta = \sqrt{1 - \cos^2\theta},\ \cos\phi = \frac{S_1}{S\sin\theta}\ \text{ and }\ \sin\phi = \frac{S_2}{S\sin\theta}\,.$$

Unit vectors of a coordinate frame K', with $\hat{z}'$ in the direction of $\hat{S}$ and $\hat{y}'$ in the $x - y$ plane, are then defined by

$$\hat{z}' = \begin{pmatrix} \sin\theta\,\cos\phi \\ \sin\theta\,\sin\phi \\ \cos\theta \end{pmatrix},\ \ \hat{x}' = \begin{pmatrix} \cos\theta\,\cos\phi \\ \cos\theta\,\sin\phi \\ -\sin\theta \end{pmatrix}\ \text{ and }\ \hat{y}' = \begin{pmatrix} -\sin\phi \\ \cos\phi \\ 0 \end{pmatrix}.$$

Expanding $\vec{s}'$ in these units vectors, we have

$$\vec{s}' = \sin\alpha^r\,\cos\beta^r\,\hat{x}' + \sin\alpha^r\,\sin\beta^r\,\hat{y}' + \cos\alpha^r\,\hat{z}'$$

$$= \begin{pmatrix} \sin\alpha^r\,\cos\beta^r\,\cos\theta\,\cos\phi - \sin\alpha^r\,\sin\beta^r\,\sin\phi + \cos\alpha^r\,\sin\theta\,\cos\phi \\ \sin\alpha^r\,\cos\beta^r\,\cos\theta\,\sin\phi + \sin\alpha^r\,\cos\beta^r\,\cos\phi + \cos\alpha^r\,\sin\theta\,\sin\phi \\ -\sin\alpha^r\,\cos\beta^r\,\sin\theta + \cos\alpha^r\,\cos\theta \end{pmatrix}$$

$$(3.62)$$

and the three numbers of the column vector (3.62) are stored in the computer. The Fortran implementation of this heat bath algorithm is given in section 3.4.3 of this chapter.

3.2.3   *Assignments for section 3.2*

(1) Write a Metropolis program to generate the normal distribution through the Markov process

$$x \rightarrow x' = x + 2\,a\,(x^r - 0.5)$$

where x^r is a uniformly distributed random number in the range $[0, 1)$ and a is a real constant. Use $a = 3.0$ and the initial value $x = 0.0$ to test your program. (i) Generate a chain of $2\,000$ events. Plot the empirical peaked distribution function for these event in comparison with the exact peaked normal distribution function. Monitor also the acceptance rate. (ii) Repeat the above for a chain of $20\,000$ events.

3.3   Potts Model Monte Carlo Simulations

We introduce generalized Potts models in an external magnetic field on d-dimensional hypercubic lattices with periodic boundary conditions. Without being overly complicated, these models are general enough to illustrate the essential features we are interested in. Various special cases of these models are by themselves of physical interest, see references in [Blöte and Nightingale (1982); Wu (1982)]. Generalizations of the algorithmic concepts to other models are straightforward, although technical complications may arise.

We define the energy function of the system by

$$-\beta\,E^{(k)} = -\beta\,E_0^{(k)} + H\,M^{(k)} \tag{3.63}$$

where

$$E_0^{(k)} = -2 \sum_{<ij>} J_{ij}(q_i^{(k)}, q_j^{(k)})\,\delta(q_i^{(k)}, q_j^{(k)}) + \frac{2\,d\,N}{q} \tag{3.64}$$

$$\text{with}\quad \delta(q_i, q_j) = \begin{cases} 1 \text{ for } q_i = q_j \\ 0 \text{ for } q_i \neq q_j \end{cases}$$

and

$$M^{(k)} = 2 \sum_{i=1}^{N} \delta(1, q_i^{(k)})\,, \tag{3.65}$$

so that the Boltzmann factor is $\exp(-\beta\,E^{(k)}) = \exp(-\beta\,E_0^{(k)} + H\,M^{(k)})$. In equation (3.64) the sum $<ij>$ is over the nearest neighbor lattice sites and $q_i^{(k)}$ is called the **Potts spin** or **Potts state** at site i. This should not be confused with the microstate which is the configuration k. Occasionally the notation q-**state** is used for the Potts states at some site. The q-state $q_i^{(k)}$ takes on the values[5] $1, \ldots, q$. The sum in equation (3.65) goes over the N sites of the lattice. The external magnetic field is chosen to interact with the state $q_i = 1$ at each site i, but not with the other states $q_i \neq 1$. The $J_{ij}(q_i, q_j)$, $(q_i = 1, \ldots, q; q_j = 1, \ldots, q)$ functions define the exchange coupling constants between the states at site i and site j.

The energy function (3.64) describes a number of physically interesting situations. With

$$J_{ij}(q_i, q_j) \equiv J > 0 \quad \text{(conventionally } J = 1) \tag{3.66}$$

the original model [Potts (1952)] is recovered[6] and $q = 2$ becomes equivalent to the Ising ferromagnet, see [Wu (1982)] for a detailed review. The Ising case of [Edwards and Anderson (1975)] spin glasses and quadrupolar Potts glasses [Binder (1997)] are obtained when the exchange constants are quenched random variables. Other choices of the J_{ij} include antiferromagnets and the fully frustrated Ising model [Villain (1977)].

For the **energy per spin** the notations

$$e_{0s} = E_0/N \quad \text{and} \quad e_s = E/N \tag{3.67}$$

are used for the Potts model, depending on whether the external magnetic field H is zero or non-zero (the configuration indices are often often suppressed). The factor of two in front of the sum in equation (3.64) for $E_0^{(k)}$ is introduced, so that e_{0s} agrees for $q = 2$ with our Ising model definition (3.28) with the energy (3.4). For sampling with respect to the partition function (3.3), our definition of β agrees also with the one of the Ising model. This convention is commonly used in the Ising model literature, *e.g.* [Huang (1987)]. But, it disagrees by a factor of two with the β definition used in the Potts model literature [Baxter (1973); Borgs and Janke (1992)]:

$$\beta = \beta^{\text{Ising}} = \frac{1}{2} \beta^{\text{Potts}}. \tag{3.68}$$

[5]In our Fortran programs this is changed to $\mathtt{iq} = 0, \ldots, \mathtt{nq} - 1$ due to a technicality: The instruction $\mathtt{int(nq*x)}$, where $\mathtt{x}$ is a uniformly distributed random number, generates integer random numbers in the range $\mathtt{iq} = 0, \ldots, \mathtt{nq} - 1$. So the addition of an extra 1 is avoided.

[6]Actually, the article by Potts deals mainly with the clock model (which C. Domb proposed to him as his Ph.D. thesis) and introduces the model (3.64) only in a footnote.

Again by reasons of consistency with Ising model notation, a factor of two multiplies the magnetic field in equation (3.65).

The expectation value

$$M_{q_0} = \langle \delta_{q_i,q_0} \rangle \tag{3.69}$$

to find the Potts state at site i to be q_0 is called **Potts magnetization** with respect to q_0. For the 2-state Potts model the usual Ising magnetization is then found as $m = \langle \delta_{q_i,1} \rangle - \langle \delta_{q_i,2} \rangle$. For $\beta = 0$ it is rather trivial to calculate the dependence of the Potts magnetization on the external magnetic field (3.65). One finds (this an exercise to this section in appendix B)

$$\langle \delta_{q_i,1} \rangle = \frac{\exp(2H)}{\exp(2H) + q - 1} \tag{3.70}$$

at any site i independently of the lattice dimension and size.

For the $2d$ Potts models a number of exact results are known in the infinite volume limit, mainly due to work by [Baxter (1973)]. The critical temperature (3.6) of the $2d$ Ising model generalizes to the transition temperatures

$$\frac{1}{2} \beta_c^{\text{Potts}} = \beta_c = \frac{1}{T_c} = \frac{1}{2} \ln(1 + \sqrt{q}), \quad q = 2, 3, \dots . \tag{3.71}$$

At β_c the average energy per state is given by

$$e_{0s}^c = E_0^c/N = \frac{4}{q} - 2 - 2/\sqrt{q} . \tag{3.72}$$

Again, by reasons of consistency with the Ising model notation (3.28), our definition of $e_{0s} = E_0/N$ differs by factor of two and an additive constant from the one used in most Potts model literature. The phase transition is second order for $q \le 4$ and first order for $q \ge 5$. The exact infinite volume **latent heats** Δe_{0s} and **entropy jumps** Δs were also derived by [Baxter (1973)]. The knowledge of e_{0s}^c and Δe_{0s} at the critical point allows to calculate the limiting energies for $\beta \to \beta_c$ from the disordered and ordered phase, e_{0d} and e_{0o} respectively,

$$e_{0s}^c = (e_{0d} + e_{0o})/2 \quad \text{and} \quad \Delta e_{0s} = e_{0d} - e_{0o} . \tag{3.73}$$

For the first order transitions the interface tensions f_s are known too [Borgs and Janke (1992)] (and citations therein). For reference and convenience of the reader these quantities are compiled in table 3.3 for some q values.

Table 3.3 Exact, analytical results for the 2d q-state Potts model: β_c, e_{0s}, Δe_{0s} and Δs from [Baxter (1973)] and f_s from [Borgs and Janke (1992)].

	$q = 2$	$q = 5$	$q = 8$	$q = 10$
$\beta_c = \ln(1 + \sqrt{q})/2$	0.4406868	0.5871795	0.6712270	0.7130312
$e_{0s} = 4/q - 2 - 2/\sqrt{q}$	-1.4142136	-2.0944272	-2.2071068	-2.2324555
Δe_{0s}	0	0.105838	0.972716	1.3921
$\Delta s = 2\beta_c \Delta e_{0s}$	0	0.124292	1.305828	1.98522
$2f_s$	N/A	0.000398	0.041879	0.094701

Let us return to our goal of performing MC simulations of the model. Each configuration k defines a particular arrangements of Potts spins and, vice versa, each arrangement of the Potts spins determines uniquely a configuration:

$$k = \{q_1^{(k)}, \ldots, q_N^{(k)}\} . \tag{3.74}$$

As there are q possible states at each site, the total number of microstates is

$$Z(0) = q^N , \tag{3.75}$$

where we have used the definition (3.3) of Z. Already for moderately sized systems q^N is an enormously large number.

We would like to implement the Metropolis algorithm. Given a configuration k, new configurations l are proposed with explicitly known transition probabilities (3.52). The symmetry condition (3.56) implies detailed balance (3.51), which can be replaced by the somewhat weaker condition of balance (3.40). The newly proposed configuration l is accepted with the probability (3.53), otherwise rejected. For generalized Potts models it is straightforward to realize these steps. The transition probabilities $f(l, k)$ can be defined by the following procedure:

(1) Pick one site i at random, *i.e.* with probability $1/N$.
(2) Assign one of the states $1, \ldots, q$ to $q_i^{(l)}$, each with probability $1/q$.

These rules amount to

$$f(l, k) = \begin{cases} 1/q \text{ for } l = k, \\ 1/(qN) \text{ for each configuration } l \text{ with } q_i^{(l)} \neq q_i^{(k)} \text{ at one site } i, \\ 0 \text{ for all other configurations } l. \end{cases}$$

$$(3.76)$$

The new configuration l is then accepted or rejected according to equation (3.53). Ergodicity is fulfilled, because with (3.76) every configuration can be reached in N update steps. The random choice of a site ensures detailed balance. A procedure where the Potts spins are updated in the systematic order of a permutation $\pi_1, \ldots, \pi_N$ of the site numbers leads still to the desired distribution. Detailed balance is then violated but the weaker condition of balance still holds (see appendix B, exercise 1 to section 3.2).

3.3.1   *The Metropolis code*

The Fortran code presented here is for the $J_{ij}(q_i, q_j) \equiv 1$ original Potts model (3.66) in arbitrary dimensions d with $H = 0$. Later on code which allows for $H \neq 0$ will be introduced. Generalization to arbitrary $J_{ij}(q_i, q_j)$ arrays is straightforward, but to use it for the situation where all elements are equal would slow down the computational speed and occupy unnecessarily RAM resources.

Instead of the energy function (3.64) the **action variable**[7]

$$\texttt{iact}^{(k)} = \sum_{<ij>} \delta(q_i^{(k)}, q_j^{(k)}) \qquad (3.77)$$

is used in the following, and

$$E_0^{(k)} = \frac{2 d N}{q} - d \, \texttt{iact}^{(k)} \qquad (3.78)$$

holds. The action variable $\texttt{iact}$ is introduced because it takes on integer values which are positive or zero

$$\texttt{iact} = 0, 1, 2, \ldots, \texttt{nlink}$$

where $\texttt{nlink}$ is the number of links of the lattice, defined by the subroutine $\texttt{lat\_init.f}$ discussed in the previous section. Note that for the 2-state Potts model, *i.e.* the Ising model, $\texttt{iact}$ increases in steps of two: $\texttt{iact} = 0, 2, \ldots, \texttt{nlink}$. This happens, because in the Ising model a flip

---

[7]It differs from the definition of the action in field theory by the factor β.

of a single spin changes always the contribution of a even number of links to equation (3.77), whereas for the $q \geq 3$ Potts models changes of an odd number of links become possible. The **action per link** is defined by

$$\text{act} = \frac{\text{iact}}{\text{nlink}} \tag{3.79}$$

so that it is conveniently normalized in the range $0 \leq \text{act} \leq 1$.

The estimator of the `act` expectation values is denoted by `actm`. (3.80)

There are now two steps: First we initialize the system, then the Metropolis simulations are performed. Programs and some of the major subroutines are provided in

$$\texttt{ForProg/MC\_Potts} . \tag{3.81}$$

Let us discuss why there are also subroutines in `ForProg/MC_Potts` and not only in `ForLib`, where they appear to belong. The subroutines listed in `ForProg/MC_Potts` all use statements like

$$\texttt{include 'lat.par'}$$

to transfer parameter files like `lat.par` (3.26) into the code. These statements are needed to dimension arrays in common blocks, like `lat.com` and `potts.com`, which these routines share with other routines and/or the main program. To include these parameter statements properly, the routines have to be moved to the directory where the code is compiled. The rationale for this undesired arrangement is that it appears not wise to avoid common blocks in MC code, see our discussion in chapter 1.3. The structure of the common blocks is not supposed to change frequently, while the dimensions of the involved arrays have to stay flexible. Therefore, routines involving these common blocks need to be moved around with the main code. This implies the danger that they evolve, over course of time, into distinct versions with identical names. Care has to be taken that the most up-to-date version is always moved to the directory `ForProg/MC_Potts`. Fortran 90 allows for more elegant solutions, but it appears that Fortran 90 compilers are not widespread enough to make it a useful alternative for our purposes.

In the following we will be content with introducing Metropolis programs suitable to generate the energy histograms and time series. Large scale simulations require a serious error analysis, which is discussed in the next chapter. One may also consider to speed up the code. For instance, one may replace the Marsaglia random number generator by a faster one.

The CPU time used for the random number generation is substantial in the Potts models, while it would be almost negligible for systems, which require a lot of number crunching to calculate the energy.

3.3.1.1 *Initialization*

Initialization is done by the subroutine

$$\texttt{potts\_init0.f} \quad \text{of} \quad \texttt{ForProg/MC\_Potts} \qquad (3.82)$$

which is (up to some print statements) reproduced in table 3.4. It relies mainly on calls to other subroutines, which are part of `ForLib`. In detail `potts_init0.f` does the following:

(1) `rmaset` initializes the Marsaglia random number (chapter 1). Of its arguments the input/output units `iud`, `iud1` and the seeds `iseed1`, `iseed2` are defined in the parameter block `mc.par` of which an example is reproduced in table 3.5.

(2) `ranmar` is called for the first random number after initialization of the random number generator. This allows to print `xr` for check purposes.

(3) `acpt=0` is the initial value of the variable `acpt`, which is part of the common block `potts.com` and used to monitor the acceptance rate of the Metropolis updates (3.53). The common block `potts.com` is reproduced in table 3.6.

(4) `potts_act_tab` initializes `idel(0:nqm1,0:nqm1)` of `potts.com` (table 3.6). This is the delta function needed to calculate the Potts model action (3.77). In the computer code `nq` denotes the number of Potts states q and `nqm1 = nq - 1`. As we explained earlier in connection with the Potts model energy definition (3.63), the states take the values `iq = 0,...,nq - 1` in the Fortran programs. The user has to define `nq` in the parameter statement `potts.par`, for example

$$\texttt{parameter (nq = 2, nqm1 = nq - 1, nqm2 = nq - 2)} \qquad (3.83)$$

for an Ising model simulation.

(5) `potts_ran` initializes the array of Potts spins, `ista` in `potts.com`, with uniformly distributed integer random numbers in the range `0,...,nqm1`.

(6) `potts_act` calculates, after initialization of the spin array `ista`, the initial action value `iact`, which is chosen to be part of the `links` common block of `potts.com` (table 3.6), as its calculations involves the sum over all links. To illustrate the lattice structure, we have included an

Table 3.4 Initialization of the Potts model for Metropolis simulations: The relevant parts (omitting the print statements) of the subroutine `potts_init0` are listed.

```
      subroutine potts_init0(lpr)
      include '../../ForLib/implicit.sta'
      include '../../ForLib/constants.par'
C Copyright, Bernd Berg, Nov 14 2000.
C Initializes arrays needed for generalized Potts Model MC calculation.
      include 'lat.par'
      include 'mc.par'
      include 'potts.par'
      include '../../ForLib/lat.com'
      include '../../ForLib/potts.com'
      dimension ix(nd)
C
C Initialize:
      if(lpr) call rmaset(iuo,iud1,iseed1,iseed2,'nexiste.pa')
      if(.not.lpr) call rmaset(-iuo,iud1,iseed1,iseed2,'nexiste.pa')
      call ranmar(xr)                        ! first random number
      acpt=zero                              ! acceptance rate
      ia_min=mlink
      ia_max=0
      call lat_init(ns,nd,ipf,ipb,nla,ix,nlink)        ! lattice
      call potts_act_tab(idel,nqm1)               ! potts action table
      call potts_ran(ista,ns,nq)        ! potts initial states (random)
      call potts_act(ista,ipf,idel,ns,nqm1,nd,iact)      ! potts action
      call potts_wght(wrat,beta,nlink,nd)               ! potts weights
      call razero(ha,0,nlink)              ! initialize action histogram
C
      return
      end
```

array `ialink`, which allows to assign an integer variable to each link, although it is not needed for the programs used in this section. If memory space matters, `ialink` may be removed, as is done in the common block `potts1.com` discussed below.

(7) `potts_wght` initializes `wrat(-2*nd:2*nd,0:mlink+2*nd)`, the weight factor array of `potts.com` with Boltzmann (3.33) weights. The use of an array has speed advantages as array element look ups are normally faster than calculations of the exponential function. However, this observation is platform dependent and may vary in course of time. The reader is advised to perform his own benchmarks whenever per-

Table 3.5 Parameter block mc.par: User defined parameters for a MC calculations.

```
c
c mc.par: Definition of parameters for MC calculations
c          ================================================
c
c Output units and random generator seeds:
      parameter(iuo=6,iud1=11,iud2=12,iseed1=1,iseed2=0)
c
c Parameters for equilibrium and data production sweeps:
c beta:   k/T for for canonical MC, or as MUCA start value, k=1.
c H:      Magnetic field (only implemented for the heat bath updating).
c nequi:  Number of equilibrium sweeps.
c nmeas:  Number of measurement sweeps per repetition.
c nrpt:   >=1: Number of measurement repetitions in one job.
c         =0: Only MUCA (if chosen) and equilibrium run(s) are done.
      parameter(beta=0.40d00,H0=ZERO, nhit=10,nequi=200)
      parameter(nrpt=2, nmeas=1 000, nsw=1)
```

formance of the code is important. The array wrat is sized up to mlink+2*nd to allow for multicanonical simulations discussed in chapter 5.1. For canonical Metropolis simulations one can reduce the array size from (-2*nd:2*nd,0:mlink+2*nd) to (-2*nd:2*nd). This matters for simulations on large lattices. The user can switch to this version of the code by including the common statement potts1.com instead of potts.com and using the appropriate "version 1" subroutines discussed in section 3.3.1.4.

(8) razero (razero.f of ForLib) initializes the array of its arguments to zero for the array range defined by its other two arguments. The call in potts_init0.f initializes the array ha of potts.com to zero. The array ha is used to histogram the action variable iact.

3.3.1.2 *Updating routines*

The subroutine

$$\text{potts\_met.f} \qquad (3.84)$$

is reproduced in table 3.7. It implements the Metropolis procedure (3.53) using **sequential updating**, *i.e.*, sites are updated in their order is = $1, 2, \ldots, N$. **Random updating**, for which sites are chosen with a uniform

Table 3.6 Common block `potts.com`: Arrays and variables used for Metropolis simulations of the q-state Potts model.

```
      common /sites/  ista(ms)
c iamin,iamax:   action minimum and maximum of the MC sweep.
      common /links/ nlink, ialink(nd,ms),iact,iamin,iamax
      common /delfct/  idel(0:nqm1,0:nqm1)
c acpt:          counts accepted updates.
      common /weights/ wrat(-2*nd:2*nd,0:mlink+2*nd),acpt
c ia_min,ia_max: action minimum and maximum for a series of sweeps.
      common /haction/ ha(0:mlink),ia_min,ia_max
```

distribution, is performed by the subroutine

$$\texttt{potts\_met\_r.f} \tag{3.85}$$

which is otherwise identical with `potts_met.f`. Both routines can be found in the folder `ForProg/MC_Potts`.

A call to `potts_met` performs what is called a **sweep**:

$$\text{Each spin of the lattice is updated once.} \tag{3.86}$$

For a call to `potts_met_r` this modifies to:

$$\text{Each spin is updated once in the average.} \tag{3.87}$$

The notion sweep is also used for this situation. Besides performing Metropolis updates the subroutines (3.84) and (3.85) increment also the **histogram**

$$\texttt{ha} \text{ for each visited action value } \texttt{iact} \tag{3.88}$$

and keep track of the **acceptance rate acpt**. Further, the subroutines compute the variables

$$\texttt{iamin} \text{ and } \texttt{iamax} \tag{3.89}$$

which are the minimum and the maximum `iact` values visited during the sweep. The variables `ia_min` and `ia_max` track the `iact` minima and maxima over many sweeps. All these variables are part of the common block `potts.com`.

Table 3.7 Metropolis MC updating for the Potts models: **potts_met.f** of ForLib.

```
      subroutine potts_met
      include '../../ForLib/implicit.sta'
      include '../../ForLib/constants.par'
C Copyright Bernd Berg, Nov 14 2000.
C Metropolis updating with sequential spin choice.
      include 'lat.par'
      include 'potts.par'
      include '../../ForLib/lat.com'
      include '../../ForLib/potts.com'
      q=nq*ONE
      iamin=iact
      iamax=iact
      do is=1,ns
        istaold=ista(is)
        call ranmar(xr)
        istanew=int(q*xr)
        if(istanew.ne.istaold) then
          idact=0
          do id=1,nd
            ista2=ista(ipf(id,is))
            idact=idact+idel(ista2,istanew)-idel(ista2,istaold)
            ista2=ista(ipb(id,is))
            idact=idact+idel(ista2,istanew)-idel(ista2,istaold)
          end do
          if(idact.ne.0) then
            call ranmar(xr)
            if(xr.lt.wrat(idact,iact)) then
              ista(is)=istanew
              iact=iact+idact
              acpt=acpt+one
              iamin=min(iamin,iact)
              iamax=max(iamax,iact)
            end if
          else
            ista(is)=istanew
            acpt=acpt+one
          end if
        end if
        ha(iact)=ha(iact)+one
      end do
      ia_min=min(ia_min,iamin)
      ia_max=max(ia_max,iamax)
      return
      end
```

3.3.1.3 *Start and equilibration*

Under repeated application of[8] potts_met the probability of states will approach the Boltzmann distribution (3.34). However, initially we have to start with a microstate which may be far off the Boltzmann distribution. Far off means, that the Boltzmann probability (at temperature T) for a typical state of the initially generated distribution can be very, very small. Suppression factors like 10^{-10000} are well possible. Although the weight of states decreases with $1/n$ as the Markov process proceeds, $1/n$ will never reach a value as small as 10^{-10000}. Therefore, we should exclude the initial states from our equilibrium statistics. In practice this means we should allow for a certain number of sweeps

$$\text{nequi} \quad \text{to equilibrate the system.} \tag{3.90}$$

Statistical analysis may determine the appropriate value for nequi, see the next chapter.

Many ways to generate start configurations exist. Two natural and easy to implement choices are:

(1) Generate a random configuration with potts_ran.f of ForLib. Such a configuration corresponds to $\beta = 0$ and is called **disordered**. This defines a **random** or **disordered start** of a MC simulation.
(2) Generate a completely **ordered** configuration, for which all Potts spins take on the same q-value. The subroutine potts_order.f of ForLib is called for an **ordered start** of a MC simulation.

Figure 3.4 shows two Metropolis time series of 200 sweeps each for an Ising model on an 80×80 lattice at $\beta = 0.4$, which may be reproduced using the program

$$\text{potts\_ts.f} \quad \text{of} \quad \text{ForProg/MC\_Potts.} \tag{3.91}$$

The time series starting off at the minimum value of the e_{0s} scale corresponds to an ordered start ($e_{0s} = -2$ is the minimum energy per spin (3.28) of the model). The other time series corresponds to a disordered start ($\langle e_{0s} \rangle = 0$ is the expectation value for the energy per spin at $\beta = 0$). After a number of sweeps the two time series mix and we may argue that configurations may be counted after neglecting the first 100 sweeps for reaching

---

[8] The same considerations are valid for the subroutine potts_met_r and other updating procedures.

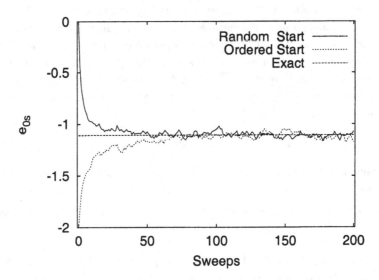

Fig. 3.4 Ising energy per spin (3.28): Two Metropolis time series of 200 sweeps each for an Ising model on an 80×80 lattice at $\beta = 0.4$ are shown. Ordered and disordered starts are used. The exact mean value $\hat{e}_{0s} = -1.10608$, calculated with **ferdinand.f** (3.15), is also indicated, see assignment 1.

equilibrium. A more careful analysis is recommended and carried out in the next chapter (see table 4.1).

To reproduce the Metropolis histograms of figure 3.3 (the $2d$ Ising model at $\beta = 0.2$ and $\beta = 0.4$) the program potts_hist.f (3.102), which is discussed in subsection 3.3.4 below, can be used, see assignment 2.

3.3.1.4 *More updating routines*

In this subsection we describe a variant and an extension of the Metropolis updating routine potts_met.f (3.84). The routines are

$$\text{potts\_metn.f,} \quad \text{potts1\_met.f} \quad \text{and} \quad \text{potts1\_metn.f} \qquad (3.92)$$

of ForProg/MC_Potts. An extension of potts_met.f is potts_metn.f: It allows to the hit the same spin n times with the Metropolis updating procedure. In this way the distribution of the spin approaches the Gibbs ensemble created by its neighbors. To optimize a simulation, one has to balance the improved performance of the spin updates against the additional CPU time needs, an issue addressed in the next chapter.

The routines potts1_met.f and potts1_metn.f agree in their function with their counterparts without the "1", but they include the common block potts1.com instead of potts.com. The common block potts1.com reduces the size of the weight array wrat (table 3.6) to its minimum dimension (-2*nd:2*nd). To save further memory space, the array ialink of potts.com is also removed in potts1.com. The inclusion of potts1.com instead of potts.com implies that we have to change the main program and other routines accordingly. For instance, the properly modified version of the program potts_hist.f of ForProg/MC_Potts is potts1_hyst.f, which includes potts1_init.f from ForProg/MC_Potts and potts1_wght.f from ForLib (instead of potts_init0.f and potts_wght.f). Similarly, for the use of potts1_metn.f a properly modified time series program, pottsn1_ts.f, is provided. It is left as assignment 13 to the reader to write a similar modification of the updating routine potts_met_r.f (3.85).

3.3.2 Heat bath code

The heat bath algorithm chooses a state q_i directly with the local Boltzmann distribution defined by its nearest neighbors, as we discussed for the $O(3)$ σ-model in section 3.2.2. For the Potts model the state q_i can take on one of the values $1, \ldots, q$ and, with all other states set, determines a value of the energy function (3.63). We denote this energy by $E(q_i)$ and the Boltzmann probabilities are

$$P_B(q_i) = \text{const } e^{-\beta E(q_i)} \qquad (3.93)$$

where the constant is determined by the normalization condition

$$\sum_{q_i=1}^{q} P_B(q_i) = 1 . \qquad (3.94)$$

In equation (3.93) we can define $E(q_i)$ to be just the energy of the interaction of q_i with its nearest neighbors. The other contributions to the total energy are the same for all q_i and absorbed into the overall constant. With this choice the $E(q_i)$ values depend only on the partition of the nearest neighbors into the values $1, \ldots, q$. For low values of q and the dimension d

the most efficient implementation of the heatbath algorithm is to tabulate all possibilities[9]. However, we do not do so now, because we want a generic code which works for arbitrary values of q and d.

The heat bath updating routine in ForProg/MC_Potts is

$$\texttt{potts\_mchb.f} \qquad (3.95)$$

and listed in table 3.8. It has to be used with the initialization routine

$$\texttt{potts\_inithb.f} \text{ of } \texttt{ForProg/MC\_Potts} . \qquad (3.96)$$

Both routines include the common block `../../ForLib/potts_hb.com`. The major change of `potts_inithb.f`, compared to `potts_init0.f` of table 3.4, is that heat bath weights `WHB_TAB(0:2*ND,0:NQM1)` are initialized with `potts_wghb.f`, which is shown in table 3.9. The weight table includes an external field `H0`, whose initial value is set in the `mc.par` parameter block (table 3.5). The use of the heat bath weights is quite different from the use of weights in Metropolis updating, as a comparison of the updating subroutines `potts_met.f` (table 3.7) and `potts_mchb.f` (table 3.8) reveals. Whereas the Metropolis weights determine directly the updating decision, the heat bath weight table is used to calculate the cumulative distribution function for the heat bath probabilities $P_B(q_i)$ of equation (3.93). This is the array `prob_hb(-1:nqm1)` of the `potts_mchb.f` code, whose final values are

$$\texttt{prob\_hb(iq)} = \sum_{q_i=1}^{iq+1} P_B(q_i) , \ \ \texttt{iq} = 0, \ldots, \texttt{nqm1} . \qquad (3.97)$$

The normalization condition (3.94) implies `prob_hb(nqm1)=1`. Comparison of these cumulative probabilities with the uniform random number `xr` yields the heat bath update $q_i \to q'_i$. Note that in the heat bath procedure the original value q_i does not influence the selection of q'_i.

The variable **achg** counts the updates, which lead to a change of the spin variable at hand. It is similar to **acpt**, which monitors the accepted changes of the Metropolis updating.

Our heat bath program calculates also **magnetization histograms**. For the Potts model each of the q distinct states defines a direction of the Potts magnetization. Hence, q histograms are kept in one array. See in

---

[9] Such tables are discussed in chapter 5.2.

Table 3.8 Heat bath MC updating for the Potts model in an external magnetic field H.

```
      subroutine potts_mchb ! Copyright Bernd Berg, Oct 20 2002.
      include '../../ForLib/implicit.sta'
      include '../../ForLib/constants.par'
C MC Heat Bath updating with sequential spin choice.
C Measurements: the action ha and the magnetization hm histograms.
C Faster when the magnetization measurements are eliminated!
      include 'lat.par'
      include 'potts.par'
      include '../../ForLib/lat.com'
      include '../../ForLib/potts_hb.com'
      dimension prob_hb(-1:nqm1),iact_a(0:nqm1)
      iamin=iact
      iamax=iact
      prob_hb(-1)=zero
      do is=1,ns
        call ranmar(xr)
        do iq=0,nqm1
          iact0=0 ! Calculates the action contribution for the state iq.
          do id=1,nd
            iact0=iact0+idel(iq,ista(ipf(id,is)))
     &                 +idel(iq,ista(ipb(id,is)))
          end do
          prob_hb(iq)=whb_tab(iact0,iq)+prob_hb(iq-1)
          iact_a(iq)=iact0                        ! Defines the array iact_a.
        end do
        do iq=0,nqm2
          ptest=prob_hb(iq)/prob_hb(nqm1) ! Heat bath probabilities.
        if(xr.lt.ptest) go to 1
        end do
        iq=nqm1
1       iqold=ista(is)
        if(iq.ne.iqold) then
          achg=achg+one
          iqnew=iq
          ista(is)=iqnew
          iact=iact+iact_a(iqnew)-iact_a(iqold)
          iamin=min(iamin,iact)
          iamax=max(iamax,iact)
          nstate(iqold)=nstate(iqold)-1 ! For magnetization measurements.
          nstate(iqnew)=nstate(iqnew)+1 ! For magnetization measurements.
        end if
        ha(iact)=ha(iact)+one
        do iq=0,nqm1              ! Do loop for magnetization measurements.
          hm(nstate(iq),iq)=hm(nstate(iq),iq)+one
        end do
      end do
      ia_min=min(ia_min,iamin)
      ia_max=max(ia_max,iamax)
      return
      end
```

Table 3.9 `potts_wghb.f` of ForLib, which generates the weight table used for our Potts model heat bath updating.

```
      SUBROUTINE POTTS_WGHB(WHB_TAB,BETA,HO,ND,NQM1)
C Copyright, Bernd Berg, Nov 22 2000.
C Table for Potts model (with magnetic field) heat bath update.
      include 'implicit.sta'
      include 'constants.par'
      DIMENSION WHB_TAB(0:2*ND,0:NQM1)
      DO IQ=0,NQM1
        H=ZERO
        IF(IQ.EQ.0) H=HO
        DO IACT=0,2*ND
          WHB_TAB(IACT,IQ)=EXP(TWO*(BETA*IACT+H))
        END DO
      END DO
      RETURN
      END
```

table 3.8 the lines commented:

$$\text{For magnetization measurements.} \qquad (3.98)$$

3.3.3 *Timing and time series comparison of the routines*

We have to balance the updating performance of our subroutines versus their CPU time consumptions. The results will depend on the hardware and compilers used and evaluations have to be done by the final user. Here an illustrative example is discussed. A Pentium II 400 MHz processor under Linux with the gnu $g77$ Fortran compiler version 2.91.66 was used. The version number of the $g77$ Fortran compiler under Linux is identified by typing

$$g77 \ - v \ . \qquad (3.99)$$

Compilations were done with the -0 optimization.

Initialization and 10 000 sweeps were performed for the 10-state Potts model on an 80×80 lattice. The number of sweeps is chosen relatively large, to make the initialization time (in which we are not really interested) negligible. The different updating routines consumed CPU time as follows (see (1.27) and (1.28) for CPU time measurements under Linux):

$$39.1 \text{ s (potts\_met.f)}, \quad 83.5 \text{ s } (2 - \text{hit potts\_metn.f})$$
$$\text{and} \quad 170.2 \text{ s (potts\_mchb.f)} . \tag{3.100}$$

Of the three routines used the heat bath updating has the best dynamics, but one heat bath updating needs more than four times the CPU time of one 1-hit Metropolis updating. Multi-hit Metropolis generates the single spin heat bath distribution for nhit $\to \infty$. Therefore, it can only beat the performance of the heat bath updating when its CPU time consumption is less than that of heat bath routine. For nhit=2 the multi-hit Metropolis CPU time consumption[10] is approximately half of that of our heat bath updating. It is assignment 4 to repeat the CPU timing on your computer.

How can we compare the performance of the different updating procedures? As we see in the next chapter, this requires to measure autocorrelation times. At the present state we are content with a look at the different time series and find that the performance can indeed be quite distinct. For the 10-state model on an 80×80 lattice at $\beta = 0.62$ we compare in figure 3.5 times series of 200 sweeps each. The simulation temperature is in the disordered phase, but not too far away from the phase transition point β_c (3.71), see table 3.3. Ordered and disordered starts were performed. While for the disordered starts there is relatively little visible difference in the approach of the simulations to the equilibrium process, this is not the case for the ordered starts. The 1-hit Metropolis process has considerable difficulties to find its way to the equilibrium chain. The obvious interpretation is that this is due to the finite lattice remnant of the first order phase transition. The convergence speed is improved for the 2-hit Metropolis process, while the heat bath algorithm performs clearly best. The question, whether the improvement due to the heat bath updating does, for the equilibrium process, (more than) compensate for the increased CPU time consumption (3.100) is addressed in the next chapter, see table 4.4.

3.3.4 *Energy references, data production and analysis code*

Using our MC subroutines, we present some high precision calculations of the average internal energy per Potts spin. The purpose is to test the

---

[10]With the gnu g77 compiler our n-hit Metropolis updating is found to scale in CPU time with the surprisingly bad factor of n and this does not improve when the neighbor spins are first stored in a small intermediate array (to convince the compiler to keep these numbers in cache during the multi-hits).

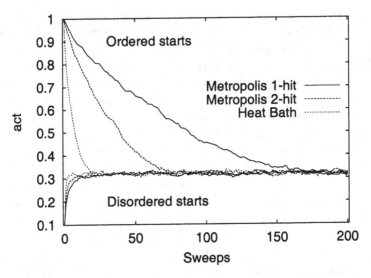

Fig. 3.5 $q = 10$ Potts model time series of 200 sweeps on an 80 × 80 lattice at $\beta = 0.62$. Measurements of the action variable (3.79) after every sweep are plotted for ordered and disordered starts. The updating was performed with the following routines: `potts_met.f`, nhit=2 `potts_metn.f` and `potts_mchb.f`, see assignment 5.

correctness of the routines versus exact results, versus numerical results from the literature, and against one another. This is immediately hit by a snag: The MC data of a Markov chain are correlated and we do not yet know how to calculate error bars reliably in this situation. At our present level of expertise we resort on a heuristic way out of the dilemma: We use the binning procedure of chapter 1.8.3. The bins become (almost) independent when the binsize is chosen large enough. How large is large enough? This is a topic of chapter 4. Here we just choose a number which is (very) large compared to the scale on which convergence of the time series ala figure 3.4 is found and hope that the correlation between bins are then negligible. Due to the central limit theorem the distribution of the bins will be approximately Gaussian and we can use Gaussian error analysis tools. Kolmogorov tests allow to check the Gaussian assumption for consistency, see assignment 11.

3.3.4.1 *2d Ising model*

We begin with the $2d$ Ising model and test against the exact results of our program ferdinand.f (3.15). As for the time series from an 80^2 lattice in figure 3.4, we use the value $\beta = 0.4$, but simulate on an 20^2 lattice (as for the histograms in figure 3.3). The convergence of the Markov chain to its equilibrium process is expected to be even faster than in figure 3.4, where the first 200 sweeps are depicted. Now we use a statistics of 10 000 sweeps for reaching equilibrium and assume that this is an overkill. The statistics for measurement is chosen 32 times larger: 64 bins of 5 000 sweeps each. The number 64 is taken, because according to the student distribution (tables 2.6 and 2.7) the approximation to the Gaussian approximation is then already excellent, while the binsize of 5 000 ($\gg 200$) is argued to be large enough to neglect correlations between the bins. With this statistics we find (assignment 6)

$$\bar{e}_{0s} = -1.1172 \, (14) \ \text{(Metropolis)} \quad \text{versus} \quad \hat{e}_s = -1.117834 \ \text{(exact)} \, .$$
$$(3.101)$$

Performing the Gaussian difference test of chapter 2.1.3 gives $Q = 0.66$, a perfectly admissible value.

In table 3.10 we list the data production program

$$\text{potts\_hist.f of ForProg/MC\_Potts} \qquad (3.102)$$

which is also suitable for long production runs. After performing **nequi** sweeps for reaching equilibrium, the program creates a filename which includes the dimension d, the number of states q, and the lattice size n^1 in x^1 direction (3.17). For the frequently encountered symmetric case n^1 determines the entire lattice. Using the same programming technique, one can extend or alter this name list to fit specific purposes. The format of the data file is chosen to be "unformatted", which is the efficient way of handling large data sets under Fortran. (For small data sets readable "formatted" files have, besides being readable, the advantage of being easily exchangeable between different computer architectures.) The program proceeds with writing general information into the first block of the data file. Subsequently, in the **nrpt** do loop, it writes one action variable histogram **ha**, plus other information, per block. Each block corresponds to **nmeas** updates with measurements (the parameter **nsw** of mc.par is only used in later time series production programs). In the production part the

Table 3.10 The data production program `potts_hist.f` performs `nequi` equilibration sweeps followed by `nrpt` bins of `nmeas` measurements. For each bin the assembled histogram `ha` is written to an unformatted data file, which is named to reflect the parameters of the run.

```
      program potts_hist
C Copyright, Bernd Berg, Dec 13 2000.
C MC production run to create energy (action variable) histograms.
      include '../../ForLib/implicit.sta'
      include '../../ForLib/constants.par'
      character cd*1,cq*2,cl*3
      include 'lat.par'
      include 'mc.par'
      include 'potts.par'
      include '../../ForLib/lat.com'
      include '../../ForLib/potts.com'
      include 'lat.dat'
C
      call potts_init(.true.) ! Initialize Potts Metropolis MC:
C
      do iequi=1,nequi         ! Sweeps for reaching equilibrium.
        call potts_met
      end do
C
      write(cd,'(I1.1)') nd
      write(cq,'(I2.2)') nq
      write(cl,'(I3.3)') nla(1)
      open(iud1,file="p"//cd//"d"//cq//"q"//cl//".d",
     &    form="unformatted",status="unknown")
      write(iud1) beta,nd,nq,nla,nlink,nequi,nrpt,nmean
      do irpt=1,nrpt ! nrpt repetitions of nmeas measurement sweeps.
        acpt=zero
        call razero(ha,0,nlink)
        do imeas=1,nmeas
          call potts_met
        end do
        acpt=acpt/(nmeas*ns)
        write(iud1) ha,acpt,irpt
        call write_progress(iud2,"irpt,iact,acpt:",irpt,iact,acpt)
      end do
      close(iud1)
C
      stop
      end
```

program calls also the subroutine

$$\text{write\_progess.f of ForLib} \tag{3.103}$$

which is listed in table 3.11. This subroutine opens the formatted data file progress.d, writes information about the progress of the calculation on the data file and closes it immediately afterwards. The contents of the file progress.d can be looked up at runtime and thus allows to monitor the simulation. This is of importance for long runs, normally performed in the background. Without opening and closing the file before and after each write, the computer may store the data in some buffer, instead of writing them immediately into the file.

Table 3.11 The subroutine write_progress.f of ForLib allows the user to to monitor the progress of a (long) MC calculation at run time by looking up the content of the file progress.d.

```
      subroutine write_progress(iud,ctxt,irpt,iact,acpt)
C Copyright, Bernd Berg, Dec 13 2000.
C Write subroutine to monitor progress.
      include 'implicit.sta'
      character*(*) ctxt
      open(iud,file="progress.d",form="formatted",status="unknown")
      rewind iud
      write(iud,'(A24,2I12,F10.3)') ctxt,irpt,iact,acpt
      close(iud)
      return
      end
```

3.3.4.2 Data analysis

For the **analysis of the histogram data** the program

$$\text{ana\_hist.f of ForProg/MC\_Potts} \tag{3.104}$$

is provided. After initializing the data filename and reading its first block, the analysis program relies to a large extent on the subroutine

$$\text{read\_steb0.f of ForLib} \tag{3.105}$$

which is reproduced in table 3.12. This subroutine adapts our basic statistical error bar routine steb0.f (2.17) to the handling of arrays, which are possibly too big to keep nrpt of them into the computer RAM. Read instructions are used to get away with just three equally sized arrays, ha, ham and hae, in the RAM. Optionally, we can normalize the incoming histograms ha, so that for each of them the sum of its entries becomes nlink. When setting the input argument norm $\neq$ 0 the ha arrays are normalized, for norm $=$ 0 they are used as read.

For each of the nrpt arrays ha the subroutine read_steb0.f calculates also the average action per link and returns these nrpt numbers in the array act. This calculation relies on calls to the Fortran function potts_actm.f of ForLib, which is listed in table 3.13. It is better to calculate the average action from the histograms than by using the iact action values after each sweep, because the updating routine generates small fluctuations during a sweep and the histograms store this information.

Our analysis program returns the mean action per link actm, and also converts it to the average energy per spin $\bar{e}_{0s}$, which is according to equations (3.78), (3.79), (3.80)

$$\bar{e}_{0s} = \frac{2\,d}{q} - 2\,d\,(\text{actm})\,. \tag{3.106}$$

As explained, to achieve consistency with Ising model conventions of the literature, our definition of $\bar{e}_{0s}$ differs from the one used in most of the Potts model literature.

3.3.4.3 *2d 4-state and 10-state Potts models*

Our next example is the 2d 4-state Potts model. We compare with a high statistics result on a 120 × 120 lattice: From table 1 of [Caselle *et al.* (1999)] we pick $\beta = 1.06722/2$ (their $\beta^{\text{Potts}} = 1.06722$) and repeat the calculation with our updating routine potts_met.f (Caselle *et al.* used Swendsen-Wang cluster updating, which we introduce in chapter 5). Our statistics is 400 000 sweeps for reaching equilibrium and 64 × 200 000 sweeps for measurement. The results are

$$\text{actm} = 0.646057\,(24)\ \text{(Metropolis)}\quad \text{versus}$$
$$\text{actm} = 0.64604\,(6)\ \text{[Caselle *et al.* (1999)]} \tag{3.107}$$

and $Q = 0.79$ for the Gaussian difference test, see assignment 7.

For the 2d 10-state Potts model at $\beta = 0.62$ we test our Metropolis routine potts_met.f versus our heat bath routine potts_mchb.f on a small 20 × 20 lattice. For the heat bath updating we use the same statistics

Table 3.12 The subroutine `read_steb0.f` of ForLib adapts the error bar subroutine `steb0.f` (2.17) to handle sets of possibly large arrays using `read` instructions.

```
      SUBROUTINE READ_STEB0(nrpt,iud,norm,nlink,ha,ham,hae,act)
C Copyright, Bernd Berg, Jan 13, 2002.
C Reads nrpt times the array ha() from unit iud and calculates the
C       histogram mean values ham() and their error bars hae().
C Further, the nrpt action variable averages are calculated.
C norm=0:  no normalization of the ha histogram, otherwise normalization.
      include 'implicit.sta'
      include 'constants.par'
      dimension ha(0:nlink),ham(0:nlink),hae(0:nlink),act(nrpt)
C
      call razero(ham,0,nlink)
      do irpt=1,nrpt
        read(iud) ha
        if(norm.ne.0) then ! Normalization of ha() to hasum-nlink.
          hasum=zero
          do ilink=0,nlink
            hasum=hasum+ha(ilink)
          end do
          do ilink=0,nlink ! Normalization of ha() to hasum=nlink.
            ha(ilink)=(ha(ilink)*nlink)/hasum
          end do
        end if
        act(irpt)=potts_actm(nlink,ha) ! Mean action variable.
        do ilink=0,nlink
          ham(ilink)=ham(ilink)+ha(ilink)
        end do
      end do
      do ilink=0,nlink
        ham(ilink)=ham(ilink)/nrpt
      end do
C
      do irpt=1,nrpt
        backspace iud
      end do
      call razero(hae,0,nlink)
      do irpt=1,nrpt
        read(iud) ha
        do ilink=0,nlink
          hae(ilink)=hae(ilink)+(ha(ilink)-ham(ilink))**2
        end do
      end do
      do ilink=0,nlink
        hae(ilink)=sqrt(hae(ilink)/(nrpt*(nrpt-1)))
      end do
      return
      end
```

Table 3.13 Fortran function `potts_actm.f` of Forlib: The average action per link is calculated from a histogram of the action variable (3.77) `iact`. A variant of this routine, which is for some applications much faster, is `potts_actm2.f` of ForLib.

```
      FUNCTION  potts_actm(nlink,ha)
C Copyright, Bernd Berg, Dec 12 2000.
C Calculation of the mean action variable from its histogram.
      include 'implicit.sta'
      dimension ha(0:nlink)
      potts_actm=0
      hsum=0
      do ilink=0,nlink
        potts_actm=potts_actm+ilink*ha(ilink)
        hsum=hsum+ha(ilink)
      end do
      potts_actm=potts_actm/(hsum*nlink) ! iact per link.
      return
      end
```

as previously for the $2d$ Ising model (3.101), 10 000 sweeps for reaching equilibrium and $64 \times 5\,000$ sweeps for measurements. For the Metropolis updating we increase these numbers by a factor of four to 40 000 sweeps for reaching equilibrium and $64 \times 20\,000$ sweeps for measurements. This increase is done, because we expect the performance of Metropolis updating for the 10-state model to be far worse than for the 2-state model. At low temperature in the ordered phase the likelihood to propose the most probable (aligned) Potts state is $1/2$ for the 2-state model, but only $1/10$ for the 10-state model. Although $\beta = 0.62$ is not yet in the ordered phase, it is sufficiently close to it, so that this effect is expected to be of relevance. The results of our simulations are

$$\text{actm} = 0.321772 \,(75) \quad \text{(Metropolis)} \quad \text{versus}$$
$$\text{actm} = 0.321661 \,(70) \quad \text{(heat bath)} \tag{3.108}$$

and $Q = 0.28$ for the Gaussian difference test, see assignment 8. The heat bath data production program is a variant of `potts_hist.f`, called `potts_hihb.f`. It stores besides the energy **also magnetization histograms**. For the energy histograms the analysis program `ana_hist.f` (3.104) still works.

3.3.4.4 *3d Ising model*

Let us move to three dimensions. In [Alves *et al.* (1990)] a precise numerical estimate of the internal energy is given for the Ising model on an 14^3 lattice at $\beta = 0.22165$. This β value is close to the infinite volume critical point. In the same reference the Curie temperature is estimated to be at $\beta_c = 0.22157$ (3). We repeat the calculation of [Alves *et al.* (1990)] with the statistics of 20 000 sweeps for reaching equilibrium and $64 \times 10\,000$ sweeps with measurements. The results compare as follows[11]

$$\bar{e}_{0s} = -1.0427\ (14)\quad \text{(Metropolis here)}\quad \text{versus}$$
$$\bar{e}_{0s} = -1.04369\ (9)\quad \text{[Alves \textit{et al.} (1990)]}\tag{3.109}$$

and the Gaussian difference test gives $Q = 0.48$, see assignment 9.

3.3.4.5 *3d 3-state Potts model*

To illustrate features of a first order phase transition, we perform a run for the $3d$ 3-state Potts model. We use the 1-hit Metropolis algorithm and simulate at $\beta = 0.275229525$. On a 24^3 lattice we perform 20 000 sweeps for reaching equilibrium and subsequently $64 \times 10\,000$ sweeps with measurements. From the latter statistics we calculate the action variable histogram and its error bars, which are depicted in figure 3.6 (see assignment 10). The histogram exhibits a **double peak** structure (related to the latent heat), which is typically obtained when systems with first order transitions are simulated on finite lattices in the neighborhood of so called pseudo-transition temperatures. **Pseudo-transition temperatures** are finite lattice temperature definitions, which converge with increasing system size towards the infinite volume transition temperature. For first order transitions a definition of a pseudo-transition temperature is the value which gives equal heights to both peaks of the histogram.

Our simulation on the 24^3 lattice is at the pseudo-transition β value from table v of [Alves *et al.* (1991)][12]. Due to statistical fluctuations, our thus obtained histogram needs to be re-weighted to a slightly higher β value to arrange for equal heights of the maxima. To find this β value for the data of figure 3.6 is left as part of assignment 10 to the reader. For reference,

---

[11][Alves *et al.* (1990)] gives the result $E_A = -715.97$ (6) in a notation for which E_A is related to our of equation (3.63) by $E = 4E_a$. Therefore, $\bar{e}_{0s} = 4\bar{E}_a/N = -2863.88/14^2 = -1.04369$.

[12]The β value given in table v of [Alves *et al.* (1991)] is $\beta_A = 0.3669727$ and related to our notation by $\beta = 3\beta_A/4$.

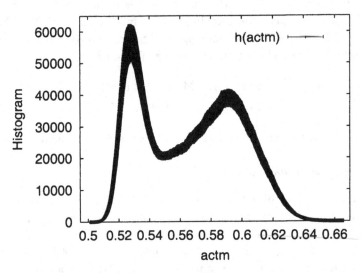

Fig. 3.6 Histogram of the action variable **actm** for the $3d$ 3-state Potts model on a 24^3 lattice at $\beta = 0.275229525$, see assignment 10.

our mean action value, corresponding to the histogram of figure 3.6 is

$$\texttt{actm} = 0.5662 \ (21) \ . \tag{3.110}$$

Due to the double peak structure of the histogram the error bar is relatively large. Still, the central limit theorem works and the Kolmogorov test of assignment 11 shows that our statistics is large enough to create an approximately Gaussian distribution for the binned data.

3.3.4.6   *4d Ising model with non-zero magnetic field*

Finally, let us test the heat bath code for a non-zero value of the magnetic field. Equation (3.70) provides for simple tests at $\beta = 0$. This equation does not depend on the dimension d of the system, and we use the opportunity to run the code in $d = 4$. In assignment 12 we choose the external magnetic field to correspond to $\langle \delta_{q_i,1} \rangle = 0.5$ for the 10-state Potts model and run 32×10 sweeps with measurements on an 10^4 lattice. The numerical result is

$$\langle \delta_{q_i,1} \rangle = 0.49961\,(71) \tag{3.111}$$

which yields $Q = 0.58$ when compared to 0.5 by means of the Gaussian difference test.

3.3.5 Assignments for section 3.3

In the following, use always Marsaglia random number with their default seed (table 1.1 of chapter 1) in the MC calculations. This makes the results easily reproducible. In research investigations the random numbers should differ from simulation to simulation, to be on the statistically save side (as long as the same random numbers are used in different settings the actually produced correlations are presumably negligible).

(1) Reproduce the energy per spin time series of figure 3.4. Initialize the parameter files to the requested values and run the program potts_ts.f from ForProg/MC_Potts. Use ferdinand.f (3.15) to calculate the exact mean value. Repeat the simulation using potts_ts_r.f.

(2) Use the program potts_hist.f of ForProg/MC_Potts to reproduce the $\beta = 0.2$ and $\beta = 0.4$ histograms of figure 3.3. Start each run with the default seed of the Marsaglia random number generator and allow 10 000 sweeps to reach equilibrium. The error bars are with respect to 20 bins of 5 000 sweeps each. Combine the results with those of assignment 2 of subsection 3.1.3 to reproduce figure 3.3.

(3) Run potts_hist.f of the previous assignment at $\beta = 0.2$ with 1 000 sweeps for reaching equilibrium and $10 \times 1\,000$ sweeps for data production. Repeat the same calculation with potts1_hist.f and compare the generated data files (with diff fln1.d fln2.d under Linux). Include the nhit Metropolis routines potts_metn.f and potts1_metn.f with nhit=1 into this test.

(4) Repeat the CPU timing (3.100) of potts_met.f, 2-hit potts_metn.f and potts_mchb.f on your computer using your favorite Fortran compiler. Try different optimization levels of your compiler. Further, include the routines potts1_met.f and potts1_metn.f in the competition.

(5) Use potts_ts.f of ForProg/MC_Potts to generate the Metropolis time series of figure 3.5 for the 10-state Potts model at $\beta = 0.62$ on an 80×80 lattice. Repeat the calculation using (a) n-hit Metropolis updating with

$n = 2$ (the program pottsn_ts.f of ForProg/MC_Potts) and (b) heat bath updating (the program pottshb_ts.f of ForProg/MC_Potts).

(6) Use potts_met.f to reproduce the $2d$ Ising model $\overline{e}_{0s}$ Metropolis result of equation (3.101). A suitable production program is potts_hist.f of ForProg/MC_Potts and the data analysis can be done with ana_hist.f of ForProg/MC_Potts. Note that the exact value from ferdinand.f differs from that of assignment 1 due to finite size effects. Is the difference detectable within the accuracy of the MC run?

(7) Use potts_met.f to reproduce the $2d$ 4-state Potts model actm Metropolis result of equation (3.107). Make an estimate of the CPU time needed (compare assignment 4) before starting this simulation.

(8) Use potts_met.f versus potts_mchb.f to reproduce the $2d$ 10-state Potts model actm results of equation (3.108).

(9) Use potts_met.f to reproduce our $3d$ Ising model $\overline{e}_{0s}$ Metropolis result of equation (3.109).

(10) Use potts_met.f to reproduce our $3d$ 3-state Potts model histogram of figure 3.6 as well as the actm Metropolis result of equation (3.110). Find the β value for which the two maxima of the double peaks take on equal heights.

(11) Check for the previous five assignments whether the assumption that the binned data are Gaussian distributed passes the upper and lower one-sided Kolmogorov test (2.142). This can be done by running the program ana_kol.f of ForProg/MC_Potts.

(12) Consider the 10-state Potts model at $\beta = 0$ and determine the magnetic field H so that $\langle \delta_{q_i,1} \rangle = 0.5$ holds according to equation (3.70). Subsequently, use this H value and potts_hihb.f of ForProg/MC_Potts to calculate $\langle \delta_{q_i,1} \rangle$ numerically on a 10^4 lattice. Use 32 bins, each of 10 sweeps with measurements. No equilibrium sweeps are necessary (why?). Estimate $\langle \delta_{q_i,1} \rangle$ from the data (the appropriate analysis program is ana_mag.f of ForProg/MC_Potts) and perform the Gaussian difference test with respect to 0.5. Calculate also the exact mean action per link and perform the Gaussian difference test with respect to its estimate from the data.

(13) Modify the updating routine potts_met_r.f (3.85) to potts1_met_r.f, to use potts1.com instead of potts.com. Check that the potts_ts_r.f time series of assignment 1 stays unchanged.

3.4 Continuous Systems

MC simulations of systems with continuous variables face additional difficulties. Histogramming of the observables leads to discretization errors and it is better to keep the time series of the data. However, dealing inappropriately with large amounts of data may slow down the code and can exhaust the available storage facilities. At some point one may also become concerned that the discretization errors of the random number generator could matter. Any resolution smaller than $\epsilon_{\text{Marsaglia}} = 2^{-24} = 1/16,777,216$ discussed at the end of section 1.2 needs special attention, compare equation (3.123). Another complication is that rounding errors may limit the bit by bit reproducibility of our numerical results, see below.

To give an impression about dealing with continuous systems, we consider simulations of $O(n)$, $n \geq 2$ spin systems in $d \geq 2$ dimensions. The systems are defined by equations (3.57) to (3.60), only that we allow now for **n-component spins**, instead of just for $n = 3$. Of particular physical interest are the cases $n = 3$, which we introduced already in subsection 3.2.2, and the $n = 2$ **planar rotator**, also called XY **model**, in $d = 2$. According to [Kosterlitz and Thouless (1973); Kosterlitz (1974)] there exists a low temperature phase of the $d = 2$ planar rotator which is characterized by a power law decay in the pair correlation function, modified by the presence of pairs of vortices. The vortex pairs unbind at the critical temperature, to create a high temperature phase where the correlations decay exponentially. In this way the transition does not violate the *Mermin-Wagner theorem* [Mermin and Wagner (1966)], which forbids the spontaneous breaking of a continuous symmetry in two dimensions[13].

It is straightforward to write a simple $O(n)$ Metropolis code for generic $n \geq 2$ values, see the next subsection. The code illustrates the simplicity of the Metropolis approach, but is too slow for serious simulations. For the planar rotator and the $n = 3$ Heisenberg model, specialized subroutines are provided in subsections 3.4.2 and 3.4.3. For the XY model this is an improved Metropolis codes, whereas for the Heisenberg model it is the heat bath algorithm which we discussed in subsection 3.2.2. It should be remarked that in both cases the cluster code of chapter 5 is highly efficient and the reader should not consider the discussion in the present chapter as

---

[13][Coleman (1973)] explains the rationale behind the Mermin-Wagner theorem as follows: The breaking of a continuous symmetry leads to massless Goldstone bosons, but this is not possible in two dimensions where the massless scalar field fails to exist.

the final word on the subject of $O(n)$ model MC simulations.

In physics MC simulations of continuous systems span a wide range of applications from particle and astrophysics to chemical and biophysics. In particle physics MC code for SU(n) *lattice gauge theories*, in particular for the physically important cases SU(2) and SU(3), can be written along similar lines [Creutz (1980); Cabibbo and Marinari (1982)] as the algorithms discussed in the present section. An implementation of SU(2) and SU(3) lattice gauge theory code, which is consistent with the STMC structure, is planned for Volume II of this book. To mention an example from biophysics, *all atom* Metropolis simulations of *peptides* (small proteins) have gained popularity in recent years, *e.g.* [Bouzida *et al.* (1992); Mitsutake *et al.* (2001)]. Several MC simulation packages are available[14]. Due to its (relative) modularity, the Fortran code SMMP [Eisenmenger *et al.* (2001)] (*Simple Molecular Mechanics for Proteins*) is recommendable to the extent that the not yet implemented atom endgroups are not crucial for the project at hand.

Over the wide range of applications our data analysis methods are rather universally applicable to MC simulations.

Rounding errors

Arithmetic in the floating point representation is not exact. See section 1.2 of *Numerical Recipes* [Press *et al.* (1992)] for a short discussion. Still one will get identical results on distinct platforms as long as all operation are covered by the IEEE (1.18) standard. However, there is no IEEE standard for the basic functions, like the exponential function or logarithms. The rounding errors become compiler dependent and may limit the reproducibility of our numerical results to staying within a certain computing platform. Only agreement in the statistical sense remains then between results obtained on distinct platforms. Whenever this happens, we shall give the values obtained on a Linux PC with the version 2.91.66 of the g77 Fortran compiler. Unfortunately, it has turned out that even the version number (3.99) of the g77 compiler matters.

3.4.1   *Simple Metropolis code for the $O(n)$ spin models*

Table 3.14 depicts our Metropolis code on_met.f for simulating the $O(n)$ model for general n $\geq$ 1 in arbitrary dimensions nd $\geq$ 1. The value of n is

---

[14]Note, however, that molecular dynamics simulations are the main stream approach in this field, see chapter 7 for more information.

Table 3.14 The Metropolis updating subroutine on_met.f of ForProg/MC_On for $O(n)$ spin models.

```fortran
      subroutine On_met
      include '../../ForLib/implicit.sta'
      include '../../ForLib/constants.par'
C Copyright Bernd Berg, Feb 10 2002.
C O(n) model: Sequential Metropolis updating (inefficient code).
      include 'lat.par'
      include 'mc.par'
      include 'on.par'
      include '../../ForLib/lat.com'
      include '../../ForLib/on.com'
      dimension stanew(n)
      amin=act
      amax=act
      do is=1,ns
1       sum=zero ! Propose new spin stanew().
        do i=1,n
          stanew(i)=two*(rmafun()-half)
          sum=sum+stanew(i)**2
        end do
        if(sum.gt.one) go to 1 ! stanew() inside the unit sphere.
        dact=zero
        fn=one/sqrt(sum)
        do id=1,nd
          do i=1,n
            dact=dact+sta(i,ipf(id,is))*(fn*stanew(i)-sta(i,is))
            dact=dact+sta(i,ipb(id,is))*(fn*stanew(i)-sta(i,is))
          end do
        end do
        if(rmafun().lt.exp(beta*dact)) then
          do i=1,n
            sta(i,is)=fn*stanew(i)
          end do
          act=act+dact
          acpt=acpt+one
          amin=min(amin,act)
          amax=max(amax,act)
        end if
      end do
      a_min=min(a_min,amin)
      a_max=max(a_max,amax)
      return
      end
```

Table 3.15 The common block on.com of ForLib for simulations of the $O(n)$ models.

```
      common /sites/  sta(n,ms)
c amin,amax:     action  act  minimum and maximum of the MC sweep.
c acpt:            counts accepted updates.
      common /links/ alink(nd,ms),act,amin,amax,acpt,nlink
c tsa: time series action array.
c a_min,a_max:  action minimum and maximum for a series of sweeps.
      common /ts_action/ tsa(nmeas),a_min,a_max
```

set in the parameter file on.par. All other parameters are set in the files lat.par and mc.par, just as in the Potts model simulations. The n spin components are stored in the array sta(n,ms) of the common block on.com, which is reproduced in table 3.15. This common block is, as always, kept in ForLib. It replaces the potts.com block used by the Potts model routines, while the lat.com common blocks stays the same. The array alink(nd,ms) is not used by the present routine and can be eliminated when storage space matters. Should one be interested in measuring energy–energy correlations, one would store the link energies in this array.

Instead of using histograms we rely now on the **time series** and introduce appropriate data analysis techniques. The time series approach avoids discretization errors, which are encountered when the histogramming approach is used for continuous observables. During a production run the time series data are intermediately kept in the array tsa(nmeas) of on.com. Whenever the array is filled up, it is written out to a file. To store time series data intermediately in reasonably sized arrays and to write the arrays out is much faster than writing many small data sets separately to disk.

We provide

$$\text{on\_ts.f \quad of \quad ForProg/MC\_On} \qquad (3.112)$$

as the main program which performs the equilibration as well as the data production. In its simulation run this program writes the array tsa nrpt times out. *I.e.*, a time series of length

$$\text{nrpt} \times \text{nmeas} \qquad (3.113)$$

is kept on file. For initialization the program on_ts.f calls the subroutine on_init.f, which invokes on_ran.f and on_act.f.

$O(n)$ spins are proposed in on_met.f between the line with the label 1 and the go to 1 statement. They are subsequently stored in the array stanew(n). The rationale of their calculation is a follows: Spin components are generated which are uniformly distributed in the n-dimensional $[-1, +1]$ box of volume $B_n = 2^n$. Spins which do not fit into the unit sphere are rejected, until a spin is found which fits. This spin is subsequently normalized to one. In the code the products fn*stanew(i) are the components of the normalized spin. This method of proposing spins works reasonably well for small n, but deteriorates quickly for large n. The volume V_n of the unit sphere is $V_n = \pi^{n/2}/\Gamma(1 + n/2)$, see appendix C of [Pathria (1972)]. For large n the acceptance ratio of the selection process, (V_n/B_n), approaches zero. In our following considerations only the cases $n = 2$ and $n = 3$ are of importance. With $\Gamma(n + 1) = n\Gamma(n)$, $\Gamma(1) = 1$ and $\Gamma(1/2) = \pi^{1/2}$ the familiar equations for the area of the unit circle and the volume of the unit sphere in three dimensions are obtained, and the acceptance ratios become

$$\frac{V_2}{B_2} = \frac{\pi}{2^2} = 0.785\ldots \quad \text{and} \quad \frac{V_3}{B_3} = \frac{4\pi/3}{2^3} = 0.523\ldots. \tag{3.114}$$

The Metropolis criterium (3.53) is implemented by the Fortran statement

$$\text{if}(\text{rmafun}().\text{lt}.\exp(\text{beta} * \text{dact})) \text{ then} \tag{3.115}$$

of the on_met.f subroutine. Note that on_met.f is written as a short routine at the price of performance. If a similar Metropolis code has to be used for long simulations, it is recommended to test possible improvements. For instance, the calculation of the exponential function is quite CPU time consuming and a code which by an additional if statement accepts the proposals with dact ≥ 0 before using (3.115) may be faster. Such conjectures need to be tested in the actual computing environment, because they depend on the CPU unit used as well as on the Fortran compiler. As a general programming rule, it is a good idea to assume working with a dumb compiler, who will miss out on simple optimization opportunities. With this rule one will gain on computing speed with the dumb compilers without loosing on the others. In the code of on_met.f a precaution would be to modify the most inner do loop by introducing an extra Fortran statement which assigns a variable to the product fn * stanew(i). This would insure that the multiplication is not done twice. Of course, a reasonable compiler will figure that out anyway. Illustrations and comparisons of the actual performance of our code are given next.

3.4.2   *Metropolis code for the XY model*

The subroutines

o2_met.f, xy_met.f, xy_met0.f and xy_met1.f of ForProg/MC_On   (3.116)

are all implementing the Metropolis algorithm for the XY model. The corresponding main programs for initialization, equilibration and data (time series) production are, respectively,

o2_ts.f, xy_ts.f, xy_ts0.f and xy_ts1.f of ForProg/MC_On .   (3.117)

For initialization the programs xy_ts.f and xy_ts0.f use xy_init.f, relying on xy_ranf.f and xy_act.f, whereas xy_ts1.f needs its own initialization subroutine, xy_init1.f, relying on xy_ran1.f and xy_act1.f. The program o2_ts.f is a specialized version of on_ts.f (3.112) and uses the same routines as on_ts.f for initialization. The analysis program

ana_ts_on.f of ForProg/MC_On   (3.118)

can be used on the output files of all the programs (3.117). It allows also to analyze the time series for autocorrelations as needed in chapter 4 (compare table 4.3).

The main deviation of the Metropolis routine xy_met0.f from on_met.f is that it specializes to $n = 2$ and uses a direct calculation of the newly proposed coordinates,

$$x = \cos(\phi), \quad y = \sin(\phi) \quad \text{and} \quad \phi = 2\,\pi\,x^r \,, \tag{3.119}$$

where $x^r \in [0, 1)$ is a uniformly distributed random number. As the calculation of the sine and cosine functions is notoriously slow on some computers, it requires some testing to find out, whether the random procedure with the 0.785 acceptance rate (3.114) or the direct calculation (3.119) is faster. If the random procedure turns out to be faster, a code which is specialized to $n = 2$ should still be used but the lines for the direct calculation should be replaced by those between label 1 and go to 1 in table 3.14.

The acceptance rate for the Metropolis update of xy_met0.f goes to zero for large β. The code of xy_met.f adds an additional twist by adjusting the new spin proposals so that the acceptance rate becomes never smaller than a preset value acpt_rate. Instead of defining the angle ϕ of equation (3.119) with respect to the x-axis and proposing it uniformly in the range $[0, 2\pi)$, ϕ is now defined with respect to the direction of the sum of the neighboring spins and proposed in a more sophisticated way: Whenever the acceptance rate falls below the value acpt_rate, which is set to 0.5 by default, the range $[-\phi_{max}, \phi_{max}]$ of the ϕ proposals is reduced to an interval smaller than $[-\pi, \pi)$ until acpt_rate is reached. The other way round, if the acceptance rate increases above acpt_rate the range is extended until either the acpt_rate acceptance rate value or the maximum ϕ range $[-\pi, \pi)$ is reached. At small β acceptance rates are, of course, already larger than 50% with the spin proposals (3.119).

To test our code, we compare two energy values with data reported in a paper by [Tobochnik and Chester (1979)][15]. In their tables I and II they give energy values for a 60×60 lattice at different temperatures, each data point relying on 1 000 sweeps for reaching equilibrium followed by 2 000 sweeps with measurements. For their data in table I they cool the initially random system and for table II they warm the initially ordered system. For comparison, we pick the high temperature value $1/T = \beta = 0.5$, where the system moves quite freely, and $1/T = \beta = 1$, a value in the critical region, slightly below the peak in the specific heat. We combine the values given in the two tables of Tobochnik and Chester assuming equal weights. This gives (assignment 1).

$$e_s = -0.54775\,(43) \quad \text{at} \quad \beta = 0.5 \quad \text{and} \quad e_s = -1.3213\,(15) \quad \text{at} \quad \beta = 1 \,.$$
(3.120)

Nowadays it is easy to produce a large statistics. We perform 10 000 sweeps for reaching equilibrium followed by $32 \times 2\,000$ sweeps with measurements. We use the general $O(n)$ code on_ts.f (3.112) as well as xy_ts0.f (3.117) and obtain at $\beta = 0.5$

$$e_s = -0.54756\,(13) \quad \text{with on_ts.f} \quad \text{and} \quad e_s = -0.54755\,(13) \quad \text{with xy_ts0.f} \,,$$
(3.121)

see assignment 2. We use the Gaussian difference test (2.33) to compare these estimates with one another as well as with $\beta = 0.5$ the estimate (3.120) and find the satisfactory Q values 0.96, 0.67 and 0.66.

3.4.2.1 *Timing, discretization and rounding errors*

Timing the on_ts.f and xy_ts0.f simulations against one another gives machine dependent results. On a Linux PC with the g77 compiler the on_ts.f simulation took longer than the xy_ts0.f simulation: 514 s versus 409 s. On a Sun workstation the results were more tight and in the

---

[15]This work was one of the first MC simulations of the XY model and investigated the Kosterlitz-Thouless scenario in some details.

reverse order: 313 s with on_ts.f and 336 s with xy_ts0.f. The routine o2_met.f illustrates that we can speed up the methods of the general on_met.f Metropolis code by specializing to $n = 2$. In particular, o2_met.f avoids the calculation of exp(beta*dact) and one random number for the acceptance case dact $\geq$ 0. The corresponding Fortran statement

$$\text{if(dact.lt.zero) then} \qquad (3.122)$$

of o2_met.f is also used in xy_met0.f. The driving program for the simulations with o2_met.f is o2_ts.f. With o2_ts.f the simulations for equation (3.121) took 459 s on the Linux PC and 260 s on the Sun workstation.

Out of this competition is the program xy_ts1.f for which the same run took 960 s on the Linux PC and 1 199 s on the Sun workstation. Instead of using the x, y coordinates, this program stores only the angles ϕ. Storage space requirements are reduced by a factor of two, but the calculation of the energy differences from the angles is far slower than from the coordinates. To use such a program makes only sense when RAM is a serious problem.

The energy values obtained after the simulation with on_ts.f agree on the Linux PC and the Sun workstation to all digits. That is not the case for the simulation with xy_ts0.f. The result on the Sun workstation was $e_s = -0.54739\,(13)$, agreeing in the statistical sense of $Q = 0.38$ for the Gauss difference test with the value (3.121) from the Linux PC. At the first look the difference is a bit surprising. Due to the portability of the Marsaglia's random number generator, the proposed XY coordinates differ only by the eventually distinct rounding errors of the sine and cosine functions in equation (3.119). Thus the energy differences encountered by the Metropolis acceptance steps are expected to be governed by the machine rounding error, which is for double precision about 2.2×10^{-16}, compare paragraph 20.1 of *Numerical Recipes* [Press *et al.* (1992)]. In our MC calculations we use 2.304×10^8 random numbers per run for acceptance/rejection decisions and the likelihood that the two calculations diverge on different machines appears to be less than 10^{-7}. So, what is going on?

What happens is uncovered in assignment 3. Due to the fact that there are only $2^{24} = 16,777,216$ distinct Marsaglia random numbers (assignment 5 of section 1.2), we introduce a discretization of the angle ϕ, and, hence, the XY coordinates (3.119). Therefore, it will occasionally happen that precisely the present (old) Potts spin is proposed at some site. The energy difference is then zero up to a $\pm\epsilon_{\text{machine}}$ rounding error, which turns out to be exactly zero on the Sun, but around $\pm 10^{-18}$ on the Linux PC. This leads to different decisions in the if statement (3.122) and, hence, an additional random number is or is not used in a machine dependent way. Once this happens, the calculations on the two machines diverge.

If one is unhappy with the accuracy of this discretization of the angle ϕ in equation (3.119), one can greatly improve on it by using two random numbers. The proposed angles are then defined by

$$\phi = 2\pi\, x_1^r + 2\pi\, \epsilon_{\text{Marsaglia}}\, x_2^r \tag{3.123}$$

where $\epsilon_{\text{Marsaglia}} = 2^{-24}$ is the discretization step (1.20) of Marsaglia's generator.

3.4.2.2 Acceptance rate

Repeating our simulation (3.121) at $\beta = 1$ gives (assignment 4)

$$e_s = -1.32194\,(87) \text{ with } \texttt{on\_ts.f} \quad \text{and} \quad e_s = -1.3213\,(10) \text{ with } \texttt{xy\_ts0.f}\,. \tag{3.124}$$

The Gaussian difference tests yield $Q = 0.12$, 0.71 and 0.36.

Whereas the Metropolis acceptance rate for the high temperature simulations (3.121) at $\beta = 0.5$ was about 0.64, it is now down to 0.32, and it will go further down with decreasing temperature. A very low acceptance rate appears to be bad. The routine $\texttt{xx\_met.f}$ allows to increase the acceptance rate by proposing an angle ϕ in the range

$$0 \leq |\phi| < \Delta\Phi \text{ with } 0 < \Delta\Phi < \pi \tag{3.125}$$

with respect to the direction of the sum of the neighboring spins. The proposed new spin at site i is then given by the equations

$$x_i^{\text{new}} = \cos(\phi)\, x_i^{\text{old}} - \sin(\phi)\, y_i^{\text{old}} \quad \text{and} \quad y_i^{\text{new}} = \sin(\phi)\, x_i^{\text{old}} + \cos(\phi)\, y_i^{\text{old}}\,. \tag{3.126}$$

By decreasing the value of $\Delta\Phi$, one increases the acceptance rate, because the proposed energy differences become smaller. A large acceptance rate is not good too: In the limit where the acceptance rate goes to one, we find $\Delta\phi = 0$, and the system does not move anymore. On the other side, the system does not move either when the acceptance rate is zero. What is the optimal acceptance rate? In essence, this question can only be answered by a careful study of autocorrelation times (see the next chapter). In practice one may not care to be somewhat off the optimum. As a rule of thumb: An acceptance rate in the range 30% to 50% works well.

We use the notation **natural acceptance rate** for the value obtained with $\Delta\Phi = \pi$ in equation (3.125). The Metropolis routine $\texttt{xy\_met.f}$ allows for adjustments of its acceptance rate to the value of its parameter acpt_rate, unless acpt_rate is smaller than the natural acceptance rate, which is its lower bound. The preset default value of actp_rate is 0.5 and can be changed in the routine. Whenever the argument lacpt of xy_met.f is set to .true., the routine adjusts $\Delta\Phi$ to attain the desired acceptance rate. For this task it relies on its so far assembled statistics to estimate the appropriate value of $\Delta\Phi$. The driving program xy_ts.f introduces a schedule for lacpt=.true. versus lacpt=.false. during the equilibration part of the run.

For the simulation of assignment 4, which leads to the xy_ts0.f result (3.124), the natural acceptance rate is about 32%. In assignment 5 no improvement of the error bar is found by producing the same statistics with xy_ts.f at an acceptance rate of 50%. On the Linux PC the result was

$$e_s = -1.3211\,(12) \text{ with } \text{xy\_ts.f at 50\% acceptance rate}. \quad (3.127)$$

This is well compatible with the estimates (3.124), while the slightly larger error bar is within the limits one may attribute to statistical fluctuations. With much higher statistics the integrated autocorrelation time is estimated in assignment 5 of chapter 4.2.2 and found to be slightly lower with 32% acceptance rate than with 50%. Altogether, the Metropolis algorithm appears to work reasonably well for acceptance rates in the entire range of 30% – 50%.

Really high β values are needed to find a significant improvement of error bars due to the adjustment of the acceptance rate to 50%. In assignment 5 MC simulations are performed on 10×10 lattice, for which a high statistics is easily assembled, at $\beta = 10$ and $\beta = 20$. The energy results at $\beta = 10$ are (on the Linux PC)

$$e_s = -1.9498\,46\,(33) \text{ with } \text{xy\_ts0.f and}$$
$$e_s = -1.9498\,46\,(18) \text{ with } \text{xy\_ts.f}. \quad (3.128)$$

Comparing the error bars with the F-test (2.107) gives $Q = 10^{-8}$. The xy_ts.f simulation performs significantly better than the one with xy_ts0.f for which the acceptance rate was 0.081 (the agreement of the mean values to all digits is an accident). At $\beta = 20$ the results are

$$e_s = -1.7841\,03\,(20) \text{ with } \text{xy\_ts0.f and}$$
$$e_s = -1.9750\,82\,(10) \text{ with } \text{xy\_ts.f}. \quad (3.129)$$

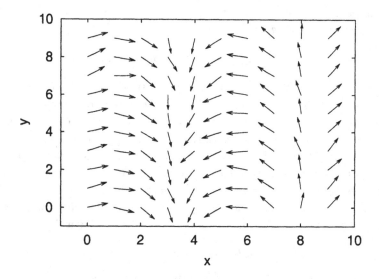

Fig. 3.7 Domain wall configuration encountered in one of the simulations when quenching the 10×10 XY system to $\beta = 20$, see assignment 5.

The acceptance rate of the xy_ts0.f simulation at $\beta = 20$ was 0.060 and the error bar of the xy_ts.f simulation is again the significantly smaller one. However, most remarkable is that the mean values do not agree at all. The value obtained with xy_ts0.f is far too high (several thousand standard deviations). It turns out that this has nothing to do with the performance of the xy_met0.f Metropolis algorithm. Figure 3.7 is taken after the 100 000 equilibration sweeps of this simulation and plots the XY spins of the $2d$ lattice with sites labeled by coordinates x and y. It shows that the system got stuck in a domain wall configuration from which there is practically no escape by small fluctuations. The spins at $x = 3$ point down, whereas the spins at at $x = 8$ point up, creating two domains in between. This can happen due to rapid quenching of the system from a disordered start to almost zero temperature. By lowering the temperature more slowly, *i.e.* annealing, one can avoid this. Or, one can simply use an ordered start configuration.

3.4.3   *Heat bath code for the $O(3)$ model*

The subroutine

$$\texttt{o3\_mchb.f} \quad \text{of} \quad \texttt{ForProg/MC\_On} \qquad (3.130)$$

is an implementation of the $O(3)$ heatbath algorithm of section 3.2.2. It is run by the driving program

$$\texttt{o3\_tshb.f} \quad \text{of} \quad \texttt{ForProg/MC\_On} \ . \qquad (3.131)$$

In the usual structure it does first initialization, then equilibration and finally data production. For the initialization the routines o3_init.f, o3_ran.f and o3_act.f are invoked. The last two routines can be replaced by the general $O(n)$ routines on_ran.f and on_act.f, if desired.

To test the $O(3)$ model code, we simulate a 100×100 lattice at $\beta = 1.1$ to compare with an energy values of table I in [Berg and Lüscher (1981)], where the topological charge of the $O(3)$ σ-model was investigated[16]. We perform 10 000 sweeps for reaching equilibrium and 20×2500 sweeps with measurement. Our results on the Linux PC, see assignment 6, are

$$e_s = -0.85230\,(14) \quad \text{with} \quad \texttt{on\_tshb.f} \quad \text{and}$$
$$e_s = -0.852122\,(96) \quad \text{with} \quad \texttt{o3\_tshb.f} \ . \qquad (3.132)$$

They are consistent with one another. The Gaussian difference test gives $Q = 0.29$. The same is true for the error bars. The F-test (2.107) gives $Q = 0.11$, a value too large to call the smaller error bar of the heatbath simulation statistically relevant. This is clarified by increasing the production part of the statistics from twenty to one hundred bins (assignment 7). The error bar of the heat bath simulation remain the smaller one, 36 to 62 is the ratio, and the Q value of the F-test decreases to 1.4×10^{-7}. The performance of the heat bath simulation is significantly better than that of our Metropolis simulation and this improvement wins over the increased CPU time consumption of the heat bath code: For the simulations of equation (3.132) the CPU time used on the Linux PC was 1 703 s for on_ts.f and 2 334 s for o3_ts.f. On the Sun workstation it was 1 109 s for on_ts.f and 1 610 s for o3_ts.f.

---

[16]In QCD the η–η' mass difference is related to fluctuations of the instanton topological charge ['t Hooft (1976)]. The $O(3)$ σ-model in $d = 2$ serves as a toy model for such investigations. To get physical results, the approach of [Berg and Lüscher (1981)] needs to be refined. See [Negele (1999)] for a review of the lattice gauge theory approach to this subject.

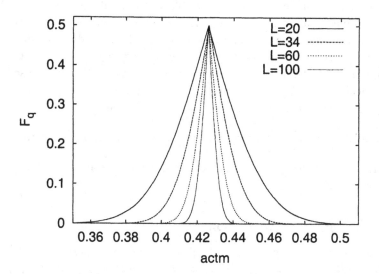

Fig. 3.8 Illustration of self-averaging: Peaked distribution functions from MC simulations of the $O(3)$ σ-model at $\beta = 1.1$, see assignments 6 and 8.

The $\beta = 1.1$ energy value of table I in [Berg and Lüscher (1981)] translates to $e_s = 0.85250\,(25)$[17]. Gaussian difference tests show that this value is consistent with those of equation (3.132): $Q = 0.49$ and $Q = 0.16$.

In figure 3.8 we compare the peaked distribution function (1.41) of the mean action per link from the simulations of assignment 6 with the corresponding peaked distribution functions from simulations on smaller lattices (assignment 8). The property of **self-averaging** is observed: The larger the lattice, the smaller the confidence range. When self-averaging works, the peaked distribution function will with increasing lattice size approach the function which is 0.5 at the median and zero otherwise. The other way round, the peaked distribution function is very well suited to investigate observables for which self-averaging does not work, as for instance

---

[17]The relation $e_s = (e_{BL} - 4)/2$ holds, where e_{BL} is the value given in table I of [Berg and Lüscher (1981)], compare their equation (29). The error bar used here assumes the maximum rounding towards their last digit given.

encountered in spin glass simulations [Berg *et al.* (2000)].

3.4.3.1 *Rounding errors*

Comparing runs from a Sun workstation with those on a Linux PC, we find that the on_ts.f values of equation (3.132) agree to all digits, whereas the o3_ts.f values differ, but remain consistent in the statistical sense. For the XY model we encountered a sudden divergence of results due to a distinct use of a random number as consequence of a rounding error. This is not the case here. In the heat bath algorithm all random numbers are used in an identical way on both platforms. We find a slow divergence of results, which can be traced to the rounding errors in the line

$$\text{snew3} = \log(\text{Sexp} + \text{rmafun}() * (\exp(-\text{Sbeta}) - \text{Sexp}))/\text{Sbeta}$$

of the heat bath routine o3_mchb.f. This line is the Fortran implementation of equation (3.61).

3.4.4 *Assignments for section 3.4*

(1) The energy per spin values in [Tobochnik and Chester (1979)] at $\beta = 0.5$ are $e_s = -0.5470\,(7)$ in table I and $e_s = -0.5485\,(5)$ in table II. At $\beta = 1$ the values are $e_s = -1.3188\,(23)$ in table I and $e_s = -1.3237\,(19)$ in table II. Use the subroutine steb2.f (2.25) to reproduce the estimates of our equation (3.120).

(2) Use Marsaglia random numbers with their default seed and the programs on_ts.f (3.112) and xy_ts0.f (3.117) to simulate the XY model at $\beta = 0.5$ on an 60×60 lattice with 10 000 sweeps for reaching equilibrium followed by $32 \times 2\,000$ sweeps with measurements. Subsequently, run the analysis program ana_ts_on.f to obtain the energy values reported in equation (3.121). Perform Gaussian difference tests to compare the results with one another as well as with the $\beta = 0.5$ estimate (3.120). Time the two simulations against one another, as well as against o2_ts.f and xy_ts.f. Explain why the energy values obtained with o2_ts.f are not to all digits identical with those from on_ts.f.

(3) Investigate the equilibrium sweep iequi = 3707 of the xy_ts0.f run from the previous assignment and print out the dact values of the subroutine xy_met0.f for $2390 \leq \text{is} \leq 2395$. Explain the value which you get for is = 2393.

(4) Repeat assignment 2 for $\beta = 1$ to reproduce the energy values reported in equation (3.124).

(5) A. Repeat the previous assignment using xy_ts.f. B. Use xy_ts0.f and xy_ts.f to perform high statistics simulations with nequi = 100 000, nrpt = 100 and nmeas = 10 000 on a 10×10 lattice at $\beta = 10$ and $\beta = 20$. Perform the F-test (2.107) to compare the error bars of the two simulations. Plot the spins after equilibration for the $\beta = 20$ run with xy_ts0.f.

(6) Use Marsaglia random numbers with their default seed and the programs on_ts.f (3.112) and o3_tshb.f (3.131) to simulate the $O(3)$ model at $\beta = 1.1$ on a 100×100 lattice with 10 000 sweeps for reaching equilibrium followed by $20 \times 2,500$ sweeps with measurements. Subsequently, run the analysis program ana_ts_on.f to obtain the energy values reported in equation (3.132). Perform Gaussian difference tests to compare the results with one another as well as with the estimate from [Berg and Lüscher (1981)]. Perform the F-test (2.107) to compare the two error bars in equation (3.132).

(7) Increase the production statistics of the previous assignment to $100 \times 2,500$ sweeps and repeat the other parts of that assignment.

(8) Repeat the simulations of assignment 6 (*i.e.*, the same statistics and β value) on lattices of size $L = 20$, 34 and 60. Reproduce figure 3.8.

Chapter 4

Error Analysis for Markov Chain Data

In large scale MC simulation it may take months, possibly years, of computer time to collect the necessary statistics. For such data a thorough error analysis is a must and it cannot hurt to apply the appropriate methods also to shorter simulations. A typical MC simulation falls into two parts:

(1) **Equilibration:** Initial sweeps are performed to reach the equilibrium distribution. During these sweeps measurements are either not taken at all or they have to be discarded when calculating equilibrium expectation values.

(2) **Production:** Sweeps with measurements are performed. Equilibrium expectation values are calculated from this statistics.

Certain simple measurements, like sampling energy histograms, are often integrated into the MC updating routine. Other measurements may be performed after one or more sweeps. Performing **extensive measurements** after each sweep may consume more CPU time than the updating does. A rule of thumb is then: **Spend about 50% of your CPU time on measurements.** The reason for this rule is that, *a priori*, we may not have a good estimate how strongly subsequent sweeps are autocorrelated (autocorrelations are discussed in the next section). With our rule of thumb we can never loose more than 50% of our CPU time. Assume, subsequent sweeps are strongly correlated. Then, it may be optimal to spend, say, 99% of our CPU time on MC sweeps and only 1% on measurements, so we are off by a factor of almost two. On the other hand, if subsequent sweeps are almost uncorrelated, then it may be optimal to spend, say, 99% of the CPU time on measurements. Again, we are off by a factor of almost two, but it cannot get worse.

How many sweeps should be discarded for reaching equilibrium? In a few exceptional situations this question can be rigorously answered with the *Coupling from the Past* method [Propp and Wilson (1998)]. Otherwise, the next best thing to do is to measure the integrated autocorrelation time self-consistently and to discard, after reaching a visually satisfactory situations (compare figures 3.4 and 3.5), a number of sweeps which is larger than the integrated autocorrelation time. In practice even this can often not be achieved, as we will discuss after equation (4.44). Therefore, it is re-assuring that it is sufficient to pick the number of discarded sweeps approximately right. Assume, we count some out off equilibrium configurations as part of our production run. This means, we include data in our measurements whose generation during the measurement time has, in the equilibrium ensemble, a probability of practically zero. With an increasing statistics the contribution of the non-equilibrium data dies out like $1/N$, where N is the number of measurements. For large N the effect is eventually swallowed by the statistical error, which declines only like $1/\sqrt{N}$. The point of discarding the equilibrium configurations is that the factor in front of $1/N$ may be large. Consider, for instance, figure 3.4 of the previous chapter. The deviation of the random initial configuration from the true average is about 1.1, whereas statistical fluctuations after reaching the neighborhood of the true value do not exceed 0.1. Similarly, the deviation of the ordered initial configuration from the true average is about 0.9 and the approach to the average values is slower than for the disordered start. In essence, it is important to exclude the large deviations from the equilibrium statistics, but it is relatively unimportant whether the cut is made after, say, 50, 100 or 200 sweeps. Of course, there can be far more involved situations, like that the Markov chain ends up in a metastable configuration and never reaches equilibrium. An example for this was given in figure 3.7. Worse are metastabilities which stay unnoticed. For instance, the proper treatment of metastabilities is a serious challenge in the simulation of complex systems like spin glasses or proteins. Besides extending the computer time of the run also algorithmic changes, as discussed in chapters 5 and 6, may help to control such situations.

4.1 Autocorrelations

We like to estimate the expectation value $\hat{f}$ of some physical observable. Let us assume that the system has reached equilibrium. How many MC sweeps are needed to estimate $\widehat{f}$ with some desired accuracy? To answer this question, one has to understand the autocorrelations within the Markov chain.

Given is a **time series** of N measurements from a Markov process

$$f_i = f_i(x_i), \quad i = 1, \ldots, N , \tag{4.1}$$

where x_i are the configurations generated. The label $i = 1, \ldots, N$ runs in the temporal order of the Markov chain and the elapsed time, measured in updates or sweeps, between subsequent measurements f_i, f_{i+1} is always the same. The estimator of the expectation value $\widehat{f}$ is

$$\overline{f} = \frac{1}{N} \sum f_i . \tag{4.2}$$

With the notation

$$t = |i - j|$$

the definition of the **autocorrelation function** of the observable $\widehat{f}$ is

$$\widehat{C}(t) = \widehat{C}_{ij} = \langle (f_i - \langle f_i \rangle)(f_j - \langle f_j \rangle) \rangle = \langle f_i f_j \rangle - \langle f_i \rangle \langle f_j \rangle = \langle f_0 f_t \rangle - \widehat{f}^2 \tag{4.3}$$

where we used that translation invariance in time holds for the equilibrium ensemble. The asymptotic behavior for large t is

$$\widehat{C}(t) \sim \exp\left(-\frac{t}{\tau_{\exp}}\right) \quad \text{for} \quad t \to \infty, \tag{4.4}$$

where $\tau_{\exp}$ is called **(exponential) autocorrelation time**. It is related to the second largest[1] eigenvalue λ_1 of the transition matrix (3.37) by

$$\tau_{\exp} = -\ln \lambda_1 \tag{4.5}$$

under the assumption that f has a non-zero projection on the corresponding eigenstate. Superselection rules are possible so that different autocorrelation times may reign for different operators.

The variance (1.51) of f is a special case of the autocorrelations (4.3)

$$\widehat{C}(0) = \sigma^2(f) . \tag{4.6}$$

---

[1]The largest eigenvalue is $\lambda_0 = 1$ and has as its eigenvector the equilibrium distribution (3.41).

The variance of the estimator $\overline{f}$ (4.2) for the mean and the autocorrelation functions (4.3) are related in the following interesting way

$$\sigma^2(\overline{f}) = \langle(\overline{f} - \widehat{f})^2\rangle = \frac{1}{N^2}\sum_{i=1}^{N}\sum_{j=1}^{N}\langle(f_i - \widehat{f})(f_j - \widehat{f})\rangle$$

$$= \frac{1}{N^2}\sum_{i=1}^{N}\sum_{j=1}^{N}\langle f_i f_j - f_i\widehat{f} - f_j\widehat{f} + \widehat{f}^2\rangle$$

$$= \frac{1}{N^2}\sum_{i=1}^{N}\sum_{j=1}^{N}\left[\langle f_i f_j\rangle - \widehat{f}^2\right] = \frac{1}{N^2}\sum_{i=1}^{N}\sum_{j=1}^{N}C_{ij} . \qquad (4.7)$$

In the last sum $|i-j| = 0$ occurs N times and $|i-j| = t$ with $1 \le t \le (N-1)$ occurs $2(N - t)$ times. Therefore,

$$\sigma^2(\overline{f}) = \frac{1}{N^2}\left[N\,\widehat{C}(0) + 2\sum_{t=1}^{N-1}(N - t)\,\widehat{C}(t)\right] \qquad (4.8)$$

holds and, using (4.6),

$$\sigma^2(\overline{f}) = \frac{\sigma^2(f)}{N}\left[1 + 2\sum_{t=1}^{N-1}\left(1 - \frac{t}{N}\right)\widehat{c}(t)\right] \quad \text{with } \widehat{c}(t) = \frac{\widehat{C}(t)}{\widehat{C}(0)} . \qquad (4.9)$$

This equation ought to be compared with the corresponding equation (1.109) for uncorrelated random variables $\sigma^2(\overline{f}) = \sigma^2(f)/N$. The difference is the factor in the bracket of (4.9) which defines the **integrated autocorrelation time**

$$\tau_{\text{int}} = \left[1 + 2\sum_{t=1}^{N-1}\left(1 - \frac{t}{N}\right)\widehat{c}(t)\right] . \qquad (4.10)$$

For correlated data the variance of the mean is by the factor τ_{int} larger than the corresponding **naive variance** for uncorrelated data:

$$\tau_{\text{int}} = \frac{\sigma^2(\overline{f})}{\sigma^2_{\text{naive}}(\overline{f})} \quad \text{with } \sigma^2_{\text{naive}} = \frac{\sigma^2(f)}{N} . \qquad (4.11)$$

Here $\sigma^2_{\text{naive}}(\overline{f})$ is called naive because its definition is simply the relationship (1.109) for statistically independent data. In most simulations one is interested in the limit $N \to \infty$ and equation (4.10) becomes

$$\tau_{\text{int}} = 1 + 2\sum_{t=1}^{\infty}\widehat{c}(t) . \qquad (4.12)$$

For the discussion of additional mathematical details see [Sokal (1997)] and references given therein. The numerical estimation of the integrated autocorrelation time (4.12) faces difficulties. Namely, the variance of the $N \to \infty$ estimator of τ_{int} diverges:

$$\overline{\tau}_{\text{int}} = 1 + 2\sum_{t=1}^{\infty} \overline{c}(t) \quad \text{and} \quad \sigma^2(\overline{\tau}_{\text{int}}) \to \infty \qquad (4.13)$$

because for large t each $\overline{c}(t)$ adds a constant amount of noise, whereas the signal dies out like $\exp(-t/\tau_{\text{exp}})$. To obtain an estimate one considers the t-dependent estimator

$$\overline{\tau}_{\text{int}}(t) = 1 + 2\sum_{t'=1}^{t} \overline{c}(t') \qquad (4.14)$$

of the integrated autocorrelation time and looks out for a window in t for which $\overline{\tau}_{\text{int}}(t)$ is (almost) independent of t. The value from this window serves as the final estimate. Another approach to calculating the integrated autocorrelation time is the binning method discussed in the next subsection.

To give a **simple example**, let us assume that the autocorrelation function is governed by a single exponential autocorrelation time

$$\widehat{C}(t) = \text{const} \exp\left(-\frac{t}{\tau_{\text{exp}}}\right) . \qquad (4.15)$$

In this case we can carry out the sum (4.12) for the integrated autocorrelation function and find

$$\tau_{\text{int}} = 1 + 2\sum_{t=1}^{\infty} e^{-t/\tau_{\text{exp}}} = 1 + \frac{2\,e^{-1/\tau_{\text{exp}}}}{1 - e^{-1/\tau_{\text{exp}}}} . \qquad (4.16)$$

In particular, the difference between the asymptotic value (4.12) and the finite t definition (4.14) becomes then

$$\tau_{\text{int}} - \tau_{\text{int}}(t) = 2\,e^{-t/\tau_{\text{exp}}} \sum_{t'=1}^{\infty} e^{-t'/\tau_{\text{exp}}} = \frac{2\,e^{-(t+1)/\tau_{\text{exp}}}}{1 - e^{-1/\tau_{\text{exp}}}} . \qquad (4.17)$$

For a large exponential autocorrelation time $\tau_{\text{exp}} \gg 1$ the approximation

$$\tau_{\text{int}} = 1 + \frac{2\,e^{-1/\tau_{\text{exp}}}}{1 - e^{-1/\tau_{\text{exp}}}} \cong 1 + \frac{2 - 2/\tau_{\text{exp}}}{1/\tau_{\text{exp}}} = 2\,\tau_{\text{exp}} - 1 \cong 2\,\tau_{\text{exp}} \qquad (4.18)$$

holds. Normally the situation is more complicated and a factor distinct from two connects τ_{int} and τ_{exp}.

Computer code

The Fortran function `autcorf.f` of ForLib calculates the estimator

$$\overline{C}(t) = \text{AUTCORF(IT, NDAT, DATA, LMEAN)} \tag{4.19}$$

of the autocorrelation function (4.3). The arguments of `autcor.f` are the elapsed time $t = $ `IT`, the number of data `NDAT`, the data array `DATA` and the logical variable `LMEAN`. The data array is of dimension `NDAT`. The logical variable `LMEAN` has to be set to `LMEAN=.TRUE.` when the estimator $\overline{f}$ for the expectation value $\hat{f}$ in equation (4.3) needs to be involved and `LMEAN = .FALSE.` when $\hat{f} = 0$ holds. If `LMEAN` is true the unbiased average is with respect to `NDAT-1` degrees of freedom, similarly as discussed in chapter 2 for the sample variance (2.14).

For a given array of autocorrelations $\overline{C}(t)$, $0 \le t \le t_n$, a call to

$$\text{AC\_INT(NT, ACOR, ACINT)} \tag{4.20}$$

returns the corresponding array of integrated autocorrelation times (4.14). On input the array `ACOR(0:NT)`, containing the autocorrelation function (4.3) for $0 \le$ `IT` $\le$ `NT`, has to be provided. On output the array `ACINT(0:NT)`, containing the $0 \le$ `IT` $\le$ `NT` estimators (4.14) of the integrated autocorrelation time, is returned.

A call to the subroutine

$$\text{AUTCORJ(IT, NDAT, NBINS, DATA, WORK, ACORJ, LMEAN)} \tag{4.21}$$

returns jackknife estimators of the autocorrelation function $\hat{C}(t)$. In addition to the arguments already discussed for `autcorf.f` we have `NBINS`, `WORK` and `ACORJ`. The integer `NBINS` denotes the number of jackknife bins (and has to be much smaller than `NDAT`). The array `ACORJ` is of dimension `NBINS` and returns the jackknife estimators of the autocorrelation function (4.3) at $t = $ `IT`. The array `WORK` is a work space of dimension `NBINS`. The subroutine `stebj0.f` or, if a bias analysis is desired, `stebj1.f` has to be used to calculate from `ACORJ` its mean with error bar (`stebj1.f` needs on input also the mean value $\overline{C}(t)$ provided by `autcorf.f`). If serious bias problems are encountered, a second level jackknife analysis may be added.

Jackknife estimators for the autocorrelation function in a range $0 \le$ `IT` $\le$ `NT` are the starting point for calculating jackknife estimators of the integrated autocorrelation time (4.14) in the same range. A call to the subroutine

$$\text{AC\_INTJ(NT, NBINS, ACORJ, ACINTJ)} \tag{4.22}$$

needs on input the array ACORJ(NBINS,0:NT) of jackknife estimators of the autocorrelation function (4.3) and returns the array ACINTJ(NBINS,0:NT) of jackknife estimators of the integrated autocorrelation time (4.14).

In subsection 4.1.2 a very simple Metropolis process is considered to illustrate the use of these subroutines.

4.1.1 *Integrated autocorrelation time and binning*

The idea pursued here is to estimate the integrated autocorrelation time via the variance ratio (4.11). This method is advocated in an article by [Flyvberg and Peterson (1989)] and is known to practitioners of MC simulations at least since the early 1980s.

The subroutine bining.f (1.113) of chapter 1 allows to bin the time series (4.1) into $N_{bs} \leq N$ bins of

$$N_b = \texttt{NBIN} = \left[\frac{N}{N_{bs}}\right] = \left[\frac{\texttt{NDAT}}{\texttt{NBINS}}\right] \tag{4.23}$$

data each. Here [.] stands for Fortran integer division, *i.e.*, $N_b = \texttt{NBIN}$ is the largest integer $\leq N/N_{bs}$, implying $N_{ba} \cdot N_b \leq N$. Vice versa, the number of bins is determined by the number of data within each bin

$$N_{bs} = \texttt{NBINS} = \left[\frac{N}{N_b}\right] = \left[\frac{\texttt{NDAT}}{\texttt{NBIN}}\right] . \tag{4.24}$$

It is convenient to choose the values of N and N_{bs} so that N is a multiple of N_{bs}. The binned data are the averages

$$f_j^{N_b} = \frac{1}{N_b} \sum_{i=1+(j-1)N_b}^{jN_b} f_i \quad \text{for} \quad j = 1, \ldots, N_{bs} . \tag{4.25}$$

For increasing binsize N_b (decreasing N_{bs}) the autocorrelations between the (binned) data get weaker and weaker. For $N_b > \tau_{\exp}$, where $\tau_{\exp}$ is the exponential autocorrelation time (4.4), the autocorrelations are essentially reduced to those between nearest neighbor bins and even these approach zero under further increase of the binsize N_b.

For a set of N_{bs} binned data $f_j^{N_b}$, $(j = 1, \ldots, N_{bs})$ we may use one of our error bar subroutines to calculate the corresponding mean

$$\overline{f}_j^{N_b} = \frac{1}{N_{bs}} \sum_{j=1}^{N_{bs}} f_j^{N_b} \tag{4.26}$$

with its **naive error bar**. Again, naive means that the data are treated as if they were uncorrelated. The variance is the error bar squared. Assuming for the moment an infinite time series, we find the integrated autocorrelation time (4.11) from the following ratio of sample variances (2.14)

$$\tau_{\text{int}} = \lim_{N_b \to \infty} \tau_{\text{int}}^{N_b} \quad \text{with} \quad \tau_{\text{int}}^{N_b} = \left(\frac{s_{\bar{f}N_b}^2}{s_{\bar{f}}^2} \right) . \tag{4.27}$$

In practice the $N_b \to \infty$ limit will be reached for a sufficiently large, finite value of N_b. The statistical error of the τ_{int} estimate (4.27) is, in the first approximation, determined by the errors of the estimators $s_{\bar{f}N_b}^2$ in the numerator. The errors of $s_{\bar{f}}^2$ in the denominator will, in the large N_b limit, be much smaller than those of the numerator. Numerically most accurate estimates of τ_{int} are obtained for the finite binsize N_b which is just large enough that the binned data (4.25) are practically uncorrelated. The typical situation is then that, due to the central limit theorem, the binned data are approximately Gaussian, so that the **error of** $s_{\bar{f}N_b}^2$ **is analytically known** from the χ^2 distribution (see table 2.9 of chapter 2). We do not have to rely on another noisy estimator, the number (4.24) of statistically independent bins N_{bs} alone determines the error. Finally, we may want to include in our error estimate the fluctuations of $s_{\bar{f}}^2$ in the denominator. For Gaussian data this leads from the χ^2 distribution to the F-ratio (2.96) distribution (2.104). The number of degrees of freedom of the numerator is the number of independent bins N_{bs}. But which number should we use for the denominator? Due to the autocorrelations the number N of data is an overestimate. The correct value to be used is the effective number of uncorrelated data

$$N_{\text{effective}} = \frac{N}{\tau_{\text{int}}} \tag{4.28}$$

as follows from the relationship (4.11) between the real and the naive variance.

For applications it is convenient to choose N and N_b to be powers of 2. In the following we assume

$$N = 2^K, \ K \geq 4 \ \text{and} \ N_b = 2^{K_b} \ \text{with} \ K_b = 0, 1, \ldots, K - 5, K - 4. \tag{4.29}$$

Choosing the maximum value of K_b to be $K - 4$ implies that the smallest number of bins is

$$N_{bs}^{\min} = 2^4 = 16 .\tag{4.30}$$

Table 2.6 for the Student distribution shows that the confidence intervals of the error bars obtained from 16 uncorrelated normal data are reasonable approximations to those of the Gaussian standard deviation. It makes sense to work with error bars from 16 binned data, but the error of the error bar, and hence a reliable estimate of τ_{int}, requires far more data: The confidence intervals of the estimator $s_{\bar{f}^{N_b}}^2$ are those collected in table 2.9. There, we learned that about 1000 independent data are needed to provide a decent estimate[2] of the corresponding variance, *i.e.* $\sigma^2(\overline{f}^{N_b})$ here.

Computer code

The subroutine `bin_inta.f` of ForLib calculates for $N = 2^K$ data the estimators $\tau_{\mathrm{int}}^{N_b}$ of equation (4.27). The bin length used, NBIN = N_b, is defined by equation (4.29) where KB = K_b runs from 0 to KBMAX. On input K, KBMAX and the data array DATA of dimension 2**K have to be provided. Further, the work array DATB of dimension 2**K and the arrays DBE, RUP, RDO and RAT, all of dimension (0:KBMAX), have to be allocated in the calling routine. A call to

$$\texttt{BIN\_INTA(K, KBMAX, DATA, DATB, DBE, RUP, RDO, RAT)}\tag{4.31}$$

returns the results in the arrays DBE, RUP, RDO and RAT. The $\tau_{\mathrm{int}}^{N_b}$ estimators (4.27) are squared ratios of error bars for binned data and contained in the *variance ratio* array RAT. The error bars themselves are returned in the array DBE. The errors of the RAT ratios are estimated in the approximation which neglects the error of the denominator in the definition (4.27) of $\tau_{\mathrm{int}}^{N_b}$. In most practical applications the denominator of these ratios is very accurate, so that this approximation is justified. If desired, its error can be taken into account once the approximate integrated autocorrelation time is known, see the discussion leading to equation (4.42). Our error estimate for $\tau_{\mathrm{int}}^{N_b}$ is thus done as follows: The number N_{bs} of bins determines the confidence limits of the error bars analytically. **Half** of the 95.54% confidence range is taken as **definition of the error of the error bar**. These limits are calculated relying on `sebar_e.f` (2.93). The obtained upper and lower one error bar limits of RAT are returned in the arrays RUP and RDO.

---

[2] At the 95% confidence level with an accuracy of slightly better than 10% .

4.1.2 *Illustration: Metropolis generation of normally distributed data*

To give a really simple illustration, we use the Metropolis process to generate normally distributed data. The subroutine gau_metro.f generates real numbers according to the Markov process

$$x' = x + 2\,a\,x^r - a \tag{4.32}$$

where x is the event at hand, x^r a uniformly distributed random number in the range $[0, 1)$, and the real number $a > 0$ is a parameter which relates to the efficiency of the algorithm. The new event x' is accepted with the Metropolis probability

$$P_{\text{accept}}(x') = \begin{cases} 1 & \text{for } x'^2 \le x^2; \\ \exp[-(x'^2 - x^2)/2] & \text{for } x'^2 > x^2. \end{cases} \tag{4.33}$$

If x' is rejected, the event x is counted again. The Metropolis process (4.33) introduces an autocorrelation time.

A call to

$$\text{GAU\_METRO(A, NDAT, DATA, ACPT)} \tag{4.34}$$

returns a time series of NDAT data in the array DATA, as well as the acceptance rate ACPT. The real number A > 0 has to be provided on input and is returned unchanged. So we have a simple Metropolis process at hand for estimating the following quantities:

(1) The autocorrelation function (4.3).
(2) The integrated autocorrelation time (4.12) by direct calculation (4.14) and by binning (4.27).

4.1.2.1 *Autocorrelation function*

For figure 4.1 we work with $K = 17$, *i.e.*, $N = 2^{17} = 131072$ data in equation (4.29). We choose $a = 3$ for the Markov process (4.32), what gives an acceptance rate of approximately 50%. Note that we can calculate the acceptance rate analytically, when the addition of $2ax^r - a$ in equation (4.32) is replaced by adding a Gaussian distributed random number. To work this out is left to the exercises of appendix B.

We compute the autocorrelation function of the Markov process with autcorj.f. Mean and error bars are subsequently calculated with

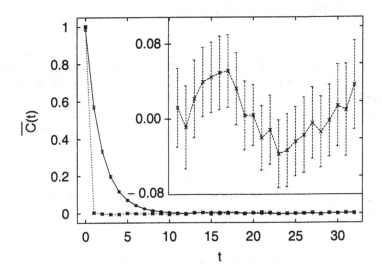

Fig. 4.1 The autocorrelation function (4.3) of a Metropolis time series for the normal distribution (upper data) in comparison with those of our Gaussian random number generator (lower data). For $t \geq 11$ the inlay shows the autocorrelations on an enlarged ordinate. The results are obtained in assignment 1. The straight lines between the data points are just to guide the eyes.

stebj0.f. The upper curve in the main part of figure 4.1 depicts the results up to $t = 32$. That the curve starts off with $\overline{C}(0) \approx 1$ comes from our choice of the random variable: $\overline{C}(0)$ is the variance of f and the variance of the normally distributed random variable is one, see (4.6) and (1.57). For $t \geq 11$ the ordinate scale of the $\overline{C}(t)$ estimates is enlarged by the inlay and it is obvious that the $\overline{C}(t)$ become consistent with zero in this range.

For comparison, the same calculations are done for Gaussian random numbers generated with rmagau.f of chapter 1, which are supposed to be uncorrelated. In this case the lower curve of figure 4.1 is obtained and, as expected, the autocorrelation function is from $t = 1$ on consistent with zero.

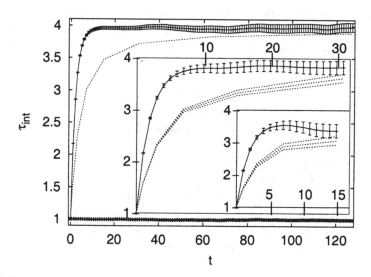

Fig. 4.2 The integrated autocorrelation time (4.12) of the Metropolis process (4.32) for normally distributed data. The upper curves in the figure and its two inlays display the estimators (4.14) as obtained by direct calculation (4.22). The lowest curve, a row of estimator compatible with one, is the integrated autocorrelation time for the Gaussian random numbers (1.33). The remaining curves are binning procedure (4.27) estimators of the integrated autocorrelation time and one standard deviation bounds of this quantity. The main figure relies on 2^{21} data and depicts estimators up to $t = 127$. The first inlay relies on 2^{17} data and depicts estimators up to $t = 31$. The second inlay relies on 2^{14} data and depicts estimators up to $t = 15$. See assignment 2 and the discussion in the text.

4.1.2.2 *Integrated autocorrelation time*

Figure 4.2 and its two inlays present a number of results for the integrated autocorrelation time τ_{int}. Feeding our jackknife estimators of the autocorrelation function into the subroutine ac_intj.f (4.22), the jackknife estimators for the integrated autocorrelation time τ_{int}, defined by equation (4.14), are obtained. These are the upper curves in figure 4.2 and in each of its two inlays. Mean values and error-bars are calculated with stebj0.f. Statistics and t-range are as follows.

(1) The full-sized figure relies on the overkill of NDAT $= 2^{21} = 2,097,152$ data and shows $\tau_{\text{int}}(t)$ up to $t = 127$. The purpose of creating this high statistics is to be sure about the true behavior of our estimators.

(2) The first inlay relies on the same statistics as our previous figure 4.1, NDAT $= 2^{17} = 131,072$ data, and follows $\tau_{\text{int}}(t)$ up to $t = 31$.

(3) The second inlay relies on NDAT $= 2^{14} = 16,384$ data and follows $\tau_{\text{int}}(t)$ up to $t = 15$. It will be used to discuss in more details the question: How many data are needed to allow for a meaningful estimate of the integrated autocorrelation time?

The lowest curve in figure 4.2 is a line of estimators for τ_{int} which is compatible with one. They are obtained by creating the 2^{21} data with our Gaussian random number generator **rmagau.f** and support, again, that these numbers are uncorrelated.

The remaining curves of figure 4.2 are the $\tau_{\text{int}}^{N_b}$ binning procedure estimates (4.27) of the integrated autocorrelation time, together with their upper and lower one standard deviation bounds. These estimators are calculated with the subroutine **bin_inta.f** (4.31) and the bounds are explained in the discussion following this equation. We compare the $\tau_{\text{int}}^{N_b}$ estimators with the direct estimators $\tau_{\text{int}}(t)$ at

$$t = N_b - 1 \, . \tag{4.35}$$

With this relation the estimators (4.14) and (4.27) agree for binsize $N_b = 1$ and for larger binsize the relation gives the range over which we combine data into either one of the estimators. Figure 4.2 shows that the approach of the binning procedure towards the asymptotic τ_{int} value is slower than that of the direct calculation of the integrated autocorrelation time.

For our large NDAT $= 2^{21}$ data set $\tau_{int}(t)$ reaches its plateau before $t = 20$. We like to convert this observation into an estimate of the integrated autocorrelation time $\tau_{\text{int}} = \tau_{\text{int}}(\infty)$. All the error bars within the plateau are strongly correlated as all $\tau_{\text{int}}(t)$ estimates rely on the same data set. Therefore, it is not recommendable to make an attempt to combine them. Instead, it is save to pick an appropriate single value and its error bar as the final estimate. With $t = 20$ we get in this way the accurate estimate

$$\tau_{\text{int}} = \tau_{\text{int}}(20) = 3.962 \pm 0.024 \ \text{ from } 2^{21} \ \text{data.} \tag{4.36}$$

The binning procedure, on the other hand, shows an increase of $\tau_{\text{int}}^{N_b}$ all the way to $N_b = 2^7 = 128$ (corresponding to $t = N_b - 1 = 127$), where the estimate is $\tau_{\mathrm{int}}^{128} = 3.89$ with the one standard deviation bounds

$$3.85 \le \tau_{\mathrm{int}}^{128} \le 3.94 \quad \text{from} \quad 2^{14} = 16,384 \text{ bins} \quad \text{of} \quad 2^{21} = 2,097,152 \text{ data.}$$
$$(4.37)$$

On the two standard deviation level this result is consistent with the estimate (4.36), but the accuracy has suffered, because we are forced to go out to a larger (effective) t-value. Even then, there appears still a slight increasing trend of $\tau_{\mathrm{int}}^{N_b}$, so that the estimate (4.36) is certainly superior. Nevertheless, the binning method remains of major importance as we discuss in the next subsection.

For the statistics of $\texttt{NDAT} = 2^{17}$ figure 4.1 shows that the autocorrelation signal disappears for $t \ge 11$ into the statistical noise. This is consistent with the inlay of figure 4.2. Still, there is clear evidence of the hoped for window of almost constant estimates. However, it is not so clear anymore which t-value one ought to pick. A conservative choice is to take $t = 20$ again, which now gives

$$\tau_{\mathrm{int}} = \tau_{\mathrm{int}}(20) = 3.86 \pm 0.11 \quad \text{from} \quad 2^{17} \text{ data.} \qquad (4.38)$$

The error bar has increased by about a factor of four compared with the error bar of equation (4.36). Worse is the binning estimate, which for the 2^{17} data is $\tau_{\mathrm{int}}^{32} = 3.63$ at $N_b = 32$ with the bounds

$$3.55 \le \tau_{\mathrm{int}}^{32} \le 3.71 \quad \text{from} \quad 2^{12} = 4,096 \text{ bins} \quad \text{of} \quad 2^{17} = 131,072 \text{ data.} \ (4.39)$$

I.e., our best value (4.36) is no longer covered by the two standard deviation zone of the binning estimate.

For the second inlay of figure 4.2 the statistics is reduced even more to $\texttt{NDAT} = 2^{14}$. With the integrated autocorrelation time (4.36) rounded to 4, this is 4096 times τ_{int}. For binsize $N_b = 2^4 = 16$ we are then down to $N_{bs} = 1024$ bins, which according to table 2.9 are needed for accurate error bars of the error, see also our discussion following equation (4.30). To work with this number we limit, in accordance with equation (4.35), our $\tau_{\mathrm{int}}(t)$ plot to the range $t \le 15$. Still, we find a quite nice window of nearly constant $\tau_{\mathrm{int}}(t)$, namely all the way from $t = 4$ to $t = 15$. Which value should constitute the final estimate and its error? By a statistical fluctuation, see **assignment 3**, $\tau_{\mathrm{int}}(t)$ takes its maximum value at $t = 7$ and this makes $\tau_{\mathrm{int}}(7) = 3.54 \pm 0.13$ a natural candidate. However, this value is inconsistent with our best estimate (4.36). The true $\tau_{\mathrm{int}}(t)$ increases monotonically as function of t, so we know that the estimators have become bad for $t > 7$. On the other hand, the error bar at $t = 7$ is obviously too small to take care of our difficulties. Hence, in a somewhat handwaving way, one may combine the $t = 15$ error bar (*i.e.*, the error bar of the $\tau_{\text{int}}(t)$ we would have liked to reach) with the $t = 7$ mean value. The result is

$$\tau_{\text{int}} = 3.54 \pm 0.21 \quad \text{for} \quad 2^{14} = 16,384 \text{ data}, \tag{4.40}$$

which achieves consistency with (4.36) in the two error bar range. For binsize $N_b = 16$ the binning estimate is $\tau_{\text{int}}^{16} = 3.06$ with the bounds

$$2.93 \leq \tau_{\text{int}}^{16} \leq 3.20 \quad \text{from} \quad 2^{10} = 1,024 \text{ bins} \quad \text{of} \quad 2^{14} \text{ data}. \tag{4.41}$$

Clearly, the binsize $N_b = 16$ is too small for an estimate of the integrated autocorrelation time. We learn from this investigation that one needs a binsize of at least ten times the integrated autocorrelation time τ_{int} to get into the right ballpark with the binning method, whereas for the direct estimate it is sufficient to have t about four times larger than τ_{int}. Later, in section 4.3, we discuss how to supplement our direct calculation of the integrated autocorrelation time with additional information by fitting the exponential autocorrelation time.

4.1.2.3   *Corrections to the confidence intervals of the binning procedure*

The confidence intervals returned by the subroutine bin_inta.f (4.31) take only the fluctuations of the numerator $s_{\bar{f}N_B}^2$ of equation (4.27) into account. Now we address shortly the corrections of the $\tau_{\text{int}}^{N_b}$ confidence intervals due to the fluctuations of the denominator $s_{\bar{f}}^2$. Besides that the fluctuations of the denominator are small, the reason for our initial definition of the error is that we need an approximate estimate of the integrated autocorrelation time (4.36) to correct for the fluctuation of $s_{\bar{f}}^2$.

The correction will be largest for our example (4.41), as it relies with 2^{14} on the smallest number of data we considered. In that case, the fluctuations of the numerator are determined by the 1024 bins, which are assumed to be uncorrelated. For the denominator we approximate the integrated autocorrelation time by $\tau_{\text{int}} = 4$ and find that according to equation (4.28) its fluctuations are determined by the effective number of 4,096 uncorrelated data. The lower and upper 95% confidence factors (2.110), for the ratio of these values are found from table 2.12: $F_{\text{min}} = 0.909$ and $F_{\text{max}} = 1.103$. This has to be compared with the corresponding factors for fluctuations of the numerator alone, given in table 2.9 to be $s_{\text{min}}^2 = 0.919$ and $s_{\text{max}}^2 =$

1.093 for 1024 independent data. The corrected version of equation (4.41) becomes[3],

$$2.92 \leq \tau_{\text{int}}^{16} \leq 3.22 \quad \text{from} \quad 2^{10} = 1,024 \text{ bins} \quad \text{for} \quad 2^{14} \text{ data}, \qquad (4.42)$$

see assignment 4. Avoiding rounding by considering more digits, the deviation (error bar) in the direction of smaller values than the estimated mean is found to increase by a good 10%, what is reasonably small for a correction of a confidence interval. In the direction of large values the agreement is even better. For our other cases of 2^{17} and 2^{21} data, the corrections to our initial estimates are considerably smaller than 10% (assignment 4). In essence, the approach of simply neglecting the fluctuations of the denominator is justified. If one likes to be on the save side, one may increase the confidence range by an ad-hoc 10%.

4.1.3   *Self-consistent versus reasonable error analysis*

In the previous chapter we discussed the usefulness of performing nequi (3.90) equilibration sweeps before starting with measurements during our MC simulation. This is not needed for the illustrative Gaussian Metropolis process of the previous subsection, because the initial value $x = 0$ is at the maximum of the probability density. The situation is quite different for simulations of the spin models of chapter 3. For most temperatures their ordered as well as their disordered start configurations are highly unlikely in the equilibrium distribution. By visual inspection of the time series, as we did for figures 3.4 and 3.5, one may get an impression about the length of the out-of-equilibrium part of the simulation. On top of this one should still choose

$$\text{nequi} \gg \tau_{\text{int}} , \qquad (4.43)$$

to allow the system to settle. That is a first reason, why it appears necessary to control the integrated autocorrelation time of a MC simulations. A second, even more important, reason is that we have to control the error bars of the equilibrium part of our simulation. Ideally the error bars are calculated from equation (4.11) as

$$\Delta \overline{f} = \sqrt{\sigma^2(\overline{f})} \quad \text{with} \quad \sigma^2(\overline{f}) = \tau_{\text{int}} \frac{\sigma^2(f)}{N} . \qquad (4.44)$$

---

[3]Note that bin_inta.f uses the 95.45% confidence interval of sebar_e.f instead of 95%. Obviously, this will not change our estimates in any relevant way.

Together with equation (4.43), where nequi $\geq 2\,\tau_{\text{int}}$ may substitute for $\gg$, this constitutes a **self-consistent error analysis** of a MC simulation.

However, the calculation of the integrated autocorrelation time can be tedious. Many more than the about twenty independent data are needed, which according to the Student distribution are sufficient to estimate mean values with reasonably reliable error bars. For computationally undemanding systems, like the Gaussian Metropolis process of our illustrations, one can just go ahead and calculate the integrated autocorrelation time accurately. But, in most large scale simulations this is utterly unrealistic. One may already be happy, if one has about twenty effectively independent data. In comparison, for the lowest statistics estimate (4.40) of the integrated autocorrelation time in our illustrations we used about 4,000 effectively independent data. Grey is all theory, what to do?

In practice, one has to be content with what can be done. Often this means to rely on the binning method. We simply calculate error bars of our ever increasing statistics with respect to a fixed number of NBINS bins, where the Student table 2.6 tells us that

$$\text{NBINS} \geq 16 \tag{4.45}$$

is reasonably large. In addition, we may put 10% of the initially planned simulation time apart for reaching equilibrium. *A-posteriori*, this can always be increased, but if measurements are CPU time extensive, one would not want to mix them up with equilibrium sweeps, unless one is focusing on non-equilibrium questions.

Once the statistics is large enough, this means at least $t_{\text{max}} > \text{NBINS}\,\tau_{\text{int}}$ for the largest t of our time series, our binned data become effectively independent and our error analysis is justified. How do we know that the statistics has become large enough? In essence, there is no other direct way, but to continue until an accurate enough estimate of the integrated autocorrelation time can be achieved. But, in practical applications there can be indirect arguments, which tell or suggest us that the integrated autocorrelation time is in fact (much) smaller than the achieved bin length. This is no longer self-consistent, as we perform no explicit measurement of τ_{int}, but it is a **reasonable error analysis** from a practitioners point of view.

Typical arguments which suggest the magnitude of τ_{int} can rely on experience with similar systems, or be finite size estimates from smaller systems of the same type. For instance, when dealing with second order phase transitions on a $V = L^d$ lattice, the integrated autocorrelation time measured in sweeps, will increase like L^{d+z}, where z is the dynamic critical exponent, which can already be estimated from the FSS behavior of rather small-sized systems. Away from critical points τ_{int} will not increase anymore with the volume size, once L has become larger than the correlation length ξ (3.12). Further, a rough, low statistics estimate of autocorrelations may indicate that the final estimate of τ_{int} will be smaller than the value needed for the error bar calculation to be valid. In the next section we study the integrated autocorrelation time for the spin systems which we introduced in chapter 3.

4.1.4 *Assignments for section 4.1*

(1) Use Marsaglia random numbers with the default seed and the subroutine gau_metro.f to reproduce the results of figure 4.1, relying on a time series of $N = 2^{17}$ events with $a = 3$. For this analysis you need the subroutines autcorf.f (4.19) and autcorj.f (4.21).

(2) Calculate the autocorrelation function in the same way as in the previous assignment, now for 2^{21} data. Use the autocorrelation function as input for the subroutine ac_intj.f to compute the integrated autocorrelation time estimator $\tau_{int}(t)$. Next, use the same data to calculate the estimator $\tau_{int}^{N_b}$ of the binning procedure. The results should reproduce the upper part of figure 4.2. Repeat this calculation for 2^{17} and 2^{14} data to obtain the integrated autocorrelation time estimators depicted in the two inlays of figure 4.2.

(3) Calculate (with successive random numbers) 128 times the integrated autocorrelation time estimators for 2^{14} data. Use steb0.f to average $\tau_{int}(15)$ with respect to the 128 values and compare with the $\tau_{int}(15)$ value obtained in the previous assignment from 10^{21} data. Explain the small difference encountered. Are the numbers identical in the statistical sense? Repeat the same, using only 2^{10} data for each estimate of $\tau_{int}(15)$.

(4) Calculate corrections to the binning estimate of the integrated autocorrelation time by taking into account fluctuations of the denominator of equation (4.27). Use $\tau_{int} = 4$ in equation (4.28) when calculating the effective number of data and compute the bounds (4.42). Find the corrected versions of the estimates (4.39) and (4.37). Hint: For the last part of this assignment the Fortran program which generates the results of table 2.12 has to be modified to deal with 4,096 and 16,384

data in the numerator. The corrections encountered will be of the order or smaller than the rounding errors in equations (4.39) and (4.37).

4.2   Analysis of Statistical Physics Data

In this section we estimate the integrated autocorrelation time (4.12) to illustrate error estimates of statistical mechanics simulations. One sweep is used as time unit. We start with the $d = 2$ Ising model. Next, we show how to compare the efficiency of competing Markov chain MC algorithms. Our first example for this is random versus sequential Metropolis updating and we find that sequential updating wins. In the second example we tune the acceptance rate of the Metropolis algorithm for the $2d$ XY model and find little dependence in the reasonable range. In the third example we find for the $2d$ 10-state Potts model at $\beta = 0.62$ that our heat bath algorithm outperforms the Metropolis algorithm. In subsequent examples similar comparisons between the Metropolis and the heat bath algorithm are performed for the $d = 3$ Ising model and for the $2d$ $O(3)$ σ model. Finally in this section, the questions of gains through averaging over small fluctuations is addressed.

The purpose of our examples is mainly to illustrate calculations of autocorrelation times for typical statistical physics applications. The results depend on temperatures and lattice sizes. When a particular algorithms is found more efficient than another, this may in some cases turn around for other parameter values or computer implementations.

4.2.1   *The $d = 2$ Ising model off and on the critical point*

We use the 1-hit Metropolis algorithm with sequential updating and show in figure 4.3 estimates $\tau_{\text{int}}(t)$ of the t-dependent integrated autocorrelation time (4.14) for simulations of the Ising model at $\beta = 0.4$ on $L \times L$ lattices. The energy, more precisely the action variable (3.77), is measured after every sweep and the times series output structure of section 3.4 is used. The relevant measurement program is

$$\texttt{p\_met\_ts.f} \quad \text{of} \quad \texttt{../ForProg/MC\_Potts} , \qquad (4.46)$$

where the name $\texttt{p\_met\_ts.f}$ is a shorthand for Potts-Metropolis-time-series.

The $L \times L$ lattice sizes used are $L = 5, 10, 20, 40, 80$ and 160. For each lattice we performed $\texttt{nequi} = 2^{19} = 524,288$ sweeps for reaching equilib-

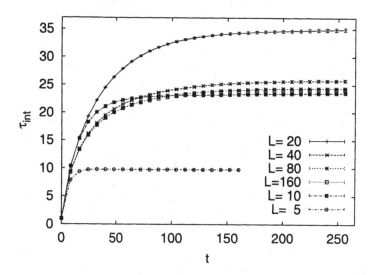

Fig. 4.3 One-hit Metropolis algorithm with sequential updating: Lattice size dependence of the integrated autocorrelation time for the $d = 2$ Ising model at $\beta = 0.4$, see assignment 1 (A). The ordering of the curves is identical with the ordering of the labels in the figure.

rium followed by $\mathtt{nrpt} = 20$ times $\mathtt{nmeas} = 2^{19}$ sweeps with energy measurements. The measurements are stored in the times series array $\mathtt{tsa(nmeas)}$ of the main program and $\mathtt{nrpt}$ times written to disk. Subsequently, the calculation of $\tau_{\mathrm{int}}(t)$ is carried out by the analysis program

$$\mathtt{ana\_ts\_p.f} \quad \text{of} \quad \mathtt{../ForProg/MC\_Potts} \,, \tag{4.47}$$

see assignment 1 (A). Instead of analyzing each of the $\mathtt{nrpt}$ times series fragments separately, one can also combine them into one long times series, from which the autocorrelations are then calculated. In fact, this is preferably, because one adds correlations connecting neighboring fragments to the statistics. This is done by the program

$$\mathtt{ana\_ts1\_p.f} \quad \text{of} \quad \mathtt{../ForProg/MC\_Potts} \,. \tag{4.48}$$

Nevertheless, in the following we use mainly the program $\mathtt{ana\_ts\_p.f}$ (4.47), because it needs about nrpt times less computer memory than the program ana_ts1_p.f, and it turns out that the final results are almost identical.

Table 4.1 Estimates of the integrated autocorrelation time for the $d = 2$ Ising model at $\beta = 0.4$ and $\beta = \beta_c = 0.44068679351$ on $L \times L$ lattices, see assignments 1 and 2. The t rows give the distances at which $\tau_{\text{int}}(t)$ was taken as the final τ_{int} estimate.

β	L	5	10	20
0.4	τ_{int}	9.772 (28)	23.58 (18)	34.95 (32)
0.4	t	32	256	256
β_c	τ_{int}	9.772 (4)	24.83 (19)	77.6 (1.3)
β_c	t	48	256	512
β	L	40	80	160
0.4	τ_{int}	25.84 (26)	24.42 (28)	24.22 (16)
0.4	t	256	256	256
β_c	τ_{int}	283.3 (10.3)	1060.0 (66.0)	3337.0 (331.0)
β_c	t	3072	8064	32128

We see from figure 4.3 that for all lattice sizes a plateau is reached, or at least almost reached. This allows for estimates of the $t \to \infty$ limiting behavior and, hence, τ_{int}. The final numbers are summarized in table 4.1. We find the remarkable behavior that τ_{int} first increases with lattice size, reaches a maximum and then decreases towards an infinite volume estimate, which is expected to be close to the $L = 160$ value.

In assignment 2 we repeat the above calculations at the critical temperature (3.6) of the $d = 2$ Ising model. By repeating the plot of figure 4.3 one finds that the $\tau_{\text{int}}(t)$ curves are now ordered according to the lattice size, the highest curve corresponding to the largest lattice. However, for lattices of size $L \geq 20$ a plateau is no longer reached in the range $t \leq 256$. For these lattices one has to follow $\tau_{\text{int}}(t)$ out to much larger values of t. This is done in assignment 2 (D) and the results are compiled in figure 4.4, where $\tau_{\text{int}}(t)$ is plotted up to $t = 2^{13} = 8192$. For the lattices up to $L = 80$ this allows the $t \to \infty$ estimates given in table 4.1. For $L = 160$ even the t range of figure 4.4 is not large enough. Following then $\tau_{\text{int}}(t)$ out to $t = 2^{15} = 32768$, a plateau is just reached and gives the value reported in the table.

The analysis program ana_ts1_p.f (4.48) has been used to generate the data for the curves of figure 4.4. Repeating the calculations with ana_ts_p.f (4.47) gives almost identical values, even at the large values

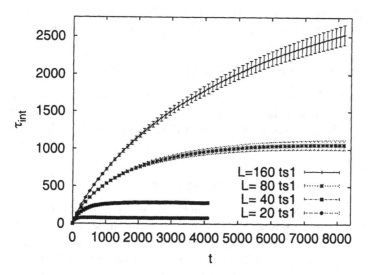

Fig. 4.4 One-hit Metropolis algorithm with sequential updating: Lattice size dependence of the integrated autocorrelation time for the $d = 2$ Ising model at its critical temperature (3.6). The label ts1 refers to the analysis program (4.48), see assignment 2 (D). The ordering of the curves is identical with the ordering of the labels in the figure.

of t, see assignment 2 (D). This is of importance when RAM restrictions prevent to run ana_ts1_p.f.

Let us investigate the finite size behavior of the integrated autocorrelation time estimates at β_c. From the literature [Blöte and Nightingale (2000)] the expectation is that τ_{int} increases like L^z with an exponent $z = 2.17$. This increase of the autocorrelation at the critical point is called **critical slowing down** and z is the **dynamic critical exponent**. More estimates of z are compiled in the book by [Landau and Binder (2000)], p.99. A 4-parameter fit with either subg_power2.f or subg_power2p.f of table 2.15 to our β_c data of table 4.1 gives as leading power

$$z = 1.893 \pm 0.059 \tag{4.49}$$

with a satisfactory goodness of fit (2.183), $Q = 0.25$, see assignment 3. Our z is somewhat lower than the best results of the literature, but our lattice sizes, as well as the precision from our largest lattices, are rather limited. Also, many modes couple to the energy-energy autocorrelations used here, so that large distances are needed to filter out the slowest mode. On the other hand, for the lattice sizes used we determine certainly the correct factors between the real and the naive energy error bars. We do not pursue the issue further, because the purpose of this book is to teach numerical methods and not to perform extensive numerical investigations.

4.2.2   Comparison of Markov chain MC algorithms

We evaluate the efficiency of a number of competing Markov chain algorithms. Note that for the second order phase transitions cluster algorithms are most efficient, which we do not evaluate in this section, because they are only introduced in chapter 5.3 (see there).

4.2.2.1   Random versus sequential updating

Is the 1-hit Metropolis algorithm more efficient with sequential updating or with random updating? To answer this question, we repeat the $L = 20$ and 40 calculations of assignments 1 (A) and 2 (A) using the updating procedure potts_met_r.f (3.85). The simulation and analysis programs for this purpose are

$$p\_metr\_ts.f \text{ and } ana\_ts\_p.f \text{ of } ../ForProg/MC\_Potts.f . \quad (4.50)$$

The results at $\beta = 0.4$ are given in figure 4.5, see assignment 1 (B). It is clear that **sequential updating is a better choice than random updating**. The same holds at the critical temperature, see assignment 2 (B).

It should be noticed that the sweep to sweep fluctuations of the energy, and therefore its naive error bars, are independent of the algorithm, as these are the physical fluctuation of the energy of the system. So, we do not have to worry about them when evaluating the efficiency of an algorithm. On the other hand, the CPU time needed depends on the algorithm as well as on the computing platform. In the present case it is obvious that random updating needs more CPU time than sequential updating, because of the calculation of an extra random number to select the lattice site.

For reference in chapter 5.2.2 we collect in table 4.2 integrated autocorrelation times from random versus sequential updating for the Ising model on an 14^3 lattice. The $\beta = 0$ result demonstrates autocorrelations between sweeps for random updating due to the fact that some spins are not hit

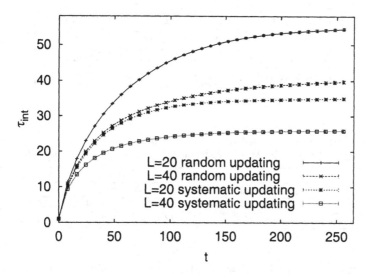

Fig. 4.5 Comparison of the integrated autocorrelation time of the Metropolis process with random updating versus sequential updating for the $d = 2$ Ising model at $\beta = 0.4$, see assignment 1 (B). The ordering of the curves is identical with the ordering of the labels in the figure.

during one sweep. In the limit of large lattices the fraction of spins missed is given by

$$\lim_{N \to \infty} (1 - 1/N)^N = e^{-1} = 0.367879\ldots. \tag{4.51}$$

4.2.2.2 *Tuning the Metropolis acceptance rate*

The Metropolis routine for the $d = 2$ planar rotator, xy_met.f (3.116), allows to tune the acceptance rate. We repeat in assignment 5 the $\beta = 1$ simulations which lead to the energy estimates of equations (3.124) and (3.127) with much higher statistics to estimate the integrated autocorrelation times τ_{int} for these simulations. The results are shown on the left-hand side in table 4.3. We find that the simulation at 32% acceptance rate is by a factor of about 1.2 more efficient than the simulation at 50% acceptance rate.

Table 4.2 Integrated autocorrelation times for random versus sequential updating: The Ising model on an 14^3 lattice. The heat bath algorithm is run except for $\beta = 0$ where the Metropolis algorithm is used, see assignment 4.

β	0	0.22165	0.3
τ_{int} random	1.313 (03)	120.5 (2.3)	4.247 (20)
τ_{int} sequential	1	61.70 (80)	2.123 (10)
β	0.4	0.5	0.6
τ_{int} random	2.702 (13)	2.384 (13)	2.250 (08)
τ_{int} sequential	1.304 (06)	1.114 (03)	1.050 (03)

Table 4.3 Estimates of integrated autocorrelation times τ_{int} at various acceptance rates acpt from simulations of the $2d$ XY model on $L \times L$ lattices, see assignment 5.

$\beta = 1$			$\beta = 2$			
acpt	$L = 60$	$L = 20$	acpt	$L = 20$	acpt	$L = 20$
0.50	314 (12)	50.1 (1.6)	0.50	7.276 (81)	0.30	7.725 (45)
0.32	262.4 (8.2)	38.97 (76)	0.19	14.69 (17)	0.24	11.376 (93)

The τ_{int} values for the $L = 60$ lattice are quite high due to the close-by critical temperature as estimated by [Tobochnik and Chester (1979)]. But they are low enough to justify the error analysis of equations (3.124) and (3.127), where error bars were calculated with respect to 32 bins of 2,000 sweeps each. This can serve as an example for which a reasonable error analysis is possible without actually calculating the integrated auto-correlation time. By repeating the $L = 60$ simulations on the far less CPU time demanding $L = 20$ lattices, we find $\tau_{\text{int}} \leq 50$ for $L = 20$. We are somewhat off the critical point, so the increase of τ_{int} with lattice size should be bounded by L^2, because a z exponent similar to (4.49) is expected at the critical point. Therefore, with nmeas $= 2,000 \gg 9 \times 50$ we are on the save side for performing the $L = 60$, nbins $= 32$ error analysis.

To see larger effects due to changing the acceptance rate, we perform in assignment 5 also simulation at $\beta = 2$, where the natural acceptance rate is lower than at $\beta = 1$, see the discussion after equation (3.125) in the previous chapter. Table 4.3 reports the results from simulations on an 20×20 lattice. We make several observations: First, the τ_{int} estimates are considerably lower than at $\beta = 1$, what is expected since we now stay away from the critical point. Second, the natural acceptance rate is down from

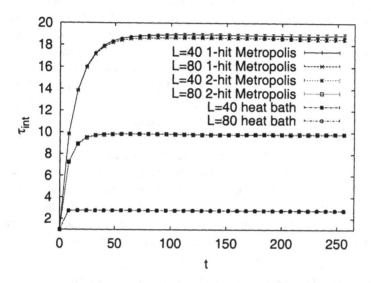

Fig. 4.6 Systematic updating: Comparison of the integrated autocorrelation times of the 1-hit and 2-hit Metropolis algorithms and the heat bath algorithm (3.95) for the 10-state Potts model on $L \times L$ lattices at $\beta = 0.62$, see assignment 6. The $L = 40$ and $L = 80$ curves lie almost on top of one another.

previously 32% to 19%. Third, enforcing a 50% acceptance rate leads to an improvement of the performance by a factor of about two. This happens in the range from 19% to about 30% acceptance, whereas the autocorrelation time is practically flat in the range from 30% to 50%.

4.2.2.3 *Metropolis versus heat bath: 2d $q = 10$ Potts*

In figure 3.5 of chapter 3 we considered the approach to equilibrium for the 10-state $d = 2$ Potts model on an 80×80 lattice, using n-hit Metropolis and heat bath algorithms (3.95) with sequential updating. Now we study the performance of the algorithms when generating equilibrium configurations, including also an 40×40 lattice, see assignment 6. The results are shown in figure 4.6. As expected, the highest curves correspond to the 1-hit Metropolis algorithm, followed by the 2-hit Metropolis algorithm and at the bottom the heat bath algorithm. In all cases a clear plateau is reached, for the heat bath algorithm almost immediately. The finite size effects due to going from the $L = 40$ to the $L = 80$ lattice are practically negligible, because of the non-critical behavior of the system. The statistics used for assignment 6 is certainly an overkill and can be reduced if desired.

Our estimates of τ_{int} are collected in table 4.4. The ratio of the 1-hit Metropolis τ_{int} over the 2-hit Metropolis τ_{int} is about 1.9. This is a large enhancement. Notably, the acceptance rate (on both lattice sizes) increases only from about 0.42 to 0.53. Nevertheless, use of the 2-hit algorithm brings no gain, as long as the CPU time increases by a factor of about two, which was measured on a Linux PC, see (3.100). For the 1-hit Metropolis algorithm versus the heat bath algorithm the ratio of integrated autocorrelation times is about 6.9, which is higher than the CPU time ratio of about 4.4 from (3.100). At $\beta = 0.62$ **it pays to use the heat bath algorithm.** For more disordered configurations at lower β values the efficiency of the Metropolis algorithm increases, so that it will then beat our heat bath method. Note that the design of a more CPU time efficient Potts model heat bath program appears possible by tabulating nearest neighbor configurations, as is done in chapter 5.2 for other purposes.

Table 4.4 Systematic updating: Estimates of the integrated autocorrelation time (4.12) for the 10-state Potts model on $L \times L$ lattices at $\beta = 0.62$, see assignment 6.

1-hit Metropolis		2-hit Metropolis		Heat bath	
$L = 40$	$L = 80$	$L = 40$	$L = 80$	$L = 40$	$L = 80$
18.96 (13)	18.65 (09)	9.854 (48)	9.837 (44)	2.767 (18)	2.781 (20)

4.2.2.4 *Metropolis versus heat bath: 3d Ising*

For the Ising model on an 14^3 lattice table 4.5 compares the efficiency

$$\frac{\tau_{int}(\text{Metropolis})}{\tau_{int}(\text{Heat bath})} \times \frac{\text{CPU time (Metropolis)}}{\text{CPU time (heat bath)}} \tag{4.52}$$

of the heat bath versus the Metropolis algorithm on a specific computing platform. The simulations were performed at $\beta = 0.22165$, close to the Curie temperature as already used for the simulations (3.109), and below this temperature. In particular at low temperatures the heat bath algorithm outperforms the Metropolis updating even when its increased demands on CPU time are taken into account. It is notable that the heat bath autocorrelation time at low temperature approaches one, when measurements are only taken after each sweep. This means that its improvement over the Metropolis algorithm will be even larger when measurements are taken within one sweep or, more practical, are averaged within each sweep, see subsection 4.2.3 below.

Table 4.5 Integrated autocorrelation times and efficiencies (4.52) of the heat bath versus the 1-hit Metropolis algorithm for 14^3 lattice Ising model simulations on an 400 MHz Pentium PC, see assignment 7.

β	0.22165	0.3	0.4	0.5
τ_{int} Metropolis	98.15 (76)	5.552 (17)	3.780 (10)	3.3062 (54)
τ_{int} Heat bath	61.70 (80)	2.123 (10)	1.304 (06)	1.1140 (28)
Efficiency (4.52)	1.18	1.93	2.21	2.24

4.2.2.5 Metropolis versus heat bath: 2d $O(3)$ σ model

Table 4.6 Systematic updating: Estimates of the integrated autocorrelation time (4.12) for the 2d $O(3)$ σ model on an 100×100 lattice at $\beta = 1.1$ and 1.5, see assignment 8. The Metropolis over heat bath ratios of the integrated autocorrelation times are also given.

β	$O(3)$ heat bath (3.130)	Metropolis (table 3.14)	Ratio
1.1	2.968 (57) at $t = 44$	12.16 (11) at $t = 78$	4.10 (09)
1.5	13.09 (44) at $t = 286$	75.6 (3.8) at $t = 1822$	5.78 (36)

By comparing error bars, we found in chapter 3 evidence that the $O(3)$ heat bath simulations for the results (3.132) are more efficient than the corresponding Metropolis simulations, even when CPU time losses were taken into account. Now we calculate in assignment 8 the integrated autocorrelation times for this situation. The results are given in table 4.6. The heat bath algorithm is found to be more than four times as efficient as the Metropolis algorithm, while the CPU time increase which we found in chapter 3 is less than a factor of two.

For the $O(3)$ σ model in $d = 2$ critical behavior is expected for $\beta \to \infty$, and one may wonder whether at higher β values the heat bath algorithm will become even more efficient when compared with the Metropolis approach. Therefore, we repeat the comparison at $\beta = 1.5$ and collect the results also in table 4.6. Indeed, we find a strong increase of the integrated autocorrelation times and a more modest increase of the relative efficiency of the heat bath over the Metropolis algorithm. As for other systems with second order phase transitions, cluster algorithms are most efficient for this situation, see chapter 5.3 and [Wolff (1989)].

4.2.3   Small fluctuations

Did we really do the best we can, when we calculated the integrated autocorrelations times? We measured the energy after each sweep, but in fact it fluctuates with each update. What will be the effect of averaging these small fluctuation out? Will this reduce the error bars of the final energy estimates considerably? Or, will it simply be irrelevant as long as the autocorrelations are (much) larger than one sweep?

First, we address the issue of small fluctuation for our Ising model examples of figures 4.3 and 4.4, see assignments 1 (C) and 2 (C). The relevant simulation and analysis programs are

$$\text{p\_met\_tsm.f} \text{ and } \text{ana\_ts\_pm.f} \text{ of } \text{../ForProg/MC\_Potts.f} . \quad (4.53)$$

In contrast to p_met_ts.f (4.46) the program p_met_tsm.f uses, after every sweep, the action variable histogram ha to calculate mean values over the updates within the sweep (remember, this histogram is collected by the Metropolis updating routine). To achieve this in a speedy way, the Fortran function

$$\text{potts\_actm2(iamin, imax, ha)} \text{ of } \text{ForLib} \quad (4.54)$$

is used, which is a variant of potts_actm.f (table 3.13), summing only histogram entries in the range iamin $\leq$ iact $\leq$ iamax. Instead of the single iact measurements, the averages are stored in the time series array tsa and written to disk. The analysis program ana_ts_pm.f is then used to calculate the autocorrelations.

At the critical temperature figure 4.7 compares for $L = 20$ and 40 the $\tau_{int}(t)$ results which we obtained in figure 4.4 with those we get from the update-averaged action variable. The difference is rather small and, at a first thought perhaps somewhat surprising, the $\tau_{int}(t)$ values for the averaged quantities are slightly higher than those from the single measurements after every sweep. At a second thought, this is what one expects. The single measurements are separated by an entire sweep, whereas for the averages

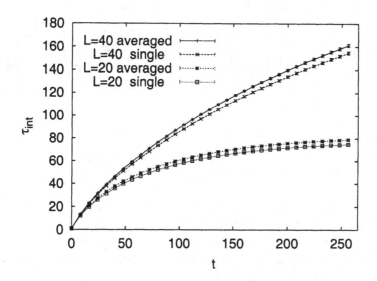

Fig. 4.7 Comparison of the integrated autocorrelation times of the $d = 2$ Ising model at its critical temperature (3.6) for the 1-hit Metropolis algorithm with sequential updating: Averaged (measured after every update) versus single (measured after every sweep) random variables, see assignment 2 (C). The ordering of the curves is identical with the ordering of the labels in the figure.

the separation is one sweep in the average. In the latter case autocorrelations of the close-by measurements (separated only by a few updates) bias the autocorrelation times towards a slightly higher value.

What about the error bars? **The error bars corresponding to averages of update-averaged quantities have to be smaller or equal to those of the averages of single measurements.** There is no contradiction with the autocorrelation times, because it is the product of the variance with the integrated autocorrelation time which determines the final error bar (4.44). The variance of the update-averaged data is small enough to ensure that its product with τ_{int} is smaller than for the non-averaged quantities.

In table 4.7 we give the integrated autocorrelation times together with

Table 4.7 Integrated autocorrelation times with (τ_{int}^{ua}) and without (τ_{int}) update-averaging and the improvement factors (4.55) due to update-averaging. Error bars for the improvement factors (Ip) are smaller than half of the last given digit. Error bars for the integrated autocorrelation times can be looked up in the assignments. Left: Ising model on $L \times L$ lattices, see assignments 1 (C) and 2 (C). Center: Ising model on an 14^3 lattice, see assignment 9. Right: $O(3)$ σ model on an 100×100 lattice, see assignment 10.

β	0.4	0.4	0.4	β_c	β_c	0.02	0.5	0.01	1.1	1.5
L	10	20	40	20	40	14	14	100	100	100
τ_{int}	23.6	35.0	25.8	74.9	155	1.02	1.12	1.01	2.97	13.1
τ_{int}^{ua}	25.3	37.4	28.0	78.9	161	1.66	1.79	1.51	1.002	16.0
Ip	1.12	1.15	1.18	1.09	1.08	2.47	1.83	1.01	1.002	1.001

the ratios

$$\text{Improvement (Ip)} = \frac{\sigma^2(\overline{\text{actm}})}{\sigma_{ua}^2(\overline{\text{actm}})} = \frac{\tau_{int}\,\sigma^2(\text{iact})}{\tau_{int}^{ua}\,\sigma^2\text{actm}} \,. \qquad (4.55)$$

Here the integrated autocorrelation time and the variance of the mean for the update-averaged action variable are labeled by ua. The variances on the left-hand side are calculated using equation (4.44). While $\tau_{int}^{ua} > \tau_{int}$ holds, update-averaging implies that the opposite is true for the variances of the mean: $\sigma^2(\overline{\text{actm}}) > \sigma_{ua}^2(\overline{\text{actm}})$.

For the $2d$ Ising model we see from table 4.7 that the gain due to update-averaging is in the range of 8% to 18%. As one would expect, it is larger at the more disordered $\beta = 0.4$ value than at the critical temperature. For the $2d$ Ising results $\tau_{int}(256)$ was used to estimate the integrated autocorrelation time. Although $t = 256$ is not a good $t \to \infty$ estimate for the lattices with really large autocorrelation, it serves the present purpose reasonably well, as corrections tend to drop out in the ratio (4.55).

Apparently, one can only expect substantial improvements due to update-averaging when the autocorrelations are small. In assignment 9 the improvement factor is calculated for the Ising model on an 14^3 lattice at $\beta = 0.02$ and 0.5 using the heat bath algorithm. We have already seen that at $\beta = 0.5$ in the ordered phase the heat-bath autocorrelations are small (table 4.5) and in the disordered phase, close the $\beta = 0$, it is obvious that the autocorrelations will be small. The improvement due to update-averaging is 2.47 at $\beta = 0.02$ and 1.83 at $\beta = 0.5$ (table 4.7).

For the $O(3)$ σ model in $d = 2$ improvement ratios for the heat bath algorithms are also included in table 4.7 for the same parameter values which where previously used when comparing with the Metropolis algorithm (table 4.6). In addition, we investigate update-averaging with the Metropolis algorithm at $\beta = 0.01$, where the system is disordered. In neither case update-averaging helps, see assignment 10.

Even when the improvement is small, should we not take it and only record update-averaged variables? There is one argument of caution: **Re-weighting of the time series to other temperatures is not (rigorously) possible from averaged quantities.** Re-weighting techniques [Ferrenberg and Swendsen (1988)] are quite useful for the analysis of canonical MC simulation data, as we discussed in chapter 3.1.2. Therefore, one better records single measurements in any case, at least in large scale simulations.

4.2.4 *Assignments for section 4.2*

(1) (A) Use the 1-hit Metropolis algorithm with sequential updating and investigate the lattice size dependence for the integrated autocorrelation time of the Ising model on $L \times L$ lattices at $\beta = 0.4$ for $L = 5, 10, 20, 40, 80$ and 160. For each lattice size use Marsaglia random numbers with their default seed and the program (4.46) to perform a simulation with nequi $= 2^{19}$ sweeps for reaching equilibrium and nrpt $= 20$ repetitions of nmeas $= 2^{19}$ sweeps with measurements. Subsequently, the analysis program (4.47) creates the data files, which allow to reproduce figure 4.3. Omit the $L = 80$ and 160 lattice, if you have CPU time constraints. (B) Compare the 1-hit Metropolis algorithms with random and sequential updating for $L = 20$ and 40. The result is depicted in figure 4.5. (C) Use the programs (4.53) to average the small fluctuations over each sweep and compare for $L = 20$ and 40 with the results obtained before.

(2) Repeat the previous exercise at the Curie temperature (3.6). Part (C) reproduces figure 4.7. (D) Extend the analysis of the integrated autocorrelation function $\tau_{int}(t)$ of part (A) to values of t up to $2^{15} = 32,768$ to reproduce figure 4.4 as well as the estimates of table 4.1. If your RAM allows it, use the analysis program ana_ts1_p.f (4.48) and compare with the results from ana_ts_p.f (4.47).

(3) Reproduce the estimate (4.49) of the dynamical critical exponent of the $d = 2$ Ising model from the β_c data of table 4.1. Use the main program gfit.f together with either subg_power2.f or subg_power2p.f from the program package of the folder ForProg/fit_g. (See table 2.15 for the functions fitted by these routines.)

(4) Test random versus sequential updating for the Ising model on an 14^3 lattice to confirm the results of table 4.2. Each data point of this table relies on 2^{17} sweeps for reaching equilibrium followed by 20×2^{17} sweeps with measurements (a smaller statistics would be sufficient, except close to the critical β).

(5) *XY* model acceptance rates: Use the program xy_ts.f together with the Metropolis routine xy_met.f to repeat the $\beta = 1$ simulation of assignment 4 of section 3.4.4 with nequi $= 2^{19}$ sweeps for reaching equilibrium and nrpt $= 20$ repetitions of nmeas $= 2^{19}$ sweeps with measurements. Analyze with ana_ts1_on.f for the integrated autocorrelation time τ_{int}. Compare with the result obtained for the natural acceptance rate, which you get by running xy_met.f with lacpt=.false.. Perform the same simulation on an 20×20 lattice (or just on this lattice, if your CPU time limitations do not allow for larger calculations). Finally, perform similar simulations on an 20×20 lattice at $\beta = 2$ to reproduce the results of table 4.3. (Note, that you have to adjust the acpt_rate parameter in xy_met.f for two of the $\beta = 2$ calculations.)

(6) Simulate the 10-state Potts model at $\beta = 0.62$ on $L \times L$ lattices of size $L = 40$ and 80. For sequential updating compare the performance of the 1-hit and 2-hit Metropolis and the heat bath algorithms (3.95). Use the statistics as well as the general approach of assignment 1. (Omit the $L = 80$ lattice, if you have CPU time constraints). The results are given in figure 4.6 and table 4.4.

(7) Use the 1-hit Metropolis and the heat bath algorithm to simulate the Ising model on an 14^3 lattice to produce table 4.5 for you favorite computer.

(8) Simulate the $O(3)$ σ model on 100×100 lattices with the heat bath (3.130) as well as with the Metropolis code (table 3.14) to calculate the integrated autocorrelation times at $\beta = 1.1$ and 1.5. Use Marsaglia random numbers with the default seed for each simulation and generate the following statistics: nmeas=nequi=2**15 for the heat bath and nmeas=nequi=2**17 for the Metropolis code, in each case with nrpt=20 repetitions of the measurement cycle.

(9) Use the heat bath algorithm and compute the improvement ratio (4.55) due to update-averaging for the Ising model on an 14^3 lattice at $\beta = 0.02$ and $\beta = 0.5$. A statistics of 2^{15} sweeps for reaching equilibrium and 20×2^{15} sweeps with measurements is sufficient as the autocorrelations are small.

(10) Repeat the $O(3)$ σ model simulation of assignment 8 and calculate the

improvements due to update-averaging. Afterward use the Metropolis code to perform the same calculation on an 100×100 lattice at $\beta = 0.02$ (2^{12} sweeps for reaching equilibrium and 20×2^{15} sweeps with measurements is sufficient at this high temperature).

4.3   Fitting of Markov Chain Monte Carlo Data

In chapter 2 we discussed the fitting of statistically independent data. While the data from MC simulations on different lattices are statistically independent, this is not the case for data from a single Markov chain. We have to deal with autocorrelations between the data. Fitting of autocorrelation times serves us as an example for this situation. The same method applies to the fitting of other observables from a Markov chain.

If you have already some background in statistics, you may think that calculating and diagonalizing the covariance matrix (1.87) will be the right way to proceed. Surprise! In practice this approach fails for most data from MC simulations. This was recognized by parts of the simulation community in the mid 1980s and [Michael (1994)] wrote down some reasons. In essence, an $n \times n$ covariance matrix can only be diagonalized if the number of statistically independent data exceeds $n \times n$ considerably, *i.e.*

$$\mathtt{ndat} \gg \tau_{\mathrm{int}}\, n \times n \,. \tag{4.56}$$

For MC simulations this is rarely the case. Fortunately, there is some remedy. It is quite sufficient to make the fitting curve look smooth, when plotted together with the data. This is simply achieved by fitting the data as if they were statistically independent, ignoring their off-diagonal covariances. The resulting fits look good, but we do not know the errors of the estimated parameters, because all the error bars of the Markov chain MC data are (strongly) correlated. What to do?

It is the jackknife method, which comes to our rescue. The error bars of the Markov chain data are only needed to get smooth fits. To get the error bars of the estimated parameters, we can bin the original data into n_j jackknife bins, repeat the entire fitting n_j times and, finally, calculate jackknife error bars for our fit parameters.

In the following examples this procedure is illustrated by fitting the autocorrelations of some of our MC time series.

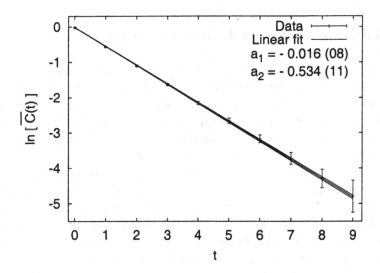

Fig. 4.8 Logarithm of the autocorrelation times of the Metropolis time series of figure 4.1 together with a linear jackknife fit to the exponential autocorrelation time, see assignment 1. Upper, lower and central fit lines are defined as in figure 2.6.

4.3.1   *One exponential autocorrelation time*

In figure 4.1 we plotted the estimated autocorrelation function $\overline{C}(t)$ for a Metropolis time series of Gaussian random numbers. In figure 4.8 we show the logarithms of these autocorrelations up to $t = 9$. The largest value of t is chosen to be $t = 9$, because the one error bar range overlaps from $t = 10$ on with zero. The visual impression from figure 4.8 is that the data are consistent with a straight line, $i.e.$ the fall-off behavior of the autocorrelation function is already at short distances dominated by the exponential autocorrelation time $\tau_{\exp}$ defined in equation (4.4). To calculate $\tau_{\exp}$ we would like to perform the fit

$$\ln[\overline{C}(t)] \; = \; a_1 + a_2\, t \tag{4.57}$$

and encounter the difficulty introduced in this section: All the (jackknife) error bars shown in figure 4.8 are correlated, as they rely on the same time series. Nevertheless, we use them in the same way as we use statistically independent error bars, because in accordance with the arguments of [Michael (1994)] it would be unstable to diagonalize the entire covariant matrix with the intention to use the error bars of the diagonal random variables (1.98) for the fit.

Due to the correlations of the error bars the χ^2 value of our fit (and correspondingly the goodness of fit value) looses its quantitative statistical interpretation, while it remains a useful entity in the sense that the fit with the smallest χ^2 values is regarded to be the best. To calculate error bars of our fit parameters, we perform the fit for jackknife sets of our data, using in each case the same (jackknife) error bars of our mean values. This is based on the following observations:

(1) An overall constant factor of the error bar does not affect the fit results.
(2) *A-priori* all jackknife data sets have identical statistical weights and the standard deviations of the jackknife data agree, up to an overall constant factor, with the standard deviations of the mean values.

Once the NBINS fits for our jackknife data sets are performed, we calculate the jackknife error bars of the fits parameters in the usual way with the routine stebj0.f (2.164). The result of the linear fit as well as the fit parameters and their error bars are given in figure 4.8, see assignment 1 to learn the technical details.

We have thus succeeded to calculate the fit parameters and their error bars, but nothing has yet been said about the consistency of the fit with the data. Visually the fit of figure 4.8 looks fine, but is there a way to quantify this? Although the use of χ^2 or of the goodness of fit Q makes no sense anymore, the error bar of each single data point has still its usual statistical meaning, which is assumed to be Gaussian. Suppose that the fit is replaced by the (normally unknown) exact curve. Then the Gaussian difference test (2.33) between the fit (entered with zero error bar) and one data point has an uniformly distributed Q. If the form of the fit includes the exact curve as a special case, the data will bias the parameters away from the exact values towards an even better agreement between the fit curve and the data, so that Q becomes biased towards a larger average than 0.5. This bias is small when the number of data points is much larger than the number of fit parameters. In any case, we see that the function $Q(t)$ of these Gaussian difference tests gives useful information in the sense that the occurrence of small values signals an inconsistent fit. In table 4.8

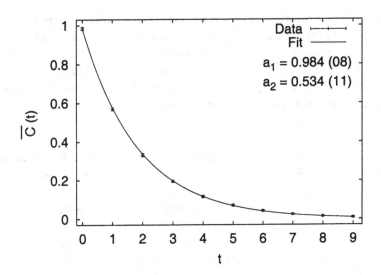

Fig. 4.9 Autocorrelation times of the Metropolis time series of figure 4.1 together with an exponential jackknife fit to the autocorrelation time, see assignment 3.

the $Q(t)$ values for the fit of figure 4.8 are given in the first Q-row. As expected from the visual inspection of the fit, these $Q(t)$ values are quite satisfactory.

Table 4.8 $Q(t)$-values of Gaussian difference tests between fit values and data points for the fits of assignments 1 to 3. The assignment numbers are given in the first column.

t	0	1	2	3	4	5	6	7	8	9
1. Q	0.92	0.39	0.46	0.71	0.91	0.76	0.69	0.86	0.98	0.96
2. Q	0.68	0.51	0.65	0.94	0.87	0.59	0.54	0.70	0.79	0.79
3. Q	0.89	0.33	0.45	0.72	0.93	0.75	0.69	0.86	0.99	0.96

Alternatively to the linearized fit (4.57), we may fit the data directly to the form

$$\overline{C}(t) = a_1 \exp(a_2 t) . \tag{4.58}$$

This can be done by using our linearized jackknife fit program `lfitj.f` of `ForProg/fit_1` together with its subroutine `subl_exp.f`, or by using our jackknife version of Levenberg-Marquardt fitting, `gfitj.f` of `ForProg/fit_g` together with its subroutine `subg_exp.f`. These fits are performed in assignments 2 and 3. The graph of the Levenberg-Marquardt fit is given in figure 4.9 and the $Q(t)$ values of both fits are included in table 4.8. All three fits are satisfactory and give quite similar results.

Estimates of the integrated autocorrelations times τ_{int} follow by using the first of the equations (4.18). In the order of assignments 1, 2 and 3 the results are

$$1.\ 3.833\,(75)\,, \qquad 2.\ 3.829\,(79) \qquad \text{and} \qquad 3.\ 3.832\,(79)\,. \qquad (4.59)$$

Although we have used only autocorrelations up to $t = 9$, the results are in reasonably good agreement with our full estimate (4.36), which relies on the $\tau_{int}(t)$ value at $t = 20$ and sixteen times more data than used for our fits. This confirms that the autocorrelations of this Metropolis chain are indeed dominated by a single exponential autocorrelation time. As we show next, this is no longer the case for many of the autocorrelations times of the physical systems, which we investigated in the previous section.

4.3.2 More than one exponential autocorrelation time

To give one simple example for this situation, we pick the $L = 10$ lattice from our $d = 2$ Ising model simulations at the critical point, see assignment 2 of the previous section. Instead of estimating the integrated autocorrelation time directly from the time series, we now generate data files with the jackknife autocorrelations, which allow for jackknife fitting (assignment 4). The data are cut off at $t = 30$, because the autocorrelations become consistent with zero for larger t values.

Our fits are depicted in figure 4.10. The two-parameter fits (4.57) and (4.58) are now of unsatisfactory quality (the slight difference of the fit curves is due to the different handling of the error bars). This is already obvious by visual inspection of figure 4.10 and confirmed by a plot of the corresponding Gaussian difference tests in figure 4.11: The $Q(t)$ values are practically zero with exceptions where the fits cross the data region. From the unacceptable quality of the two-parameter fits we conclude that there have to be relevant contributions from one or more subleading autocorrelations times $\tau_2 > \tau_1 = \tau_{exp}$. Consequently, we try the four-parameter

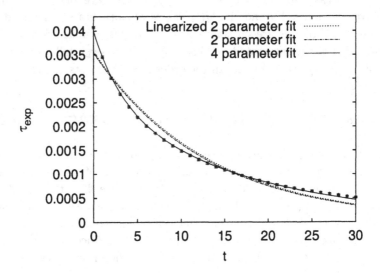

Fig. 4.10 Autocorrelation times of the Metropolis time series of assignment 2 of the previous section together with 2– and 4–parameter exponential jackknife fits, see assignment 4.

fit

$$\overline{C}(t) = a_1 \exp(a_2 t) + a_3 \exp(a_4 t), \quad (a_2 = -1/\tau_1 \text{ and } a_4 = -1/\tau_2). \quad (4.60)$$

This fit is implemented by the subroutine `subg_exp2.f` of `ForProg/fit_g` and used in assignment 4 by the program `gfitj2.f`. From figures 4.10 and 4.11 we observe that the quality of this fit is satisfactory, although drops of the Q-values for the smallest ($t = 0$) and largest ($t = 30$) t-values are a bit worrying. The fit (4.60) can then be used to calculate the integrated autocorrelation time by feeding the fit into its definition (4.12). With jackknife errors the result is

$$\overline{\tau}_{\text{int}} = 24.61 \pm 0.55, \quad (4.61)$$

which is in good agreement with the 24.83 (19) estimate of table 4.1.

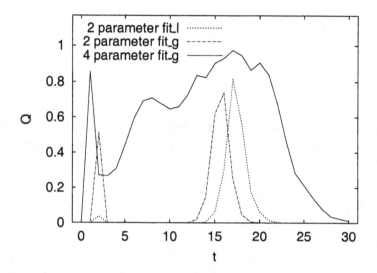

Fig. 4.11 Gaussian difference tests for the plots of figure 4.10, see assignment 4.

4.3.3 Assignments for section 4.3

(1) Perform a linear jackknife fit to the logarithms of the jackknife data of figure 4.1 to obtain figure 4.8 and the corresponding row of table 4.8.

(2) Use the linear jackknife fit routine together with subl_exp.f to perform the fit of assignment 1. Obtain the corresponding row of table 4.8.

(3) Use the general jackknife fit routine together with subg_exp.f to perform the fit of assignment 1. Obtain figure 4.9 and the corresponding row of table 4.8.

(4) Repeat the data production of assignment 2 of the previous section for $L = 10$ (or use the data generated there). Subsequently, use the program ana_ts_ac.f to generate data files with the jackknife autocorrelation times. Perform the following jackknife fits as shown in figure 4.10: (A) The linearized two-parameter exponential fit as in assignment 2. (B) The two-parameter exponential fit as in assignment 3. (C) The four-parameter exponential fit (4.60). For each of the fits reproduce the Gaussian difference tests depicted in figure 4.11.

Chapter 5

Advanced Monte Carlo

In this chapter we discuss a number of topics, which go beyond a basic understanding of Markov chain MC algorithms. Many of them have increased in importance over the recent years.

5.1 Multicanonical Simulations

One of the questions which ought to be addressed before performing a large scale computer simulation is "What are suitable weight factors for the problem at hand?" So far we used the Boltzmann weights (3.33) as this appears natural for simulating the Gibbs ensemble. However, a broader view of the issue is appropriate. With the work of [Berg and Neuhaus (1992)] it became an accepted idea that simulations with a-priori unknown weight factors are feasible and deserve to be considered. Similar ideas were already put forward in the 1970th under the name *umbrella sampling* [Torrie and Valleau (1977)]. Nowadays such simulation methods are often summarized under the header **generalized ensembles** and employed for an increasing number of applications in Physics, Chemistry, Structural Biology and other areas, for a reviews see [Mitsutake *et al.* (2001); Berg (2002); Hansmann and Okamoto (1999)]. Conventional canonical simulation calculate expectation values at a fixed temperature T and can, by re-weighting techniques, only be extrapolated to a vicinity of this temperature [Ferrenberg and Swendsen (1988)], compare figure 3.3. For multicanonical simulations this is different. A single simulation allows to obtain equilibrium properties of the Gibbs ensemble over a range of temperatures which would require many canonical simulations. Canonical simulations do not provide the appropriate implementation of importance sampling when the physically important configurations are rare in the canonical ensemble or when a rugged free energy landscape makes it difficult to reach them. In both situations multicanonical simulations can improve the performance remarkably. Transition Matrix Methods, which to some extent are reviewed in [Wang and Swendsen (2002)], can be understood as a further extension of the generalized ensemble approach.

MC calculation of the interface tension of a first order phase transition, discussed now, provide an example where canonical MC simulation miss the important configurations. Canonical simulations sample configurations k with the Boltzmann weights (3.33) and the resulting probability density, $P_\beta(E)$, is given by equation (3.32). Let L characterize the lattice size (for instance consider $N = L^d$ spins). For first order phase transition pseudo-transition temperatures $\beta^c(L)$ exist, so that the energy distributions $P(E) = P(E; L)$ become double peaked and the maxima at $E^1_{\max} < E^2_{\max}$ are of equal height $P_{\max} = P(E^1_{\max}) = P(E^2_{\max})$. Figure 3.6 gives an example for which canonical simulations still work, because the first order transition is relatively weak and the lattice size small enough. In-between the maxima a minimum is located at some energy $E_{\min}$. Configurations at $E_{\min}$ are exponentially suppressed like

$$P_{\min} = P(E_{\min}) = c_f \, L^p \, \exp(-f^s A) \tag{5.1}$$

where f^s is the interface tension, A is the minimal area between the phases ($A = 2L^{d-1}$ for an L^d lattice), c_f and p are constants (computations of p have been done in the capillary-wave approximation, see [Caselle et al. (1994)] and references given therein). The interface tension can be determined by Binder's histogram method [Binder (1982)]. One has to calculate the quantities

$$f^s(L) = -\frac{1}{A(L)} \ln R(L) \quad \text{with} \quad R(L) = \frac{P_{\min}(L)}{P_{\max}(L)} \tag{5.2}$$

to make a finite size scaling (FSS) extrapolation of $f^s(L)$ for $L \to \infty$. However, for large systems a canonical MC simulation will practically never visit configurations of energy $E = E_{\min}$ and estimates of the ratio $R(L)$ will be very inaccurate. The terminology **supercritical slowing down** was coined to characterize such an exponential deterioration of simulation results with lattice size. [Berg and Neuhaus (1992)] approached this problem by sampling, in an appropriate energy range, with an **approximation**

$$\widehat{w}_{mu}(k) = w_{mu}(E^{(k)}) = e^{-b(E^{(k)}) \, E^{(k)} + a(E^{(k)})} \tag{5.3}$$

to the weights

$$\widehat{w}_{1/n}(k) = w_{1/n}(E^{(k)}) = \frac{1}{n(E^{(k)})} \qquad (5.4)$$

where $n(E)$ is the spectral density (3.31). Here the function $b(E)$ defines the inverse **microcanonical temperature** at energy E and $a(E)$ the **dimensionless, microcanonical free energy** (see the next subsection). The function $b(E)$ has a relatively smooth dependence on its arguments, which makes it a useful quantity when dealing with the weight factors. Instead of the canonical energy distribution $P(E)$, one samples a new multicanonical distribution

$$P_{mu}(E) = c_{mu}\, n(E)\, w_{mu}(E) \approx c_{mu} . \qquad (5.5)$$

The desired canonical probability density (3.32) is obtained by re-weighting

$$P(E) = \frac{c_\beta}{c_{mu}}\, \frac{P_{mu}(E)}{w_{mu}(E)}\, e^{-\beta E}. \qquad (5.6)$$

This relation is rigorous, because the weights $w_{mu}(E)$ used in the actual simulation are exactly known. With the approximate relation (5.5) the average number of configurations sampled does not longer depend strongly on the energy and accurate estimates of the ratio $R(L) = P_{\min}/P_{\max}$ become possible. Namely, for $i = 1, 2$ the equation

$$R(L) = R_{mu}(L)\, \frac{w_{mu}(E^i_{\max})\, \exp(-\beta^c E_{\min})}{w_{mu}(E_{\min})\, \exp(-\beta^c E_{\max})} \qquad (5.7)$$

$$\text{with} \quad R_{mu}(L) = \frac{P_{mu}(E_{\min})}{P_{mu}(E^i_{\max})}$$

holds and the statistical errors are those of the ratio R_{mu} times the exactly known factor.

The multicanonical method requires two steps:

(1) Obtain a **working estimate** $\widehat{w}_{mu}(k)$ of the weights $\widehat{w}_{1/n}(k)$. Working estimate means that the approximation to (5.4) has to be good enough to ensure movement of the Markov process over the desired energy range, while deviations of $P_{mu}(E)$ from the constant behavior (5.5) are tolerable.

(2) Perform a MC simulation with the **fixed** weights $\widehat{w}_{mu}(k)$. The thus generated configurations constitute the **multicanonical ensemble**.

Canonical expectation values are found by re-weighting to the Gibbs ensemble and standard jackknife methods allow reliable error estimates.

A condensed version of the following presentation was published in [Berg (2003)] (see also references given therein).

5.1.1 *Recursion for the weights*

In this subsection a recursion is presented which allows to obtain a working estimates $w_{mu}(k)$ of the weight factors (5.4).

We first discuss the relationship of the weights (5.3) with the microcanonical quantities $b(E)$ and $a(E)$, because it turns out to be advantageous to formulate the recursion in terms of them. We have

$$w(k) = e^{-S(E^{(k)})} = e^{-b(E^{(k)}) E^{(k)} + a(E^{(k)})} \tag{5.8}$$

where $S(E)$ is the **microcanonical entropy**. By definition, the **microcanonical temperature** is given by

$$b(E) = \frac{1}{T(E)} = \frac{\partial S(E)}{\partial E} \tag{5.9}$$

and the **dimensionless, microcanonical free energy** is defined by the relation

$$a(E) = \frac{F(E)}{T(E)} = \frac{E}{T(E)} - S(E) = b(E)\,E - S(E) \ . \tag{5.10}$$

We show below that it is determined by the relation (5.9) up to an additive constant, which is irrelevant for the MC simulation. We consider the case of a discrete minimal energy ϵ and choose

$$b(E) = [S(E + \epsilon) - S(E)]/\epsilon \tag{5.11}$$

as the definition of $b(E)$. The identity $S(E) = b(E)\,E - a(E)$ implies

$$S(E) - S(E - \epsilon) = b(E)E - b(E - \epsilon)(E - \epsilon) - a(E) + a(E - \epsilon) \ .$$

Inserting $\epsilon\,b(E - \epsilon) = S(E) - S(E - \epsilon)$ yields

$$a(E - \epsilon) = a(E) + [b(E - \epsilon) - b(E)]\,E \tag{5.12}$$

and for the MC simulation $a(E)$ is fixed by defining $a(E_{\max}) = 0$. Note, however, that this value is not arbitrary in the data analysis part of the investigation, if one wants to calculate the free energy and the entropy, see section 5.1.7. For the simulation, once $b(E)$ is given, $a(E)$ follows for free.

A convenient starting condition for the initial ($n = 0$) simulation is

$$b^0(E) = b_0 \geq 0 \quad \text{and} \quad a^0(E) = 0. \tag{5.13}$$

The nth simulation is carried out using $b^n(E)$. Let the energy histograms of the nth simulation be given by $H^n(E)$. To avoid $H^n(E) = 0$ we map for the moment

$$H^n(E) \rightarrow \hat{H}^n(E) = \max[h_0, H^n(E)], \tag{5.14}$$

where h_0 is a number $0 < h_0 < 1$. Our final equations will allow for the limit $h_0 \rightarrow 0$. In the following subscripts $_0$ are used to indicate quantities which are are not yet our final estimators from the nth simulation. We define

$$w_0^{n+1}(E) = e^{-S_0^{n+1}(E)} = c \, \frac{w^n(E)}{\hat{H}^n(E)},$$

where the (otherwise irrelevant) constant c is introduced to ensure that $S_0^{n+1}(E)$ can be an estimator of the microcanonical entropy

$$S_0^{n+1}(E) = -\ln c + S^n(E) + \ln \hat{H}^n(E). \tag{5.15}$$

Inserting this relation into (5.11) gives

$$b_0^{n+1}(E) = b^n(E) + [\ln \hat{H}^n(E + \epsilon) - \ln \hat{H}^n(E)]/\epsilon. \tag{5.16}$$

The estimator of the variance of $b_0^{n+1}(E)$ is obtained from

$$\sigma^2[b_0^{n+1}(E)] = \sigma^2[b^n(E)] + \sigma^2[\ln \hat{H}^n(E + \epsilon)]/\epsilon + \sigma^2[\ln \hat{H}^n(E)]/\epsilon.$$

Now $\sigma^2[b^n(E)] = 0$, as $b^n(E)$ is the fixed function used in the nth simulation. The fluctuations are governed by the sampled histogram $H^n = H^n(E)$,

$$\sigma^2[\ln(\hat{H}^n)] = \sigma^2[\ln(H^n)] = [\ln(H^n + \Delta H^n) - \ln(H^n)]^2,$$

where ΔH^n is the fluctuation of the histogram, which is known to grow with the square root of the number of entries $\Delta H^n \sim \sqrt{H^n}$. Hence, in an approximation where the autocorrelation times of neighboring histograms are assumed to be identical, the equation

$$\sigma^2[b_0^{n+1}(E)] = \frac{c'}{H^n(E + \epsilon)} + \frac{c'}{H^n(E)}, \tag{5.17}$$

holds where c' is an unknown constant.

Equation (5.17) emphasizes that the variance is infinite when there is zero statistics for either histogram, *i.e.* $H^n(E) = 0$ or $H^n(E + \epsilon) = 0$. The statistical weight for $b_0^{n+1}(E)$ is inversely proportional to its variance and the overall constant c' is irrelevant. We define

$$g_0^n(E) = \frac{H^n(E + \epsilon)\, H^n(E)}{H^n(E + \epsilon) + H^n(E)}. \tag{5.18}$$

Note that $g_0^n(E) = 0$ for $H^n(E+\epsilon) = 0$ or $H^n(E) = 0$. It is now straightforward to combine $b_0^{n+1}(E)$ and $b^n(E)$ according to their respective statistical weights into the desired estimator

$$b^{n+1}(E) = \hat{g}^n(E)\, b^n(E) + \hat{g}_0^n(E)\, b_0^{n+1}(E), \tag{5.19}$$

where the normalized weights

$$\hat{g}_0^n(E) = \frac{g_0^n(E)}{g^n(E) + g_0^n(E)} \text{ and } \hat{g}^n(E) = 1 - \hat{g}_0^n(E) \tag{5.20}$$

are determined by the recursion

$$g^{n+1}(E) = g^n(E) + g_0^n(E), \ g^0(E) = 0. \tag{5.21}$$

We can eliminate $b_0^{n+1}(E)$ from equation (5.19) by inserting its definition (5.16) and get

$$b^{n+1}(E) = b^n(E) + \hat{g}_0^n(E) \times [\ln \hat{H}^n(E + \epsilon) - \ln \hat{H}^n(E)]/\epsilon. \tag{5.22}$$

Notice that it is now save to perform the limit $h_0 \to 0$, because $\hat{g}_0^n(E) = 0$ for $h_0 = 0$. Finally, equation (5.22) can be converted into a direct recursion for the ratios of weight factor neighbors. We define

$$R^n(E) = e^{\epsilon b^n(E)} = \frac{w^n(E)}{w^n(E + \epsilon)} \tag{5.23}$$

and get

$$R^{n+1}(E) = R^n(E) \left[\frac{\hat{H}^n(E + \epsilon)}{\hat{H}^n(E)} \right]^{\hat{g}_0^n(E)}. \tag{5.24}$$

It is recommended to start the recursion in the disordered phase of a system, where it moves freely under MC updates. Typical is the choice (5.13), where β_0 is safely located in the disordered phase of the system, *e.g.* $\beta_0 = 0$.

An alternative recursion scheme is given by the *random walk algorithm* of [Wang and Landau (2001)].

Table 5.1 Multicanonical recursion routine for Potts models.

```
      subroutine p_mu_rec ! Recursion of multicanonical weights.
C Copyright: Bernd Berg, Nov 25, 2000.
      include '../../ForLib/implicit.sta'
      include '../../ForLib/constants.par'
      include 'lat.par'
      include 'mc.par'
      include 'potts.par'
      include 'muca.par'
      include '../../ForLib/lat.com'
      include '../../ForLib/potts.com'
      include '../../ForLib/potts_muca.com'
C
C Weight changes outside the [namin,namax] range forbidden:
      if(ia_min.lt.namin) then
        do ia=ia_min,(namin-iastep),iastep
          hmuca(ia)=hmuca(ia)+ha(ia)
          ha(ia)=zero
        end do
        ia_min=namin ! Re-definition of ia_min for this case.
      end if
C
      if(ia_max.gt.namax) then
        do ia=(namax+iastep),ia_max,iastep
          hmuca(ia)=hmuca(ia)+ha(ia)
          ha(ia)=zero
        end do
        ia_max=namax ! Re-definition of ia_max for this case.
      end if
C
      hmuca(ia_min)=hmuca(ia_min)+ha(ia_min)
      hup(ia_min)=hup(ia_min)+ha(ia_min)
      hdn(ia_min)=hdn(ia_min)+ha(ia_min)
      ha(ia_min)=zero
      if(ia_max.eq.ia_min) stop "p_mu_rec: ia_max.eq.ia_min."
C
      ia1=ia_min
      ia=ia_min
1     ia=ia+iastep ! tame label 1.
        hmuca(ia)=hmuca(ia)+ha(ia)
        hup(ia)=hup(ia)+ha(ia)
        hdn(ia)=hdn(ia)+ha(ia)
        ha(ia)=zero
        if(hmuca(ia).gt.half) then
          ndel_new=ia-ia1
          ndel_old=ndel_muca(ia1)
          if(ndel_old.gt.ndel_new) then
            gw(ia1)=zero
            ndel_muca(ia1)=ndel_new
            ndel_muca(ia)=ndel_old-ndel_new
          end if
          ia1=ia
        end if
      if(ia.lt.ia_max) goto 1 ! Only goto to label 1.
```

```
C
C     Calculation of the smallest action considered, ia1:
      ia1=ia_min
      iatest=ia_min
2     iatest=iatest-iastep ! Tame label 2.
      if(iatest.ge.namin) then
        if(iatest+ndel_muca(iatest).eq.ia_min) ia1=iatest
      else
        if(iatest.gt.ia_min-n2d) goto 2 ! Only goto to label 2.
      end if
C
C Weight factor recursion:
      ia2step=2*iastep
      do ia=ia1,ia_max,iastep
        idel_muca=ndel_muca(ia)
        ia_up=ia+idel_muca
        hdn_a=hdn(ia)
        hup_a=hup(ia_up)
        if(hup_a.gt.half .and. hdn_a.gt.half) then ! if 1
C         The new weight factor:
C         g-weights:
          gw0=hdn_a*hup_a/(hdn_a+hup_a)
          gw00=gw0/(gw(ia)+gw0) ! normalized.
          gw(ia)=gw(ia)+gw0       ! g-weight iteration.
          wfactor=wrat(idel_muca,ia)*(hdn_a/hup_a)**gw00
          hdn(ia)=zero
          hup(ia_up)=zero
          if(idel_muca.eq.iastep) then ! if 2
            wrat(iastep,ia)=wfactor
            wrat(-iastep,ia+iastep)=one/wfactor
            iab=ia
            do id=ia2step,n2d,iastep
              iab=iab-iastep
              wrat(id,iab)=wrat(id-iastep,iab)*wfactor
              wrat(-id,iab+id)=one/wrat(id,iab)
            end do
          else ! if 2
            if(idel_muca.lt.iastep) stop "p_mu_rec: idel_muca error."
            wfactor=wfactor**((iastep*one)/(idel_muca*one))
            nstep=(idel_muca/iastep)
            iaa=ia-iastep
            do istep=1,nstep
              iaa=iaa+iastep
              wrat(iastep,iaa)=wfactor
              wrat(-iastep,iaa+iastep)=one/wfactor
              iab=iaa
              do id=ia2step,n2d,iastep
                iab=iab-iastep
                wrat(id,iab)=wrat(id-iastep,iab)*wfactor
                wrat(-id,iab+id)=one/wrat(id,iab)
              end do
            end do
          end if ! if 2
        end if ! if 1
      end do
```

```
C
C Weight factors for the largest action values encountered:
      ia2=ia_max-iastep
      ia1=ia_max-n2d+iastep
      iastep1=iastep
      do ia=ia2,ia1,-iastep
        iastep1=iastep1+iastep
        do id=iastep1,n2d,iastep
          wrat(id,ia)=wrat(id-iastep,ia)*wrat(iastep,ia)
          wrat(-id,ia+id)=one/wrat(id,ia)
        end do
      end do
C
      ia_max=0
      ia_min=mlink
C
      return
      end
```

5.1.2 Fortran implementation

The subroutine

$$\texttt{p\_mu\_rec.f} \text{ of ForProg/MC\_Potts} \qquad (5.25)$$

implements the recursion (5.24) and is reproduced in table 5.1. In addition to the usual MC files, the routine includes muca.par and potts_muca.com, which define the parameters of the recursion and the common block where the relevant information is stored. Table 5.2 reproduces muca.par and table 5.3 muca.com. Details of the Fortran implementation are explained in the following.

As in chapter 3, the computer programs use the action variable iact (3.77) instead of the energy, because iact takes on integer values ≥ 0. Let us discuss the arrays and variables of potts_muca.com.

(1) The array hmuca(0:mlink+n2d) accumulates the histogram of the iact variable over the entire recursion time. Its main purpose is to keep track whether a certain iact value has been visited or not. The extension of the array size from mlink to mlink+n2d, where n2d = 2d, is of technical nature: It allows to write formulas with hmuca(ia+iastep) for ia up to mlink, thus avoiding tedious if statements.

(2) The arrays hup(0:mlink+n2d) and hdn(0:mlink) serve to implement the definition (5.20) of the normalized weights $\hat{g}_0^n(E)$. When two neighbor histograms are non-zero, they give a contribution to $g_0^n(E)$ in equa-

Table 5.2 Parameter file `muca.par` for the multicanonical recursion.

```
c
c muca.par:  Parameters for the MUCA recursion
c            ==================================
c
c lmuca=.true.:   MUCA recursion with  ba(.)=beta initially.
c lmuca=.false.:  NO MUCA recursion and beta used for canonical MC.
c namin<=iact<=namax is the action range of the recursion.
c bet_min=beta [from ms.par] <= b(ia) <= bet_max temperature range.
c
      parameter(lmuca=.true., namax=mlink,namin=mlink/nq)
      parameter(namaxp1=namax+1)
c
c maximum number of recursions:          nrec_max.
c update sweeps per recursion:           nmucasw.
c maximum number of tunnelings allowed:  maxtun.
c
      parameter(nrec_max=10*nq*ms,nmucasw=10*nq,maxtun=10)
```

Table 5.3 The common block `potts_muca.com` for the multicanonical recursion.

```
      common/muca/ hmuca(0:mlink+n2d),hup(0:mlink+n2d),hdn(0:mlink),
     & gw(0:mlink),ndel_muca(0:mlink),iastep
```

tion (5.18) and, after being used, they have to be reset to zero to prepare for the accumulation of the forthcoming statistics. Most iact values have upper and lower neighbor values and, after being used in connection with one of its neighbors, the histogram entry may still be needed to cope with its other neighbor. The solution adopted here is to double the number of histograms hdn (dn for down) is used for iact together with hup for iact+ndel_muca(iact), where ndel_muca(iact) is explained in 4 below.

(3) The array gw(0:mlink) is used to iterate the weights from $g^n(E)$ to $g^{n+1}(E)$ according to equation (5.21).

(4) The array ndel_muca(0:mlink) keeps track of the distance found from one iact value to the next higher. This means when hmuca entries

exist for iact, then

$$\text{iact} + \text{ndel\_muca(iact)}$$

is the next higher iact value which was covered in the simulation. The values of ndel_muca(0:mlink) are initially set to point out of the recursion range, see equation (5.27) below.

Before p_mu_rec.f can be used, the subroutine

$$\text{p\_mu\_init.f of ForProg/MC\_Potts} \qquad (5.26)$$

has to be called, which performs a number of initializations: the arrays ha, hmuca, hup, hdn and gw are set to zero and the values of the array ndel_muca are set to

$$\text{ndel\_muca(ia)} = \text{namax} + 1 - \text{ia} . \qquad (5.27)$$

Finally, iastep is initialized to iastep=2 for nq=2, *i.e.* the Ising model, and to iastep=1 otherwise.

We are now ready to explain the different parts of the subroutine p_mu_rec.f. The first section accumulates the histogram hmuca for those action values which were found outside the range [namin,namax] to which we apply the recursion. Afterwards, the array ha is set to zero for these values[1], as those entries are now accounted for. Next, the humuca, hup and hdn histograms are accumulated for ia_min, where ia_min is the maximum of namin and the minimum iact value reached in the preceding sweeps. The entries at ia_min are needed as starting values for the subsequent calculation of the ndel_muca array entries, which starts at ia1=ia_min. This is done between the label 1 and the goto 1 statement, where the accumulation of the humuca, hup and hdn histograms is also completed.

After label 1, using the label 2 and the go to 2 statement, the program assigns ia1 to the smallest iact value considered in the subsequent recursion part. In essence ia1 is determined by ia1+ndel_muca(ia1)=ia_min, but a special case is ia_min=namin, because array values ndel_muca(ia) are not calculated for ia1<namin. We then simply take ia1=namin-n2d, which is safe because n2d is the largest stepsize possible and it does not matter for the later parts of the program when the actual stepsize is smaller.

The recursion itself begins after the end if statement following label 2. With the if 1 statement it is checked that the hup_a=hup(ia_up) and

---

[1] Remember, the Metropolis updating routine (3.84) accumulates the histogram (3.88) ha.

the hdn_a=hdn(ia) histogram value are greater than 0.5, corresponding to both $H^n(E + \epsilon)$ and $H^n(E)$ in equation (5.18) being ≥ 1. The histograms hup(ia_up) and hdn(ia) are then used to calculate the new weight factor ratio wfactor according to equation (5.24) and afterwards they are set to zero. The update of the weight factor array wrat(-n2d:n2d,0:mlink+n2d) follows, which is kept in potts.com (see table 3.6 of chapter 3). Care is taken that the entire weight matrix is derived from a single weight function based on the minimum distance between iact values found during the simulation. The if 2 statement distinguishes two cases, ndel_muca(ia) agrees with iastep or it does not. The stepsize used in the wrat array is always iastep, but in the second case nstep=ndel_muca(ia)/iastep entries have to be filled, where nstep is an integer ≥ 2. It may happen that the computer lowers in a later stage of the recursion the value of ndel_muca(ia), say from 2*iastep to iastep. Then, the subsequent accumulation of gw(ia) proceeds with the lower value, i.e., in our example gw(ia+iastep) enters the game. In practice the process turns out to be convergent, although a better theoretical understanding of some steps appears desirable.

Some of the formulas used do not apply to the ia_max end of the recursion. The final part of the program deals with this problem and overwrites eventually obtained incorrect values with the correct ones.

5.1.3 *Example runs*

First, we illustrate the recursion for the 20^2 Ising model. We use the multicanonical data production program

$$\text{p\_mu.f} \quad \text{of} \quad \text{ForProg/MC\_Potts} \tag{5.28}$$

to run the recursion in the range

$$\text{namin} = 400 \leq \text{iact} \leq 800 = \text{namax} . \tag{5.29}$$

The recursion part of p_mu.f is reproduced in table 5.4 and the parameter values are preset in muca.par of table 5.2. The values of namin and namax are chosen to cover the entire range of temperatures, from the completely disordered ($\beta = 0$) region to the groundstate ($\beta \to \infty$). In many applications the actual range of physical interest is smaller. Then, one should adjust namin and namax correspondingly, because the recursion time increases quickly with the size of the energy range.

For nlink=mlink lattices namin=mlink/nq is (up to eventual decimal increments) the expectation value of iact at infinite temperature, $\beta = 0$.

Table 5.4 Recursion part of the main program p_mu.f of ForProg/MC_Potts.

```
      mu_sweep=0
      ntun_old=0
      do irec=1,nrec_max
        acpt=zero
1       continue
          mu_sweep=mu_sweep+1
          call potts_met
          call tun_cnt(namin,namax,iamin,iamax,ntun)
          if(ntun.gt.ntun_old) then
            write(iuo,'(" irec,ntun:",2I10)') irec,ntun
            ntun_old=ntun
            if(ntun.ge.maxtun) go to 2
          end if
        if((acpt/(ns*one)).lt.(nmucasw*one)) go to 1 !
C The MUCA recursion is only done, when enough moves have been accepted:
        call p_mu_rec
      end do
      write(iuo,*) "Muca recursion  n o t   finished."
      STOP           "Muca recursion  n o t   finished."
2     continue
```

The value namax=mlink corresponds to the iact value of one of the nq groundstate. The recursion is terminated when maxtun so called tunneling events are completed. A **tunneling event** is defined as an updating process which finds its way from

$$iact = namin \quad \text{to} \quad iact = namax \quad \text{and back} . \qquad (5.30)$$

This notation comes from applications to first order phase transitions [Berg and Neuhaus (1992)], for which namin and namax are separated by a free energy barrier in the canonical ensemble. Although the multicanonical method aims at removing this barrier, the terminus tunneling was kept for the Markov process bridging this energy range. The requirement that the process tunnels also back is included in the definition, because a one way tunneling is not indicative for convergence of the recursion. An alternative notation for tunneling event is **random walk cycle**. The counting of the tunneling events (random walk cycles) is done by calls to the subroutine tun_cnt.f of ForLib which is reproduced in table 5.5. The routine makes use of the iamin and iamax values of potts.com as calculated by the MC updating routine (see tables 3.6 and 3.7 of chapter 3). An initial value for ltun0 has to be supplied, ltun0=.false. when the initial action value is in the disordered phase iact.le.namin.

Table 5.5 Subroutine to count tunneling events (5.30).

```
      subroutine tun_cnt(namin,namax,iamin,iamax,ntun,ltun0)
C Copyright, Bernd Berg, May 27, 2001.
C Tunneling count. None are missed, if this routine is called after
C every sweep and (namax-namin) is sufficiently large.
      logical ltun0
      if(ltun0) then
        if(iamin.le.namin) then
          ntun=ntun+1
          ltun0=.false.
        end if
      else
        if(iamax.ge.namax) then
          ltun0=.true.
        end if
      end if
      return
      end
```

For most applications the value maxtun=10, as preset in muca.par of table 5.2, leads to acceptable weights. The parameter nrec_max of muca.par serves to discontinue the recursion, if maxtun is not reached after a certain maximum number of recursion steps. If this happens for a reasonably large nrec_max value, you may be in trouble. But in most cases the problem will disappear when you rerun (eventually several times) with different random numbers. Otherwise, enlarge the parameter nmucasw of muca.par. The parameter nmucasw times the inverse acceptance rate equals the number of updating sweeps done between recursion steps (*i.e.* between calls to the subroutine p_mu_rec.f). The recursion formula (5.24) is only valid when the system is in equilibrium. Each update of the weights disturbs it and the recursion may become unstable when numucasw is chosen too small[2]. The disturbance of the equilibrium becomes weak when the weight function approaches its fixed point. We take the number of sweeps between recursion steps to be numcasw/acpt (up to rounding errors), because the acceptance

---

[2]One may consider to discard some sweeps after each recursion step to reach equilibrium, but experimental evidence shows that the achieved improvement (if any) does not warrant the additional CPU time.

rate acpt may vary in the [namin,namax] range and an equal number of accepted moves is a better relaxation criterium than an equal number of sweeps. The numucasw value of table 5.2 is chosen to be large enough to work for typical applications. For our illustrations numucasw=1 would be sufficient.

We use Marsaglia random numbers with their default seed for our Ising model simulation and find the requested 10 tunneling events after 787 recursions and 64,138 sweeps, corresponding to an average acceptance rate of 20*787/64138=0.245. A more detailed analysis of the recursion process shows that almost half of the sweeps are spent to achieve the first tunneling event, see assignment 1. Afterwards a working estimate of the desired weights exists apparently and the additional statistics serves to stabilize and smoothen this estimate. Assignment 2 repeats this illustration for the $2d$ 10-state Potts model on an 20^2 lattice.

5.1.4   *Performance*

We are interested in the scaling of the performance with volume size. Let us consider the second part of the simulation, when the weights are frozen and assume that the distribution is reasonably flat in the energy. In principle one could then measure autocorrelation times, but in practice it is more convenient to count the average number of updates needed for one random walk cycle (5.30) and to plot this number versus the lattice size. We denote this number by τ_{tun} and it is intuitively quite obvious that τ_{tun} is proportional to the integrated autocorrelation time (note that the random walk cycles become independent events when at least one (sequential) sweep per cycle is spent at $\beta = 0$).

If the multicanonical weighting would remove all relevant free energy barriers, the behavior of the updating process would become that of a free *random walk*. Therefore, the optimal performance of the Markov chain for the second part of the multicanonical simulation is in updates per spin

$$\tau_{\text{tun}} \sim V^2 \, , \tag{5.31}$$

where the volume V is proportional to the lattice size, *e.g.* $V = L^d$.

For the original multicanonical simulation of the 10-state Potts model [Berg and Neuhaus (1992)] $\tau_{\text{tun}} = V^{2+\epsilon}$ with $\epsilon = 0.325\,(10)$ was reported, where the correction ϵ was attributed to violations of the perfect random walk behavior, which remained at that time unaccounted for. Nowadays, one understands first order phase transitions better [Neuhaus and Hager

(2003)] and it is found that the multicanonical procedure removes only the leading free energy barrier, while at least one subleading barrier causes still a residual supercritical slowing done. Up to certain medium sized lattices the behavior $V^{2+\epsilon}$ gives a rather good effective description. For large lattices supercritical slowing down of the form

$$\tau_{\text{tun}} \sim \exp\left(+\text{const}\, L^{d-1}\right) \tag{5.32}$$

dominates again, where the constant encountered is (much) smaller than two times the interface tension, which due to equation (5.1) dominates the slowing down of a canonical simulation of the transition.

For *complex systems* with a rugged free energy landscape [Mitsutake *et al.* (2001); Berg (2002); Hansmann and Okamoto (1999)] one finds in general a multicanonical slowing down far worse than for first order phase transitions. Many barriers appear to be left, which are not flattened by the multicanonical procedure [Berg *et al.* (2000)]. In contrast to first order transitions, an explicit parameterization of the leading barriers is not available.

The slowing down of the weight recursion with volume size is expected to be even (slightly) worse than that of the second part of the simulation. The reason lies in the additional difficulty of approaching the correct weights. The situations is better when finite size scaling methods can be employed for calculating working estimates of the weights in one step, from the already simulated lattices sizes to the next larger one.

5.1.5 *Re-weighting to the canonical ensemble*

Let us assume that we want to perform a multicanonical simulation which covers the action histograms needed for a temperature range

$$\beta_{\min} \leq \beta = \frac{1}{T} \leq \beta_{\max} . \tag{5.33}$$

In practice this means that the parameters namax and namin in muca.par of table 5.2 have to be chosen so that

$$\text{namin} \ll \overline{\text{act}}(\beta_{\min}) \quad \text{and} \quad \overline{\text{act}}(\beta_{\max}) \ll \text{namax}$$

holds, where $\overline{\text{act}}(\beta)$ is the canonical expectation value of the action variable (3.77) at temperature $\beta = 1/T$. The $\ll$ conditions may be relaxed to equal signs, if $b(\text{iact}) = \beta_{\min}$ is kept for all values iact $\leq \overline{\text{act}}(\beta_{\min})$ and $b(\text{iact}) = \beta_{\max}$ for all values iact $\geq \overline{\text{act}}(\beta_{\max})$.

Given the multicanonical time series, where $i = 1, \ldots, n$ labels the generated configurations, the definition (3.2) of the canonical expectation values leads to the **estimator**

$$\overline{\mathcal{O}} = \frac{\sum_{i=1}^{n} \mathcal{O}^{(i)} \exp\left[-\beta E^{(i)} + b(E^{(i)}) E^{(i)} - a(E^{(i)})\right]}{\sum_{i=1}^{n} \exp\left[-\beta E^{(i)} + b(E^{(i)}) E^{(i)} - a(E^{(i)})\right]} . \tag{5.34}$$

This formula replaces the multicanonical weighting of the simulation by the Boltzmann factor. The denominator differs from Z by a normalization constant which drops out, because the numerator differs by the same constant from the numerator of equation (3.2). For discrete systems it is sufficient to keep histograms instead of the entire time series when only functions of the energy (in practice the action variable of our computer programs) are calculated. For an operator $\mathcal{O}^{(i)} = f(E^{(i)})$ equation (5.34) simplifies then to

$$\overline{f} = \frac{\sum_{E} f(E) \, h_{mu}(E) \, \exp\left[-\beta E + b(E) E - a(E)\right]}{\sum_{E} h_{mu}(E) \, \exp\left[-\beta E + b(E) E - a(E)\right]} , \tag{5.35}$$

where $h_{mu}(E)$ is the histogram sampled during the multicanonical production run and the sums are over all energy values for which $h_{mu}(E)$ has entries.

The computer implementation of equations (5.34) and (5.35) requires care. The differences between the largest and the smallest numbers encountered in the exponents can be really large. To give one example, for the Ising model on an 100×100 lattice and $\beta = 0.5$ the groundstate configuration contributes $-\beta E = 10^4$, whereas for a disordered configuration $E = 0$ is possible. Overflow disasters will result if we ask Fortran to calculate numbers like $\exp(10^4)$. When the large terms in the numerator and denominator take on similar orders of magnitude, one can avoid them by subtracting a suitable number in all exponents of the numerator as well as of the denominator, resulting in dividing a common factor out. Instead of overflows one encounters harmless underflows of the type $\exp(-10^4)$, which the program will set to zero. We implement the basic idea in a more general approach, which remains valid when the magnitudes of the numerator and the denominator disagree. We avoid altogether to calculate large numbers and deal only with the logarithms of sums and partial sums.

We first consider sums of positive numbers and discuss the generalization to arbitrary signs afterwards. For $C = A + B$ with $A > 0$ and $B > 0$ we calculate $\ln C = \ln(A + B)$ from the values $\ln A$ and $\ln B$, without ever storing either A or B or C. The basic observation is that

$$\ln C = \ln \left[\max(A, B) \left(1 + \frac{\min(A, B)}{\max(A, B)} \right) \right] \tag{5.36}$$
$$= \max(\ln A, \ln B) + \ln\{1 + \exp[\min(\ln A, \ln B) - \max(\ln A, \ln B)]\}$$
$$= \max(\ln A, \ln B) + \ln\{1 + \exp[-|\ln A - \ln B|]\}$$

holds, because the logarithm is a strictly monotone function. By construction the argument of the exponential function is negative, so that a harmless underflow occurs when the difference between $\min(\ln A, \ln B)$ and $\max(\ln A, \ln B)$ becomes too big, whereas it becomes calculable when this difference is small enough. The Fortran function

$$\text{addln.f} \quad \text{of} \quad \text{ForLib} \tag{5.37}$$

implements equation (5.36) and is reproduced in table 5.6.

Table 5.6 The Fortran function routine addln.f of ForLib.

```
      FUNCTION ADDLN(ALN,BLN)
C Copyright Bernd Berg, Nov 25 2000.
C Given ln(A) and ln(B), the function returns ln(C) with C=A+B.
      include 'implicit.sta'
      include 'constants.par'
      AALN=MAX(ALN,BLN)
      BBLN=MIN(ALN,BLN)
      ADDLN=AALN+LOG(ONE+EXP(BBLN-AALN))
      RETURN
      END
```

Underflow warnings at execution time can, however, be annoying. Worse, when such warnings are expected to be caused by the subroutine addln.f, we may no longer be able to discover underflows in other parts of the program, where they may be unintended. Therefore, the subroutine

$$\text{addln\_cut.f} \quad \text{of} \quad \text{ForLib} \tag{5.38}$$

introduces a cut-off on the argument of the exponential function in addln.f, so that the result zero is used whenever the argument is smaller than this cut-off. As the cut-off distracts from the main idea, we have listed the routine addln.f instead of addln_cut.f in table 5.6. In simple applications one may perfectly well live with the underflow warning errors, or use a compiler option which turns them off. For more complicated applications addln_cut is recommended. For precisions other than IEEE real*8 the cut-off value in the routine should be adjusted.

To handle alternating signs, one needs in addition to (5.36) an equation for $\ln|C| = \ln|A-B|$ where $A > 0$, $B > 0$ and $\ln A \neq \ln B$. Equation (5.36) converts into

$$\ln|C| = \max\left(\ln A, \ln B\right) + \ln\{1 - \exp\left[-|\ln A - \ln B|\right]\} \qquad (5.39)$$

and the sign of $C = A - B$ is positive for $\ln A > \ln B$ and negative for $\ln A < \ln B$. A cut-off ϵ is needed, so that $|\ln A - \ln B| < \epsilon$ terminates the calculation with the result $C = 0$.

The Fortran function addln2.f of ForLib combines the implementation of equations (5.36) and (5.39). A call to

$$\text{ADDLN2(ALN, BLN, CLN, ISGN)} \quad \text{with} \qquad (5.40)$$

$$\text{ALN} = \ln A\,, \ \ \text{BLN} = \ln B \ \ \text{and} \ \ \text{ISGN} = \pm 1 \ \ \text{on input}$$

returns $\ln C = \ln(A+B)$ for ISGN $= +1$ on input and $\ln|C| = \ln|A-B|$ for ISGN $= -1$ on input. In the latter case ISGN may change too and is returned as ISGN $= +1$ for $C = A - B > 0$ and as ISGN $= -1$ for $C = A - B < 0$. As this routine needs the cut-off $|\ln A - \ln B| < \epsilon$, a cut-off on the argument of the exponential function is introduced simultaneously, so that we have only one version of the addln2.f routine. In case of a change of precision, the cut-offs need to be adjusted.

Calculations of sums with the addln subroutines are, of course, slow. For data analysis purposes this is rather irrelevant, as we are not dealing with frequently repeated calculations like in MC simulations. The program addln_test.f of ForProg/Test illustrates the use of the addln routines for sums of random numbers.

5.1.6   *Energy and specific heat calculation*

We are now ready to produce the multicanonical data for the energy per spin of the Ising model on an 20×20 lattice, which in figure 3.1 of chapter 3 are compared with the exact results of [Ferdinand and Fisher (1969)]. Relying on the program p_mu.f of ForProg/Potts_MC with the parameters depicted in table 5.2, the multicanonical recursion is continued until ten tunnelings are performed. Afterwards 10 000 sweeps are carried out for reaching equilibrium followed by $32 \times 10\,000$ sweeps with measurements.

To create these data is part of assignment 1. Subsequently, the statistical analysis is done with the program

$$\texttt{ana\_pmu.f of ForProb/Potts\_MC} \qquad (5.41)$$

which relies heavily on calls to the subroutine

$$\texttt{potts\_zln.f of ForLib.} \qquad (5.42)$$

In part, this subroutine is reproduced in table 5.7. Its purpose is to compute a number of quantities which are functions of the logarithm of the partition function Z and its derivatives with respect to the temperature. We focus here on the calculation of functions of the energy and postpone the discussion of the other quantities calculated by potts_zln.f to the next subsection.

As for our other Potts model routines, potts_zln.f relies on using the action variable (3.77). The arguments of potts_zln.f on input are: the total number of links nlink, the minimum action of the multicanonical re-weighting range namin, the inverse temperature beta0 to which we re-weight, the multicanonical inverse temperature array b, two action histogram arrays ha and hasum, and an option variable iopt. The reason why we work with two histogram arrays is that we perform a jackknife error analysis. The subroutine potts_zln.f does this by calculations with respect to the array elements

$$\texttt{hasum(ilink)} - \texttt{ha(ilink)} . \qquad (5.43)$$

If hasum contains the accumulated statistics over all production repetitions and ha the statistics of the repetition irpt, their difference is, up to an irrelevant normalization factor, precisely the irpt jackknife histogram. If one wants to analyze a single histogram, this can be done by assigning zero to all ha entries and using hasum to enter the data.

The multicanonical array b has to be calculated from the weight factor array wrat, which is stored by the MC production run[3]. This is done by a call to the subroutine

$$\texttt{wrat\_to\_b(n2d, mlink, namin, namax, wrat, ndel\_muca, b)} \qquad (5.44)$$

---

[3]If disk space matters the production program could already store the array b instead of wrat. When the data production is split over several MC runs, one has then to re-calculate wrat from b at the beginning of each run.

Table 5.7 The first (and major) part of the subroutine potts_zln.f of ForLib.

```
      SUBROUTINE POTTS_ZLN(nlink,namin,beta0,b,ha,hasum,iopt,
     &                     Zln,Aln,A2ln)
C Copyright Bernd Berg, May 12 2002.
C Potts model ln of the partition function Z and related variables.
      include 'implicit.sta'
      include 'constants.par'
      dimension b(0:nlink),ha(0:nlink),hasum(0:nlink)
C
      iloop_ha=0 ! Counts runs through the ha(iact).gt.half loop.
      if(namin.le.0) stop "POTTS_ZLN: namin false."
      do iact=0,nlink
        ha(iact)=hasum(iact)-ha(iact) ! Jackknife histogram.
        if(ha(iact).gt.half) then
          iloop_ha=iloop_ha+1
          if(iloop_ha.eq.1) then
            a=-(beta0-b(iact))*namin ! Ferdinand-Fisher normalization.
            Zln=log(ha(iact))+two*((beta0-b(iact))*iact+a)
            Zln1_max=Zln
            Aln=Zln+log(iact*one)
            A2ln=Zln+two*log(iact*one)
          else
            a=a+(b(iact)-b(iact_old))*iact_old
            Zln1=log(ha(iact))+two*((beta0-b(iact))*iact+a)
            Zln1_max=max(Zln1_max,Zln1)
            Aln1=Zln1+log(iact*one)
            A2ln1=Zln1+two*log(iact*one)
            Zln=addln(Zln,Zln1)
            Aln=addln(Aln,Aln1)
            A2ln=addln(A2ln,A2ln1)
          end if
          iact_old=iact
        end if
      end do
C
      if(iopt.ne.1) return
C
C iact probability density at beta0: (Zln1_max used only here.)
```

of ForLib. The input arguments are n2d,mlink,namin,namax,wrat and ndel_muca. They are explained along with the discussion of p_mu_rec.f following equation (5.25).

Subsequently, the array a of the dimensionless, microcanonical free energy is calculated in the routines potts_z0ln.f and potts_zln.f. For the re-weighting to a temperature defined by $\beta_0 = 1/T_0$ the normalization of

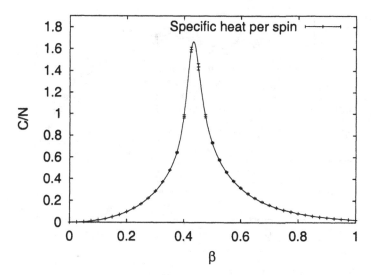

Fig. 5.1 Specific heat per spin versus β for the Ising model on an 20×20 lattice. Multicanonical data are compared with the exact result of [Ferdinand and Fisher (1969)] (full line), see assignment 3. This figure was first published in [Berg (2003)].

the partition function (*i.e.* an additive constant to the array a) is of importance when calculating the entropy or free energy, see the next subsection. A subtle point in this respect is that one has to start the recursion (5.12) with a β_0 dependent constant to reproduce the normalization used by [Ferdinand and Fisher (1969)] for the $d = 2$ Ising model, see the correspondingly commented line in table 5.7.

On output potts_zln.f returns the quantities Zln, Aln, A2ln (explained below) and overwrites the array ha. If the option variable is chosen iopt=1, the array ha contains the canonical re-weighted action density at the temperature defined by beta0, otherwise it contains the unnormalized multicanonical jackknife array (5.43).

Up to an additive, temperature independent normalization constant, discussed in the next subsection, Zln agrees with the (natural) logarithm of the partition function (3.3). The quantities Aln and A2ln serve to cal-

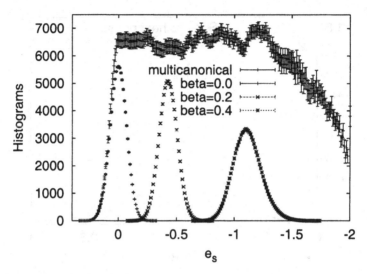

Fig. 5.2 Energy histogram from a multicanonical simulation of the Ising model on an 20×20 lattice together with the canonically re-weighted histograms at $\beta = 0$, $\beta = 0.2$ and $\beta = 0.4$, see assignment 4. Compare with figure 3.3 of chapter 3. This figure was first published in [Berg (2003)].

culate the expectation value of the action variable (3.77) and its second moment. Up to the same additive normalization constant, they agree with the logarithms of the corresponding sums in equation (3.2). Their calculation by potts_zln.f is an illustration for the use of addln.f[4]. The properly normalized expectation values are then given by (3.79), (3.80)

$$\texttt{actm} = \frac{\overline{\texttt{iact}}}{\texttt{nlink}} = \frac{\exp\left(\texttt{Aln} - \texttt{Zln}\right)}{\texttt{nlink}} \tag{5.45}$$

and

$$\texttt{a2ctm} = \frac{\overline{\texttt{iact}^2}}{\texttt{nlink}} = \frac{\exp\left(\texttt{A2ln} - \texttt{Zln}\right)}{\texttt{nlink}^2} \ . \tag{5.46}$$

---

[4] The constant Zln1_max, which is calculated along the way, is only used together with iopt=1 for normalization purposes of the re-weighted histogram in the second part of the subroutine potts_zln.f.

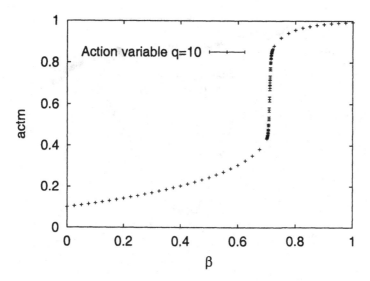

Fig. 5.3 Multicanonical mean action variable (5.45) data for the 10-state Potts model on an 20 × 20 lattice, see assignment 5. This figure was first published in [Berg (2003)].

Using the relation (3.106) for the Ising model, we reproduce the multicanonical energy data of figure 3.1. This is the first task of assignment 3. The same numerical data allow us to to calculate the **specific heat**, which is defined by

$$C = \frac{d\widehat{E}}{dT} = \beta^2 \left(\langle E^2 \rangle - \langle E \rangle^2 \right) . \tag{5.47}$$

The last equal sign follows by working out the derivative and is known as **fluctuation-dissipation theorem**. Using (3.106) again, we find

$$C = \beta^2 N d^2 \left(\texttt{a2ctm} - \texttt{actm}^2 \right) \tag{5.48}$$

for the Ising model. Figure 5.1 compares the thus obtained multicanonical data with the exact results of [Ferdinand and Fisher (1969)]. The latter are calculated using the program `ferdinand.f` (3.15). Again, see assignment 3.

Figure 5.2 shows the energy histogram of the multicanonical simulation

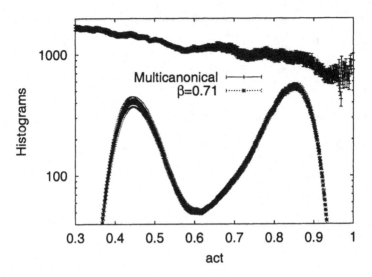

Fig. 5.4 Action histogram from a multicanonical simulation of the 10-state Potts model on an 20×20 lattice together with the canonically re-weighted histograms at $\beta = 0.71$, see assignment 6. This figure was first published in [Berg (2003)].

together with its canonically re-weighted descendants at $\beta = 0$, $\beta = 0.2$ and $\beta = 0.4$, see assignment 4. The normalization of the multicanonical histogram is adjusted so that it fits into one figure with the three re-weighted histograms. In figure 3.3 of chapter 3 the same canonical histograms are shown, relying on three simulations instead of one.

Figures 5.1 and 5.2 are based on the $2d$ Ising model data of assignment 1. In assignment 2 we produce similar data for the $2d$ 10-state Potts model. It is the first task of assignment 5 to use these data to produce the action variable `actm` data shown in figure 5.3 (via relation (3.106) `actm` is equivalent to the mean energy per site). Around $\beta = 0.71$ we observe a sharp increase of `actm` from 0.433 at $\beta = 0.70$ to 0.864 at $\beta = 0.72$, which signals the first order phase transition of the model. In figure 5.4 we plot the canonically re-weighted histogram at $\beta = 0.71$ together with the multicanonical histogram using suitable normalizations. The canonically re-weighted histogram exhibits the double peak structure which is characteristic for first order phase transitions. The ordinate is on a logarithmic scale and the minimum in-between the peaks is much deeper than it is in figure 3.6 for the $3d$ 3-state Potts model, for which the first order transition is much weaker than for the $2d$ 10-state Potts model. The multicanonical method allows to estimate the interface tension of the transition by following the minimum to maximum ratio over many orders of magnitude [Berg and Neuhaus (1992)].

It is assignment 7 to test our multicanonical code for the Ising model in $d = 1$.

5.1.7 *Free energy and entropy calculation*

At $\beta = 0$ the Potts partition function Z is given by equation (3.75). Therefore, multicanonical simulations allow for proper normalization of the partition function, if $\beta = 0$ is included in the temperature range (5.33). The analysis requires a call to the subroutine

$$\text{POTTS\_ZOLN(nq, ns, nlink, namin, b, ha, hasum, Zln\_dif)} \qquad (5.49)$$

of ForLib. Besides the number of Potts states nq and the number of lattice sites ns the arguments on input are a subset of those discussed for the subroutine potts_ln.f of table 5.7. On output Zln_dif is returned as the difference between $\ln(Z)$ from the multicanonical simulation re-weighted to $\beta = 0$ and the exact result $N \ln(q)$ at $\beta = 0$. This difference does not depend on the temperature. At all β values the properly normalized partition function logarithms are obtained by performing the replacement

$$\text{Zln} \rightarrow \text{Zln} + \text{Zln\_dif} \qquad (5.50)$$

where the left-hand side Zln is the one calculated by potts_zln.f of table 5.7. The normalization of the partition function is chosen so that it agrees for the $d = 2$ Ising model with the one used by [Ferdinand and Fisher (1969)]. A slightly tricky point is to calculate the additive constant for the dimensionless microcanonical free energies (5.10) consistently. The interested reader may look this up in the routine (5.49).

The properly normalized partition function allows to calculate the **Helmholtz free energy**

$$F = -\beta^{-1} \ln(Z) \qquad (5.51)$$

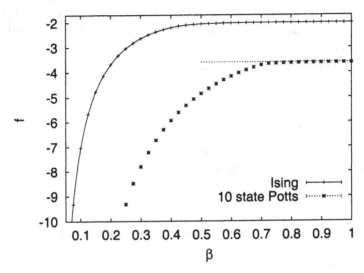

Fig. 5.5 Free energies from multicanonical simulations of the Ising and 10-state Potts models on an 20×20 lattice, see assignments 3 and 5. The full lines are the exact results of [Ferdinand and Fischer (1969)] for the Ising model and the asymptotic equation (5.54) for the 10-state Potts model. This figure was first published in [Berg (2003)].

and the **entropy**

$$S = \frac{F - \widehat{E}}{T} = \beta \, (F - \widehat{E}) \qquad (5.52)$$

of the canonical ensemble. Here $\widehat{E}$ is the expectation value of the internal energy and the last equal sign holds because of our choice of units for the temperature (3.1). Note, our proper notation for F, Z and S would be $\widehat{F}$, $\widehat{Z}$ and $\widehat{S}$. By simply using F, Z and S we follow the notation of the statistical physics literature [Landau and Lifshitz (1999); Huang (1987); Pathria (1972)] and hope that the reader will by now be able to distinguish between expectation values, estimators of expectation values and fluctuating quantities, even when they are not explicitly distinguished by notation (F, Z and S show up in all three variants).

For the $2d$ Ising model as well as for the $2d$ 10-state Potts model, we

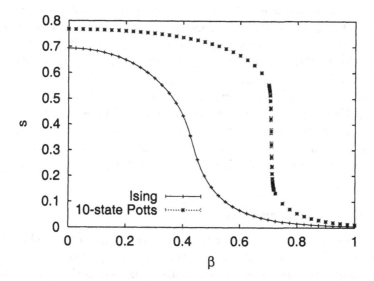

Fig. 5.6 Entropies from multicanonical simulations of the Ising and 10-state Potts models on an 20×20 lattice, see assignments 3 and 5. The full line is the exact result of [Ferdinand and Fischer (1969)] for the Ising model. This figure was first published in [Berg (2003)].

show in figure 5.5 our multicanonical estimates of the free energy density per site

$$f = F/N .\qquad(5.53)$$

As in previous figures, the Ising model data are presented together with the exact results obtained with `ferdinand.f` (3.15), whereas we compare the 10-state Potts model data with the $\beta \to \infty$ asymptotic behavior. Using our Potts energy definition (3.64), we see that for large β the partition function of our q-state Potts models approaches $q \exp(+\beta\, 2\, d\, N - \beta\, 2\, d\, N/q)$. The asymptotic behavior of the free energy is therefore

$$F_{as} = \frac{2\,d\,N}{q} - 2\,d\,N - \beta^{-1} \ln(q) \ \text{ and } \ f_{as} = \frac{2\,d}{q} - 2\,d - \beta^{-1} \frac{\ln(q)}{N} .\qquad(5.54)$$

In figure 5.6 we plot the entropy density

$$s = S/N \qquad (5.55)$$

of the 2d Ising model together with its exact curve. Further, entropy data for the 2d 10-state Potts model are included in this figure. In that case we use $s/3$ instead of s, so that both graphs fit into the same figure. In all the figures of this subsection excellent agreement between the numerical and the analytical results is found.

For the 2d Ising model one may also compare directly with the number of states. Up to medium sized lattices this integer can be calculated to all digits by analytical methods [Beale (1995)]. However, MC results are only sensitive to the first few (not more than six) digits. Therefore, one finds no real advantages over using other physical quantities.

5.1.8   *Time series analysis*

Often one would like to re-weight the time series of a multicanonical simulation to a temperature $\beta_0 = 1/T_0$. Typically, one prefers in continuous systems time series data over keeping histograms, because one avoids then dealing with the discretization errors introduced by histogramming continuous variables. Even in discrete systems time series data are of importance, as we have already seen in our investigation of autocorrelation times in chapter 4. Besides, normally one wants to measure more physical quantities than just the energy. Then the RAM limitations may require to use a time series instead of histograms. We illustrate this for the Potts magnetization (3.69), which we encountered before when we tested the Potts model heat bath algorithm.

To enable measurements of the Potts magnetization in our heat bath code of chapter 3.3.2, the array hm(0:ms,0:nqm1) of dimension (ms + 1) × nq is kept in the common block potts_hb.com of ForLib, where ms is the maximum number of lattice sites allowed in the simulation. However, for the proper re-weighting of a multicanonical simulation, the energy value has to be stored together with each measurement of the Potts magnetization. If the histogram method is used, arrays of size $\sim$ ms × ms are needed. For an increase of an array size proportional to the square of the lattice size, one runs for larger lattices quickly out of RAM. In this situation it becomes even in discrete system advantageous to keep a time series of measurements.

Our program

$$\texttt{p\_mu\_ts.f} \ \text{of} \ \texttt{ForProg/MC\_Potts} \qquad (5.56)$$

modifies the previous multicanonical data production program (5.28) for this purpose. The program (5.56) performs time series measurements in the second part of the run after finishing the unchanged multicanonical recursion in its first part. For the organization of the time series measurements we follow closely the scheme developed in section 3.4 for continuous systems. Measurements of the energy and the q Potts magnetizations are stored after every $\texttt{nsw}$ sweeps in the array $\texttt{tsa(nmeas,0:nq)}$ and $\texttt{nrpt}$ of those arrays are written to disk.

The subsequent data analysis is carried out with the program

$$\texttt{ana\_pmu\_ts.f} \ \text{of} \ \texttt{ForProg/MC\_Potts} \qquad (5.57)$$

which calls the following subroutines of $\texttt{ForLib}$

$$\texttt{p\_mu\_a.f, p\_ts\_z0ln.f, p\_ts\_z0lnj.f, p\_ts\_zln.f} \ \text{and} \ \texttt{p\_ts\_zlnj.f} \ .$$
$$(5.58)$$

This code is an all-log jackknife implementation of the time series formula (5.34). The logic is in part distinct from that of the histogram based analysis code (5.41). Whereas the jackknife binning can efficiently be coded directly for the $\texttt{nrpt}$ histograms, this is not the case for the $\texttt{nrpt}$ time series fragments. While the routines $\texttt{p\_ts\_z0ln.f}$ and $\texttt{p\_ts\_zln.f}$ perform similar tasks as the $\texttt{potts\_z0ln.f}$ and $\texttt{potts\_zln.f}$ routines of the histogram code, it is not for jackknife binned data. Instead the jackknife binning is subsequently performed by the routines $\texttt{p\_ts\_z0lnj.f}$ and $\texttt{p\_ts\_zlnj.f}$. The routine $\texttt{p\_mu\_a.f}$ calculates the dimensionless microcanonical free energies according to equation (5.12), normalized so that the conventions of [Ferdinand and Fisher (1969)] are met for the $d = 2$ Ising model.

The use of the new programs is illustrated in assignments 8 and 9. We create the same statistics on 20×20 lattices as in assignments 1 and 2, including time series measurements for the energy and also for the Potts magnetization. For energy based observables the analysis of the histogram and the time series data give consistent results.

For zero magnetic field, $H = 0$, the expectation value of the Potts magnetization on a finite lattice is not particularly interesting. Independently of the temperature it is

$$M_{q0} = \langle \delta_{q_i,q_0} \rangle = \frac{1}{q} . \qquad (5.59)$$

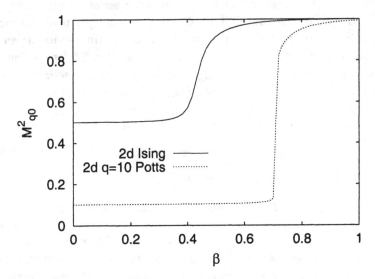

Fig. 5.7 The Potts magnetization (3.69) per lattice site squared (5.60) for the $q = 2$ and $q = 10$ Potts models on an 20×20 lattice, see assignments 8 and 9.

For a multicanonical simulation it is quite obvious that even at low temperatures each Potts state is visited with equal probability $1/q$. In contrast to this the expectation value of the magnetization squared

$$M_{q0}^2 = q \left\langle \left(\frac{1}{N} \sum_{i=1}^{N} \delta_{q_i,q_0} \right)^2 \right\rangle \qquad (5.60)$$

is a non-trivial quantity. At $\beta = 0$ its $N \to \infty$ value is $M_{q0}^2 = q\,(1/q)^2 = 1/q$, whereas it approaches one for $\beta \to \infty$, because at low temperature the Potts states of each lattice configuration collapse to one of the q possible values, *i.e.* $M_{q0}^2 = q\,(1/q) = 1$. For $q = 2$ and $q = 10$ figure 5.7 shows our numerical results on an 20×20 lattice and we see that the crossover of M_{q0}^2 from $1/q$ to 1 happens in the neighborhood of the critical temperature. A FSS analysis would reveal that a singularity develops at β_c, which is in the derivative of M_{q0}^2 for the second order phase transitions ($q \leq 4$) and in M_{q0}^2 itself for the first order transitions ($q \geq 5$).

5.1.9 *Assignments for section 5.1*

(1) Use the default initialization of the Marsaglia random number generator and run the multicanonical program p_mu.f of ForProg/Potts_MC for the Ising model on an 20 × 20 lattice with the muca.par parameter file of table 5.2. How many recursions and how many sweeps are needed for the first tunneling event? What is the average acceptance rate during this part of the run? Verify the answers given in the text to the same questions for ten tunneling events. After freezing the multicanonical parameters, perform 10 000 sweeps for reaching equilibrium and 32 × 10 000 sweeps for measurements. What is the acceptance rate during the equilibration run? How many tunneling events occur during the measurement period? (The answers to the last questions are 0.245 and 84.)

(2) Repeat the previous assignment for the 10-state Potts model on an 20 × 20 lattice, but increase the number of sweeps for equilibrium and data production by a factor of ten. Some of the answers are: 4,269 recursions with 1,211,480 sweeps for the first ten tunneling events and 45 tunneling events during the production phase.

(3) Use the analysis program ana_pmu.f of ForProg/MC_Potts together with the data set mu2d02q020.dat generated in assignment 1 to reproduce the multicanonical Ising model data plotted in figures 3.1, 5.1, 5.5 and 5.6. Use ferdinand.f (3.15) to obtain the corresponding exact results (full lines in the figures).

(4) Use the statistics of assignment 1 and plot the multicanonical histogram together with its canonically re-weighted descendants at $\beta = 0$, $\beta = 0.2$ and $\beta = 0.4$. (The analysis program ana_pmuh.f of ForProg/MC_Potts works for this task). With suitably chosen normalizations figure 5.2 will result.

(5) Use the analysis program ana_pmu.f of ForProg/MC_Potts together with the data set mu2d10q020.dat generated in assignment 2 to reproduce the multicanonical data plotted in figures 5.3, 5.5 and 5.6. Compare the free energy result with the asymptotic formula (5.54).

(6) Use the statistics of the previous assignment to plot the canonically re-weighted histogram at $\beta = 0.71$ together with the multicanonical histogram. With suitably chosen normalizations figure 5.4 will result.

(7) Test the code of this section for the $d = 1$ Ising model. Calculate the exact energy per spin on lattices with free as well as with periodic boundary conditions analytically. Then, perform multicanonical

simulations on an $L = 20$ lattice. Rely on $32 \times 100\,000$ sweeps with measurements (after determining the multicanonical parameters). Are the data accurate enough to distinguish between the free and the periodic boundary conditions? Continue to calculate as many results as you like in $d = 1$.

(8) Repeat the $2d$ Ising model simulation of assignment 1 with the program p_mu_ts.f (5.56), which records also time series data for the action variable and the magnetization. To analyze these data use the program ana_pmu_ts.f (5.57). Subsequently, plot the same quantities as in assignment 3, now relying on the time series data for the action variable. Finally, re-weight the magnetization data to reproduce the Ising part of figure 5.7.

(9) Repeat the $2d$ 10-state Potts model simulation of assignment 3 with the program p_mu_ts.f (5.56) and analyze these data with the program ana_pmu_ts.f (5.57). Plot the same quantities as in assignment 5, now relying on the time series data. Finally, re-weight the Potts magnetization data to reproduce the 10-state Potts model part of figure 5.7.

5.2   Event Driven Simulations

A Metropolis or heat bath simulation moves slowly at low temperature, because the likelihood that an update proposal is rejected becomes high, and one stays for an extended time with the configuration at hand[5] (which has to be counted again). This time is called **waiting time**. **Event driven simulations** (EDS) were introduced in a paper by Bortz, Kalos and Lebowitz [Bortz *et al.* (1975)], there called **the n-fold way**. (This name appears to be chosen as a pun to the eightfold way in Gell-Mann's quark model [Gell-Mann and Neeman (1964)].) The idea of EDS is to save the waiting time by calculating it exactly and modifying the algorithm so that the newly proposed state is always accepted. In recent years extensions of this idea have been considered [Korniss *et al.* (2003); Novotny (2001)], including to continuous systems [Munoz *et al.* (2003)]. In this section we discuss the original EDS for q-state Potts models.

Let us consider all updates of a configuration which are possible through a change of a single spin $q_i \rightarrow q_i'$, $(i = 1, \dots, N)$ where N is the number of sites. The new spin q_i' can take the values $1, \dots, q$. There are N updates

---

[5]Cluster updates discussed in section 5.3 suffer from a similar problem, although they change low temperature configurations. See the discussion at the end of section 5.3.1.

(one for each site) for which the new configuration agrees with the old one (*i.e.* $q_i' = q_i$) and $(q-1)N$ updates for which the systems moves (*i.e.* $q_i' \neq q_i$). Once a site is chosen, the Metropolis as well as the heat bath probabilities of these updates are known. By summing the probabilities of all the $q_i' = q_i$ cases, we obtain the probability P^{stay} that the system will not move in an update attempt. The main idea is now to count the configuration at hand with the weight

$$w_c = 1 + \sum_{i=1}^{\infty} \left(P^{\text{stay}} \right)^i = \frac{1}{1 - P^{\text{stay}}} \geq 1 \qquad (5.61)$$

instead of uniformly with weight 1, as for the usual Metropolis and heat bath updates. A bit of thought reveals that the thus defined w_c is precisely the average waiting time before the next update. Counting configurations with the weight w_c allows then to pick one of the $q_i' \neq q_i$ moving updates directly.

The catch of this nice idea is that it requires a global bookkeeping of the allowed updates. One has to rely on table lookups to overview **all** single spin updates of a configuration and to pick one with its correct probability. This complicates the computer program considerably and one update step in EDS becomes far slower than it is in the normal Metropolis or heat bath algorithm. Remarkably, the computer time for a single EDS update is still of order N^0, *i.e.* a volume independent constant. To make EDS efficient the increase in computer time for a single update has to be beaten by the gain on the waiting time. For Potts models this happens at sufficiently low temperatures, where $P^{\text{stay}} \to 1$ holds, implying $w_c \to \infty$. Hence, EDS are of relevance for the study of physically interesting low temperature phenomena.

In the following we use heat bath probabilities to implement EDS for q-state Potts models in d dimensions. The EDS algorithm in its Metropolis version would be similarly CPU time consuming. We define a **local neighbor configuration** at site i by the $2d$ neighbor spins of a central spin q_i. Thus there are

$$\texttt{nconf0} = q^{2d} \qquad (5.62)$$

local neighbor configurations. The contribution of a local neighbor configuration to the action variable (3.77) depends on the value of the central spin q_i. We denote these contributions by $\texttt{ia}_i(q_i)$. The values of $\texttt{ia}_i(q_i)$ lie in the range

$$\mathtt{ia}_i(q_i) \in [0, 1, \ldots, 2d] \quad \text{for} \quad q_i = 1, \ldots, q . \tag{5.63}$$

The function $\mathtt{ia}_i(q_i)$ gives the distribution of the action values, which defines the heat bath probabilities for the spins $q_i = 1, \ldots, q$ at the site i, namely

$$P_i(q_i) = \frac{1}{N} \frac{\exp[\mathtt{ia}_i(q_i)]}{\sum_{q_i=0}^{q} \exp[\mathtt{ia}_i(q_i)]} . \tag{5.64}$$

We have included the factor $1/N$ for the likelihood to choose site i, because this is convenient in the following. We are interested in the selection of a move from the set of all updates. At site i the probability not to move depends on the initial value of the spin q_i, which we denote by q_i^{in}. The total stay probability is given by the sum over the stay probabilities at all sites

$$P^{\text{stay}} = \sum_{i=1}^{N} P_i^{\text{stay}} \quad \text{with} \quad P_i^{\text{stay}} = P_i(q_i^{\text{in}}) . \tag{5.65}$$

The probability that a move will happen at site i is simply

$$P_i^{\text{move}} = \frac{1}{N} - P_i^{\text{stay}} . \tag{5.66}$$

The total move probability is $P^{\text{move}} = \sum_{i=1}^{N} P_i^{\text{move}}$, so that $P^{\text{stay}} + P^{\text{move}} = 1$ holds. By counting configurations with the weight (5.61), the P^{stay} probability becomes removed and the move probabilities are transformed according to

$$P_i^{\text{move}} \rightarrow P_i^{'\,\text{move}} = \frac{P_i^{\text{move}}}{\sum_{i=1}^{N} P_i^{\text{move}}} = \frac{P_i^{\text{move}}}{1 - P^{\text{stay}}} , \tag{5.67}$$

so that $\sum_{i=1}^{N} P_i^{'\,\text{move}} = 1$ holds, *i.e.* we move with certainty.

5.2.1   *Computer implementation*

The task is to develop a calculational scheme, which implements the general concept in a way so that the computer time for one update stays of order N^0. From equations (5.66) and (5.65) it is obvious that the move probability at site i does not only depend on the $2d$ neighbor spins, but also on q_i^{in}, the present spin at site i, which we call central spin at site i. We define the

Table 5.8 Counting distinct local action distributions for EDS: Relevant parts of the program p_e_cases.f of ForProg/MC_Potts and of the subroutine p_e_compare.f of ForLib are shown.

```
program p_e_cases ! Copyright Bernd Berg, Jan 02 2002.
...
    do id=1,n2d
      nqa(id)=nq ! Sets the nq-box dimensions for the ixcor call.
    end do
    ncase=0
    do iconf0=1,nconf0 ! Loop over the neighbor configurations.
      call ixcor(iqarray,nqa,iconf0,n2d)
      do iq_in=0,nqm1    ! Loop over the central state.
        do id=0,n2d
          nact_new(id)=0
        end do
        iconf=iconf0+iq_in*nconf0
        do iq=0,nqm1
          ia=0
          do id=1,n2d
            if(iq.eq.iqarray(id)) ia=ia+1
          end do
          if(iq.ne.iq_in) nact_new(ia)=nact_new(ia)+1
          if(iq.eq.iq_in) nact_new(n2dp1)=ia
        end do
        call p_e_compare(nact_old,nact_new,n2d,mcase,ncase,icase)
        if(ncase.gt.mcase) stop "p_e_cases: enlarge mcase!"
      end do
    end do
    write(iuo,'(" nd,nq,ncase:",2I4,I10,/)') nd,nq,ncase

subroutine p_e_compare(nact_old,nact_new,n2d,mcase,ncase,icase)
dimension nact_old(0:n2d+1,mcase),nact_new(0:n2d+1)
n2dp1=n2d+1
if(ncase.ge.1) then
  do icase=1,ncase
    lold=.true.
    do id=0,n2dp1
      if(nact_old(id,icase).ne.nact_new(id)) lold=.false.
    end do
    if(lold) return
  end do
end if
ncase=ncase+1
do id=0,n2dp1
  nact_old(id,ncase)=nact_new(id)
end do
...
```

Table 5.9 Numbers **ncase** of distinct **local cases** as explained in the text.

q	2	3	4	5	6	7	8
$d = 2$	5	9	11	12	12	12	12
$d = 3$	7	16	23	27	29	30	30

local configuration at site i by the central and its neighbor spins. There are

$$\texttt{nconf} = q^{2d+1} \tag{5.68}$$

local configurations, in contrast to only $\texttt{nconf0} = q^{2d}$ local neighbor configurations (5.62). Now it looks as if a table of size $q^{2d+1} \times N$ would be needed for our endeavor. Fortunately, this is not the case. Many of the nconf local configurations lead to identical **distributions** for the values of the **local probabilities** P_i^{stay} and $P_i(q_i)$, $q_i \neq q_i^{\text{in}}$. This allows us to get away with reasonably sized tables.

Distinct sets of $\texttt{ia}_i(q_i^{\text{in}})$ and $\texttt{ia}_i(q_i)$, $q_i \neq q_i^{\text{in}}$ values are in one to one correspondence with the local probabilities. In the following we enumerate them. Instead of doing this analytically, we let the computer count how many distinct cases there are. This is done by the program p_e_cases.f of ForProg/MC_Potts which, in turn, uses calls to the subroutine p_e_compare.f of ForLib. Relevant parts of the program p_e_cases.f and of the subroutine p_e_compare.f are reproduced in table 5.8. First, we note that running our subroutine ixcor (3.21) of chapter 3 in q dimensions allows to determine the neighboring q-states from a single configuration number iconf0 in the range (5.62)

$$\texttt{inconf0} = 1, \ldots, \texttt{nconf0} .$$

The entries of the array nact_new(0:n2d+1) define the distinct local ia distributions in the following way: The last element of the array stores the action contribution of the present central spin iq_in, while the other array elements correspond to the action contributions $\texttt{ia} = 0, \ldots, \texttt{n2d}$ and record how often a particular action contribution occurs when the central spin takes on the iq.ne.iq_in q values. Distinct nact_new arrays define distinct **local cases** of the action distributions. The subroutine p_e_compare.f counts the number of distinct local cases by testing whether a particular nact_new array agrees with an already encountered one. If this is not the case, nact_new is stored in the array nact_old(0:n2d+1,ncase_max).

For $d = 2$ and $d = 3$ Potts models with $q = 2, \ldots, 8$ we list in table 5.9 the numbers ncase of distinct local cases. It is seen that ncase becomes constant for $q \geq 2d + 1$. That this has to be so is easily proven: One of the allowed q values will be the central spin and there are $q - 1$ different q-state values left. As there are only $2d$ neighbors, the maximum number of distinct cases is reached for $q - 1 = 2d$. Further spins (encountered for $q - 1 > 2d$) lead to already covered nact_new action distributions and do not change the number of distinct cases, ncase. Therefore, independently of q the maximum number of distinct local cases stays in $d = 2$ at ncase_max=12 and in $d = 3$ at ncase_max=30. Higher values are found in higher dimensions.

Table 5.10 The parameter file p_eds.par and the common block p_eds.com for event driven simulations are depicted, respectively in the upper and lower parts of the table.

```
C Additional parameters for event driven MC simulations.
      parameter(n2dp1=n2d+1,nq2=nq**2,nconf0=nq**n2d,nconf=nq**n2dp1)
      parameter(mcase=12,mcam1=mcase-1,mcap1=mcase+1)

C Tables and variables for event driven simulations:
      common /etabs/ qmove(mcase),Peds1(0:n2d,mcase),Pstay,Pmove,
     & Pcase(mcase),Pcsum(mcase),NAcase(mcase),IAcase(nconf),
     & IScase(ms),Index(ms,mcap1),ia_array(mcase),Icase_per(mcase)
```

To convert our insights into a working updating routine, we need some bookkeeping. This is achieved through a number of new arrays which are kept in the EDS common block p_eds.com of ForLib, listed in the lower part of table 5.10. The maximum dimensions of the newly encountered arrays are set in the EDS parameter file p_eds.par, listed in the upper part of table 5.10, and in the lat.par and potts.par parameter files (3.26), (3.83) of chapter 3. The updating routine for EDS is

$$p\_e\_mc.f \quad \text{of} \quad ForProg/MC\_Potts . \tag{5.69}$$

In the following we describe the use of the arrays of p_eds.com in the order encountered in the routine p_e_mc.f.

We trace for each of the local cases (table 5.9) the probability that the next move will correspond to it. These probabilities are available in the array Pcase, which is after each move updated by the subroutine p_etabs_update discussed below. Drawing one random number allows then to pick the case, icase, to which the next move belongs. There will be a certain number of sites on the lattice, which belong to the case icase of our classification, otherwise we would have Pcase(icase)=0. This number is stored in the element NAcase(icase) of the array NAcase (Number array cases). With the draw of a second random number we pick a uniformly distributed integer

$$is1 \in [1, \ldots, NAcase(icase)] \,,$$

which determines the site is at which the update is performed, using an index array so that

$$is = Index(is1, icase) \,. \qquad (5.70)$$

We have to dimension the index array to the size ms × mcase, as we have to allow for the extreme situation that all sites contribute to the same local case[6]. For Potts models in $d \leq 3$ this can, according to table 5.9, lead to an enhancement of the original lattice storage space by a factor of up to 30 (the original allocation space needed is that of the state array ista(ms), see table 3.6 of chapter 3). For large lattices this seems to be a problematic waste of memory space, as most of the Index array elements remain unoccupied. However, presently I am not aware of a better solution. Any dynamical memory allocation would likely lead to an unacceptable slowing down of the code.

Once we have the update site is identified (5.70), we can replace its present spin, iq_in, by a new one. The relevant probabilities are stored in the arrays qmove and Peds1 of the common block p_eds.com and the actual decision relies on a third random number. After the new spin iq ≠ iq_in has been determined, the main work consists in updating all tables. This has not only to be done for the updated site, but also for all its neighbor sites, as their cases of our classification change too. We identify them relying on the array IAcase of p_eds.com, which returns for each local configuration (5.68) the case number. This introduces another large array. For instance, nconf = 10^5 for the 2d 10-state Potts model and nconf = 2187 for the 3d 3-state Potts model. In contrast to our previously encountered arrays, which scale with the volume, the size of the IAcase array stays constant and for most applications the encountered nconf values fit well into the RAM of a modern PC. On the other hand a number like nconf = 10^9, as found for the 4d 10-state Potts model, would still be prohibitively large (for PCs of the year 2003). The array size can be considerably reduced at

---

[6]The actual size of the index array Index is ms × (mcase + 1) allowing for workspace.

Table 5.11 The subroutine p_etabs_update.f which updates the EDS tables. It is attached to the subroutine p_e_mc.f (5.69).

```
      subroutine p_etabs_update(icase,jcase,is0,is1)
      include '../../ForLib/implicit.sta'
      include '../../ForLib/constants.par'
C Copyright, Bernd Berg, Jan 13 2002. Update EDS tables.
      include 'lat.par'
      include 'potts.par'
      include 'p_eds.par'
      include '../../ForLib/lat.com'
      include '../../ForLib/potts.com'
      include '../../ForLib/p_eds.com'
      is2=Index(NAcase(icase),icase)
      Index(is1,icase)=is2
      Index(is2,mcap1)=is1
      NAcase(icase)=NAcase(icase)-1
      Pstay=Pstay+Pcase(icase)                     ! Sensitive to rounding.
      Pcase(icase)=NAcase(icase)*qmove(icase)  ! Avoids rounding.
      Pstay=Pstay-Pcase(icase)
      NAp1=NAcase(jcase)+1
      NAcase(jcase)=NAp1
      Pstay=Pstay+Pcase(jcase)
      Pcase(jcase)=NAcase(jcase)*qmove(jcase)
      Pstay=Pstay-Pcase(jcase)
      Pmove=One-Pstay
      Index(Nap1,jcase)=is0
      Index(is0,mcap1)=Nap1
      IScase(is0)=jcase
      return
      end
```

the expense of spending more CPU time on calculations. It is then possible to divide nconf by the factor $q(2d)!$, see appendix B, exercise 1 for the present section.

Once the case labels of the old (in) and the new local energy distributions are calculated, it is straightforward to update all the involved arrays, as well as the Pstay probability. The workhorse for this update is the subroutine p_etabs_update.f, which is attached to the file p_e_mc.f and reproduced in table 5.11. This routine has to be used for the updated central spin as well as for each of its forward and backward nearest neighbors. In the recursive computation of the Pstay probability rounding errors would show up over an extended time. The program p_e_mc.f takes care of that by re-calculating Pstay at the end of each sweep using the integer array NAcase.

The initialization of the EDS arrays and probabilities is done by the subroutine

$$p\_e\_init.f \quad \text{of} \quad ForProg/MC\_Potts\ , \qquad (5.71)$$

which has attached the routine p_e_tabs(beta), doing most of the actual work. As the parameter mcase of p_eds.par leads to creating quite large arrays, the initialization program forces the user to impose the equation

$$mcase\ =\ ncase\ , \qquad (5.72)$$

i.e. array dimensions in excess to the actually encountered number of local cases cannot be used. The stop command which enforces the identity (5.72) should not be removed, because this identity is later on assumed to be true and used in various parts of the update routine (5.69).

Finally, the updating is somewhat faster when we sort the array Pcase of p_eds.com, so that the most often used elements come first. Other arrays of the EDS bookkeeping have then to be sorted correspondingly. The sorting is done by the subroutine Pcase_sort.f, which is included in the p_e_mc.f file. The array Pcsum of p_eds.com accumulates the Pcase probabilities used, to allow the sorting decision on a statistical basis. The tests of the next subsection are done by running the main program

$$p\_e\_hist.f \quad \text{of} \quad ForProg/MC\_Potts\ . \qquad (5.73)$$

It includes calls to Pcase_sort.f after one quarter and after one half of the equilibrium sweeps. At least the second of these calls should rely on a sufficiently equilibrated system to allow for a meaningful sorting. More calls to Pcase_sort.f should be avoided, because this routine is rather slow. Also, the calls should be made relatively early in the simulation, so that the rest of it can benefit from the proper ordering of the Pcase array. Altogether it seems that the CPU time advantage does not really justify the sorting effort.

5.2.2   *MC runs with the EDS code*

To test the code we compare with runs of the standard Metropolis and heat bath algorithms. Towards the end of this subsection we study the efficiency at low temperatures, where EDS have their strength.

For the 10-state Potts model we run at $\beta = 0.62$ on an 20×20 lattice with $10\,000$ sweeps for reaching equilibrium and $64 \times 5\,000$ sweeps for measurements, *i.e.*, the same statistics as for the heat bath run of equation (3.108) of chapter 3. Now our result is

$$\text{actm} = 0.321793\,(68)\;, \tag{5.74}$$

which gives $Q = 0.84$ for the Gaussian difference test when comparing with the Metropolis result (3.108) and $Q = 0.18$ when comparing with the heat bath result (3.108), see assignment 1. CPU timing versus the heat bath run (3.108) gives rather platform dependent results: The EDS code took 2.6 times longer on a Sun workstation, but only 1.7 times longer on a Linux PC. The user is advised to evaluate the performance in her or his own computing environment.

For the Ising model we run on an 14^3 lattice at $\beta = 0.22165$ and compare with the results listed in equation (3.109). We choose the statistics to be $10\,000$ sweeps for reaching equilibrium and $64 \times 5\,000$ sweeps for measurement, half of that of the Metropolis run of (3.109). Our EDS result is

$$e_s = -1.0456\,(11)\;, \tag{5.75}$$

see assignment 2. The Gaussian difference test gives $Q = 0.10$ for comparison with the previous Metropolis result and $Q = 0.08$ for comparison with the high statistics results of [Alves *et al.* (1990)], which is also listed in (3.109). These Q values are a bit on the low side, but the assumption that this is due to a statistical fluctuation is still reasonable.

For one more comparison we run the EDS code for the 3-state Potts model on an 24^3 lattice at $\beta = 0.275229525$. We perform $10\,000$ sweeps for reaching equilibrium and $64 \times 5\,000$ sweeps with measurement. Our result is

$$\text{actm} = 0.5669\,(19) \tag{5.76}$$

and comparison with the result (3.110) by means of the Gaussian difference test gives $Q = 0.78$, see assignment 3.

Efficiency

To compare the efficiency of EDS with the standard heat bath simulations, it is **unnecessary** to measure autocorrelation times. Instead, it is sufficient to focus on the change rate of the heat bath algorithm. The reason is that the dynamics of EDS is almost the same as that of the heat

Table 5.12 Efficiency (5.77) of EDS versus heat bath simulations for the Ising model on an 14^3 lattice using a 400 MHz Pentium PC, see assignment 4. The heat bath change rates are also given. One thousand sweeps were used for equilibration and 32×500 sweeps for measurements of the change rates.

β	0.22165	0.3	0.4	0.5	0.6
Efficiency EDS / Heat bath	0.21	0.75	3.1	11.4	39.8
Heat bath change rate	0.32	0.083	0.020	0.0054	0.0016

bath simulation with random updating. All the difference is that EDS save the waiting time, but spend more CPU time on a single updating step. Therefore, the efficiency of EDS over standard MC simulations becomes

$$\frac{\text{Heatbath CPU time}}{\text{EDS CPU time} \times \text{Heat bath change rate}} \times \frac{\tau_{\text{int}}(\text{sequential updates})}{\tau_{\text{int}}(\text{random updates})}. \tag{5.77}$$

In table 5.12 the results are compiled for Ising model simulation on an 14^3 lattice at β values which increase from the Curie value up to $\beta = 0.60$, see assignment 4. The integrated autocorrelations times needed were previously collected in table 4.2 of chapter 4.2.2. The CPU time measurements are platform dependent, whereas the change rates (also given in table 5.12) and integrated autocorrelation times are not. For the PC used we find that at the Curie temperature the standard heat bath updating beats the EDS by a factor of almost five two. Between $\beta = 0.3$ and $\beta = 0.4$ the efficiencies of the simulations cross over and for even larger β values the EDS efficiency increases rapidly over the heat bath performance, essentially because the heat bath change rate approaches zero like $\exp(-\text{const}\,\beta)$.

According to table 4.2 the autocorrelations between heat bath sweeps are small for $\beta \geq 0.4$. Therefore, to take advantage of the gain of performance in EDS, it is essential to accumulate the desired quantities after every update.

5.2.3   *Assignments for section 5.2*

(1) Use Marsaglia random numbers with the default seed and p_e_mc.f to reproduce the $2d$ 10-state Potts model e_{0s} result (5.74). Time this run versus the heat bath run of assignment 8 of chapter 3.3.

(2) Use Marsaglia random numbers with the default seed and p_e_mc.f to reproduce the $d = 3$ Ising model result of equation (5.75).

(3) Use Marsaglia random numbers with the default seed and p_e_mc.f to

reproduce the $3d$ 3-state Potts model result of equation (5.76).

(4) Compare the efficiency of EDS simulations and default heat bath simulations on your favorite computer for Ising model simulations on an 14^3 lattice at the β values listed in table 5.12.

5.3 Cluster Algorithms

The q state Potts model can be mapped onto a percolation model of spin clusters [Fortuin and Kasteleyn (1972)]. For a review of percolation theory see [Stauffer and Aharony (1994)]. Using the Fortuin-Kasteleyn representation, [Swendsen and Wang (1987)] formulated the MC cluster algorithm for Potts models. Clusters of spins are flipped in one step. For a second order phase transitions the algorithm greatly improves, or even eliminates, the critical slowing down of the MC simulation. [Blöte *et al.* (2002)] summarize further developments in the realm of cluster algorithm.

Swendsen-Wang (SW) clusters are formed by the following procedure: Bonds between neighboring sites i and j are created with probability

$$p_{\langle ij \rangle} = \max \left[0, 1 - \exp \left(-2\beta \, \delta_{q_i q_j} \right) \right] , \qquad (5.78)$$

i.e., with probability $[1 - \exp(-2\beta)]$ between identical spins (recall our convention (3.68) for β) and not at all between sites occupied by different spins. In one sweep through the lattice each site becomes assigned to precisely one cluster. Entire cluster are then flipped in the following way: A spin $q_0 \in \{1, \ldots, q\}$ is picked at random and assigned to each site of a cluster. Before coming to the SW procedure, let us consider an update procedure due to [Wolff (1989)], which flips single clusters.

The Wolff algorithm[7] fulfills detailed balance. One performs the following steps:

(1) A site i is chosen at random and added as first element to the cluster c.
(2) All links connecting i to its nearest neighbors are visited and the bond $\langle ij \rangle$ is activated with probability (5.78). The sites j of activated bonds are added to the cluster c.
(3) In the same way, continue iteratively for all links connected to newly added sites until the process stops.

---

[7]Wolff's paper deals with $2d$ $O(n)$ σ models by embedding Ising spins. For the Potts model the logic is slightly different, because more than two alternative spins exist per site.

(4) Assign to all sites of the cluster c a new spin $q_i' \neq q_i$, which is uniformly chosen from the $q - 1$ alternatives. (One could also allow $q_i' = q_i$, but the procedure would move less efficiently, in particular when q is small, *e.g.* $q = 2$.)

Ergodicity of the procedure is guaranteed by the fact that there is a non-vanishing probability that the cluster consists only of the single site i and each spin $q_i' \neq q_i$ can be reached. Repeating this $\geq N$ times (N is the number of sites of the system), the probability to cover all sites becomes non-vanishing too. Hence, all configurations can be generated.

To show detailed balance is more involved. We consider two configuration $\{q_i\}$ and $\{q_i'\}$, which differ only by a flip of the cluster c. We just have discussed how to build the cluster c in the configuration $\{q_i\}$. The probability of the cluster c is

$$P_c = \frac{|c|}{N} \prod_{\langle ij \rangle \in c} p_{\langle ij \rangle} \prod_{\langle ij \rangle, i \in c, j \notin c} \exp\left(-2\beta\, \delta_{q_i q_j}\right) , \qquad (5.79)$$

where $|c|$ is the number of sites in the cluster c, so that $|c|/N$ is the probability to pick as initial seed a spin of the cluster. The first product gives the probability $[1 - \exp(-2\beta)]^{|c|}$ to activate the bonds within the cluster and the second product is the probability to terminate at the boundary. For the same cluster in the configuration $\{q_i'\}$ we find the probability

$$P_c' = \frac{|c|}{N} \prod_{\langle ij \rangle \in c} p_{\langle ij \rangle}' \prod_{\langle ij \rangle, i \in c, j \notin c} \exp\left(-2\beta\, \delta_{q_i' q_j'}\right) . \qquad (5.80)$$

We have $p_{\langle ij \rangle}' = p_{\langle ij \rangle}$ for links within the cluster, because in either case all spins within the cluster are identical. The difference lies entirely in the last product. Outside of the cluster the configurations are the same and for the ratio of the transition probabilities one obtains the detailed balance condition (3.51)

$$\frac{W(\{q_i'\},\{q_i\})}{W(\{q_i\},\{q_i'\})} = \frac{W(\{q_i\} \to \{q_i'\})}{W(\{q_i'\} \to \{q_i\})} = \frac{\exp\left(-2\beta \sum_{\langle ij \rangle} \delta_{q_i,q_j}\right)}{\exp\left(-2\beta \sum_{\langle ij \rangle} \delta_{q_i',q_j'}\right)} = \frac{\exp\left(2\beta\, E\right)}{\exp\left(2\beta\, E'\right)}$$
$$(5.81)$$

where E and E' are the energies of the respective configurations.

Table 5.13 shows our implementation of the Wolff algorithm, the subroutine `potts_clw.f` of `ForProg/Potts_MC`. The basic idea is quite simple: A site `is` is chosen at random and its old spin `iqold` flipped to the new spin

Table 5.13 Essential parts of the Wolff cluster updating routine.

```
      subroutine potts_clw(nflip) ! Copyright, Bernd Berg, Oct 17 2001.
C Wolff cluster updating for q-state Potts models in d dimensions.
      include '...'
      dimension iacl(ms)
      qm1=one*nqm1
      boltz1=one-wrat(-1)
      ncl=1 ! Cluster size.
      icl=1 ! Presently considered cluster variable.
      call ranmar(xr)
      is=1+int(ns*xr)                ! Random selection of a site.
      iqold=ista(is)                 ! Present state.
      call ranmar(xr)
      iqnew=int(qm1*xr)
      if(iqnew.ge.iqold) iqnew=iqnew+1 ! New state.
1     ista(is)=iqnew
        do id=1,nd
          isf=ipf(id,is)             ! Forward direction.
          iqtest=ista(isf)
          if(iqold.eq.iqtest) then
            iact=iact-1
            call ranmar(xr)
            if(xr.lt.boltz1) then
              ncl=ncl+1
              iacl(ncl)=isf
              ista(isf)=-1
            end if
          else
            if(iqnew.eq.iqtest) iact=iact+1
            if(-1.eq.iqtest) iact=iact-1
          end if
          ... the same for the backward direction ...
        end do
        icl=icl+1
        if(icl.le.ncl) then
          is=iacl(icl)
      go to 1
        end if
      ha(iact)=ha(iact)+one ! Action histogram.
      return
      end
```

iqnew $\neq$ iqold. Afterwards, the program goes through the forward and
backward neighbors of the site is and, if their spins agree with iqold, they
are flipped to iqnew with the probability boltz1 defined in equation (5.78).
The sites whose spins will be flipped are entered into the cluster list iacl(),

as their nearest neighbors have later on to be treated in the same way. At the same time the flip to the new spin could be done, what would also prevent a double counting of a site, because iqnew $\neq$ iqold holds. However, there is a subtle point: We want to keep track of the change in the action variable iact. This works only properly, when a site gets the new spin assigned immediately before its nearest neighbors are treated. We resolve the problem by assigning temporarily the artificial spin -1 to the site. This serves the purpose of preventing double counting, because $-1 \neq$ iqold holds and allows to track the energy (*i.e.*, the action variable iact) properly.

The original SW cluster algorithm is more difficult to implement than the Wolff algorithm. One has to partition the entire lattice into clusters and this requires to label the clusters uniquely, a problem which was first studied and overcome by [Hoshen and Kopelman (1976)]. For details of the solution implemented by our subroutine potts_cl.f of ForProg/Potts_MC see the comments in the routine. A call to

$$\text{potts\_cl(ncl)} \tag{5.82}$$

performs one SW sweep of the lattice and returns with ncl the number of clusters found, where each site of the lattice belongs to one of the clusters. The average cluster size for one Swendsen-Wang update is given by ns/ncl, whereas the size of one Wolff cluster is simply given by the number of flipped spins (the argument nflip of potts_clw.f in table 5.13).

A CPU timing of the cluster updating routines for the $2d$ Ising model is our assignment 1. The result is that a sweep (through the entire lattice) with the SW routine potts_cl.f takes about two times longer than a Metropolis sweep. For Wolff updating this corresponds approximately to calling the subroutine potts_clw.f until the number of flipped spins equals or exceeds the lattice size ns times the number of requested sweeps.

In the following we take the timing into account by comparing two calls to potts1_met.f with one call to potts_cl.f and a (variable) number of calls to potts_clw.f (whenever the number of flipped spins nflip equals or exceeds ns we record a data point and reset nflip by -ns). For disordered and then ordered start configurations, figure 5.8 plots the first 40 events of a $2d$ Ising model times series, see assignment 2. There is some indication, in particular for the ordered start, that the Wolff updating works best. For a conclusive statement a detailed investigation of the autocorrelation times is needed (see below).

In assignment 3 the correctness of the cluster updating code is tested by

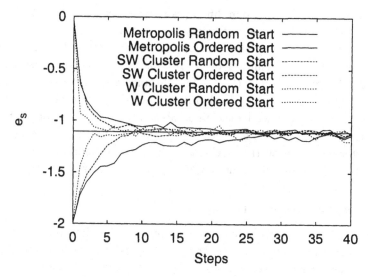

Fig. 5.8 Time series for the Ising model on an 80×80 lattice at $\beta = 0.4$. One step are: two Metropolis sweeps, one Swendsen-Wang (SW) sweep or as many Wolff (W) single cluster updates as it takes to generate $\geq 80 \times 80$ spin flips (in the average), see assignment 2. The straight line is the exact result obtained with ferdinand.f (3.15).

performing some longer runs for the Ising model on an 20×20 lattice. The parameter values are chosen to correspond to the runs for the results (3.101) of chapter 3. The statistics is adjusted so that each run takes about the same CPU time. We get

$$-1.11752\,(65)\quad \text{SW}\quad \text{and}\quad -1.11820\,(47)\quad \text{Wolff.}\qquad (5.83)$$

Comparison with the exact result of equation (3.101) by means of the Gaussian difference test (2.33) gives the satisfactory values $Q = 0.63$ (SW) and $Q = 0.44$ (Wolff). Are the differences in the error bars significant? To answer this question, we perform the F-Test (2.107) and obtain a significant better performance of the Wolff algorithm ($Q = 0.011$). In addition to the energy mean values, the average cluster size is investigated in assignment 3 and found to be larger for Wolff clusters (it is part of assignment 3 to answer "why?").

In assignment 4 similar investigations are done for the 10-state Potts model on an 20×20 lattice. The parameter values are chosen to correspond to the runs for the results (3.108). We get

$$-1.28724\,(39) \ \ \text{SW} \quad \text{and} \quad -1.28706\,(27) \ \ \text{Wolff} \qquad (5.84)$$

and $Q = 0.70$ for the Gaussian difference test between these numbers. The Wolff algorithm is again more efficient than the SW algorithm.

Whereas the improvement over the standard Metropolis algorithm is really large for the $2d$ Ising model at $\beta = 0.4$, this is not the case for the 10-state Potts model results (5.84). The obtained accuracies for the cluster, the Metropolis and heat bath algorithms are then quite similar. Their differences are outweighed by the differences in the CPU time consumption, which is a platform and code dependent quantity. These results reflect that the cluster algorithm works excellent in the neighborhood of second order phase transitions, but yields no major improvement away from those.

5.3.1 *Autocorrelation times*

We illustrate the improvements of the cluster over the Metropolis algorithm for the $d = 2$ Ising model at the critical temperature. Due to the second order phase transition the gains are really large in this case. The Metropolis performance was reported in figure 4.4 and table 4.1 of chapter 4. The corresponding cluster results are now compiled in figure 5.9 and table 5.14. Most notably, for the $L = 160$ lattice the integrated autocorrelation time of the Wolff algorithm of table 5.14 is almost 800 times smaller than the one listed in table 4.1 for the Metropolis algorithm. Compared to the Metropolis algorithm, the performance of the Swendsen-Wang cluster updating is also very good, but it is not competitive with the Wolff updating, as we already noticed for the results (5.83) and (5.84). The reason is that the Wolff algorithm flips all spins touched and, in the average, larger clusters than the SW algorithm. The latter point comes from the fact that the likelihood to hit a cluster by chance is proportional to its size.

We conclude with a few technical remarks. As the Wolff algorithm does not allow anymore for the notion of a sweep, one has to find a substitute. The solution used for assignment 5 in our program

$$\texttt{p\_tsclw\_p.f} \ \ \text{of} \ \ \text{ForProg/MC\_Potts} \qquad (5.85)$$

is to record the energy whenever the number of flips exceeds the number of lattice sites **ns** and, subsequently, to reset the **nflip** variable by $-\mathbf{ns}$. Con-

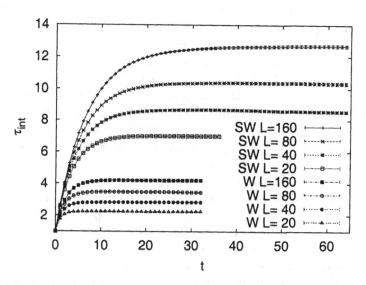

Fig. 5.9 Estimates of integrated autocorrelation times from cluster simulations of the $d = 2$ Ising model at the critical temperature $\beta_c = 0.44068679351$, see assignment 5.

Table 5.14 Estimates of the integrated autocorrelation time for the $d = 2$ Ising model at $\beta_c = 0.44068679351$ on $L \times L$ lattices from simulations with the Swendsen-Wang (SW) and Wolff (W) cluster algorithms, see assignment 5.

cluster	β	L	20	40	80	160
SW	β_c	τ_{int}	6.997 (39)	8.668 (50)	10.385 (75)	12.685 (87)
W	β_c	τ_{int}	2.245 (08)	2.820 (11)	3.472 (16)	4.199 (19)

sequently, the average distance between recorded values is **ns**, but it fluctuates. Remarkably, if the thus recorded data are used in equation (4.14) to estimate the integrated autocorrelation time, the relation (4.11) between the error bar and the naive error bar survives. This can be seen by starting all-over with equation (4.7) and is left as an exercise to the reader.

5.3.2   *Assignments for section 5.3*

(1) Use your favorite compiler with a suitable optimization turned on and simulate the Ising model on an 20×20 lattice. Measure the CPU time needed for 50 000 sweeps with (a) the SW cluster updating subroutine potts_cl.f and (b) the Metropolis subroutine potts1_met.f. The main programs potts_histcl.f and potts1_hist.f of the folder ForProg/MC_Potts are suitable for this task. One SW sweep should take about the same CPU time as two Metropolis sweeps. Next, perform updatings with the Wolff subroutine potts_clw.f until $50\,000 \times$ ns spins are flipped. A modification of the program potts_histw.f of ForProg/MC_Potts is suitable for this task. On our PC the CPU time needed for the Wolff updates was somewhat smaller than the CPU for 50 000 SW sweeps. Approximately seven calls to the subroutine potts_clw.f achieved ns flips.

(2) Consider the Ising model on an 80×80 lattice at $\beta = 0.4$ and compare for disordered and ordered starts the first 40 events of the following time series: 1. One event defined by two Metropolis sweeps. 2. One event defined by one SW sweep. 3. One event defined by a number of Wolff flips which becomes ns in the average (the factor seven found in assignment 1 holds only on an 20×20 lattice at $\beta = 0.4$). Using Marsaglia random numbers with their default seed for each disordered start and continuing after the first 40 events with the ordered start, the results depicted in figure 5.8 are obtained.

(3) Use the Marsaglia random number generator with its default seed and perform simulations of the Ising model on an 20×20 lattice at $\beta = 0.4$. Perform 5 000 sweeps for reaching equilibrium and $64 \times 2\,500$ sweeps with measurements for SW updating. With the Wolff algorithm perform 35 000 (single) cluster updates for reaching equilibrium and $64 \times 17\,500$ cluster updates with measurements. You should obtain the average energies and their error bars as listed in equation (5.83). The Gaussian difference test (2.33) shows that they are consistent with the exact value of equation (3.101). Perform the F-test (2.107) to find out whether the difference in the error bars is statistically significant. Finally calculate the average cluster sizes. You should find 4.8121 (40) for SW and 57.34 (11) for Wolff clusters. Explain the difference.

(4) Use Marsaglia random numbers as in the previous assignment and perform simulations of the 10-state model on an 20×20 lattice at $\beta = 0.62$. Perform 20 000 sweeps for reaching equilibrium and $64 \times 10\,000$ sweeps

with measurements for SW updating. With the Wolff algorithm perform 2 400 000 (single) cluster updates for reaching equilibrium and 64 × 1 200 000 cluster updates with measurements. The numbers are chosen so that the simulation takes about the same CPU time with either algorithm. You should obtain the average energies and their error bars as listed in (5.84). Use the Gaussian difference test to show that the two mean values are consistent with one another. Perform the F-test (2.107) to find out whether the difference in the error bars is statistically significant. Finally calculate the average cluster sizes. You should find 1.7769 (04) for SW and 3.6865 (17) for Wolff clusters.

(5) Use the SW and the Wolff cluster algorithms to perform simulations of the $d = 2$ Ising model at its critical temperature and reproduce the results of figure 5.9 and table 5.14.

5.4 Large Scale Simulations

Some aspects of large scale simulations were already discussed at the beginning of chapter 4. In particular, the rule of thumb was given not to spend more than 50% of the CPU time on measurements. To give a definition of large scale MC simulations: These are simulations which last longer than the average up-time of your computer system. Therefore, one needs to create regularly backups of the simulation status, which allow its continuation in case of a computer crash.

We illustrate this for a multicanonical simulation of the Potts model, which becomes rather CPU time consuming on larger lattices. The Fortran program for long multicanonical simulations is prod_mu.f of the folder ForProg/MC_Potts. It has attached the subroutine write_bak, which is reproduced in table 5.15. Calls to this subroutine create backups of the simulation on a file with its name stored in the character variable cbak. The filename has to be defined by the calling program. After a computer crash or an interruption by hand, the re-started main program calls a subroutine read_bak (also attached to prod_mu.f), which restores the status of the simulation at the last call to write_bak. Note that in the write_bak routine the file cbak is first opened by a call to rmasave, which saves the status of the Marsaglia random number generator. Subsequently, all relevant information from the ongoing run is attached to this file. In addition readable information is written to the file progress.d, which allows the user to monitor the progress of the simulation at run-time.

Table 5.15 The subroutine write_bak.f, which creates a backup of the MC simulation status on a file with name cbak.

```
      SUBROUTINE write_bak(cbak,irec,mu_swp,na_swp,ntun,ltun0,
     &                     iequi, irpt,imeas,ntu_m)
      include '../../ForLib/implicit.sta'
      character*(*) cbak
      include '../../ForLib/constants.par'
      include 'lat.par'
      include 'mc.par'
      include 'potts.par'
      include 'muca.par'
      include '../../ForLib/lat.com'
      include '../../ForLib/potts.com'
      include '../../ForLib/potts_muca.com'
      call rmasave(iud1,cbak)
      open(iud1,file=cbak,form="unformatted",status="old",
     &          access="append")
      write(iud1) wrat, ha,acpt,  hmuca,hup,hdn, gw,
     & ista,nlink,iact,iamin,iamax, ia_min,ia_max, ndel_muca,iastep,
     & irec,mu_swp,na_swp,ntun,ltun0, iequi,irpt,imeas,ntu_m
      close(iud1)
C
      open(iud1,file="progress.d",form="formatted",status="unknown")
      write(iud1,'(/," nb_swp =",I10,6X,"Last backup with:")') nb_swp
      write(iud1,'(" irec1, mu_swp,   ntun =",3I12)')irec,mu_swp,ntun
      write(iud1,'(" iequ1,irp1,imea1,ntu_m =",4I12,/)')
     &             (iequi+1),irpt,(imeas+1),ntu_m
      close(iud1)
      return
      end
```

For the $2d$ 10-state Potts model we illustrate the use of the program prod_mu.f in two assignments. In both cases the mc.par parameter file of table 5.16 is used, which allows for a multicanonical simulation of an 100×100 lattice. This is about the largest lattice size for which the recursion of subsection 5.1.1 still works. Multicanonical Potts model simulations on even larger lattices are possible, using FSS scaling estimates of the weights or, possibly, the [Wang and Landau (2001)] recursion. In addition to the usual parameters (note that H0 and nhit are not used in this simulation), the parameter file of table 5.16 introduces

nb_swp .

Table 5.16 The parameter file mc.par for the multicanonical simulations of this section.

```
c
c mc.par: Definition of parameters for MC calculations
c         ================================================
c
c Output units and random generator seeds:
      parameter(iuo=6,iud1=11,iud2=12,iseed1=1,iseed2=0)
c
c Parameters for equilibrium and data production sweeps:
c beta:   k/T for for canonical MC, or as MUCA start value, k=1.
c nequi:  Number of equilibrium sweeps.
c nmeas:  Number of measurement sweeps per repetition.
c nrpt:   >=1: Number of measurement repetitions in one job.
c         =0: Only MUCA (if chosen) and equilibrium run(s) are done.
      parameter(beta=zero,HO=ZERO, nhit=0,nequi=1 000 000)
      parameter(nrpt=64, nmeas=4 000 000, nb_swp=500 000)
```

This is the number of sweeps after which a backup is performed. Look up prod_mu.f for details of the implementation of the backup and restart procedures.

In assignment 1 a test run is performed on an 20×20 lattice, which is several times killed by hand and each time re-started. This results in an output file for which the section with the first interruption is shown in table 5.17. The interruption is done after six tunneling events are achieved. The re-start begins at a backup, which was done after four tunneling events. The program repeats tunneling events five and six at precisely the same recursion and sweep numbers where they happened in the initial run and continues to perform all ten tunneling events of the recursion part of this simulation.

During very long runs the backup file itself should occasionally be copied (by hand) on a second backup file (preferably located on another computer). This protects against system failures at precisely the times when the program writes the backup file as well as against catastrophic disk crashes.

5.4.1 *Assignments for section 5.4*

(1) Use the program prod_mu.f for a multicanonical simulation of the 10-state Potts model on an 20×20 lattice with the parameter file depicted in table 5.16 . Monitor the progress.d file at run-time and kill the

run at least once during the recursion, equilibration and production part. Continue each time and create an output file similar to the one of table 5.17.

(2) Use the program prod_mu.f for a multicanonical simulation of the 10-state Potts model on an 100×100 lattice with the parameter file depicted in table 5.16. If necessary, you may occasionally re-start the run from the last available backup.

Table 5.17 Initial part of the output file from a multicanonical simulation of the 10-state Potts model on an 20 × 20 lattice, showing the interruption and re-start during the recursion period.

```
RANMAR INITIALIZED.
Initial random number: xr =    0.116391
lat.par: nd,nq,nla = 2 10     20   20
         ms,mlink  =          400         800
mc.par:
iuo,iud1,iseed1,iseed2,     beta
  6  11      1      0   0.000000
Initial action:    iact  =           82

MUCA Recursion:
namin,namax,maxtun:              80        800         10
nrec_max,nmucasw,iastep:  10000000         20          1
ntun,irec,mu_swp,acpt:     1    5014       343129    0.292
ntun,irec,mu_swp,acpt:     2    5637       388001    0.291
ntun,irec,mu_swp,acpt:     3    6751       448618    0.301
ntun,irec,mu_swp,acpt:     4    7464       496207    0.301
ntun,irec,mu_swp,acpt:     5    8181       542717    0.301
ntun,irec,mu_swp,acpt:     6    8850       588820    0.301

Interruption ...

MARSAGLIA CONTINUATION.

MUCA Recursion:
namin,namax,maxtun:              80        800         10
nrec_max,nmucasw,iastep:  10000000         20          1
ntun,irec,mu_swp,acpt:     5    8181       542717    0.301
ntun,irec,mu_swp,acpt:     6    8850       588820    0.301
ntun,irec,mu_swp,acpt:     7    9733       641313    0.304
ntun,irec,mu_swp,acpt:     8   10180       680420    0.299
ntun,irec,mu_swp,acpt:     9   12545       824964    0.304
ntun,irec,mu_swp,acpt:    10   13916       891540    0.312
Muca recursions done.

Data file mu2d10q020.dat created.

    1000000 sweeps for reaching equilibrium.

... continued
```

Chapter 6

Parallel Computing

After briefly discussing the often neglected, but in praxis frequently encountered, issue of trivially parallel computing[1], we turn to parallel computing with information exchange. To work in a definite framework, we introduce the Fortran version of the Message Passing Interface (MPI) to the extent needed in our subsequent illustration, which is the parallel tempering MC algorithm [Swendsen and Wang (1986); Geyer (1991); Hukusima and Nemoto (1996)]. Parallel tempering has a low, but non-trivial, communication overhead between the processes. Information about the temperatures, energies and nearest neighbor locations of n systems is communicated, logical decisions have to be made and the order of the information exchange matters. This algorithm gives the newcomer not only some kind of ideal start into the world of parallel processing, but is also of major practical importance. In some applications it competes with the multicanonical algorithm of section 5.1, in others both algorithms supplement one another. For instance, combinations of them have been explored in simulations of peptides (small proteins), see the review by [Mitsutake et al. (2001)].

6.1 Trivially Parallel Computing

There are many tasks which simply can be done in parallel, for instance canonical MC simulations at different temperatures or on distinct lattices. For such applications the gain due to having many (identical) processors available scales linearly with the number of processors. In practice trivially

---

[1]An example of a trivially parallel large scale endeavor is the *Folding@Home* distributed computing project [Shirts and Pande (2000)], which achieved sustained speeds of about thirty teraflops in a molecular dynamics simulation [Snow et al. (2002)].

parallel applications are frequently encountered. One example are simulations of replica in spin glass investigations, *e.g.* [Berg *et al.* (2000)]. It would be desirable to have sufficiently many computer clusters available, because it is far more cost efficient to run trivially parallel applications on them than on dedicated parallel machines (supercomputers), which feature expensive communication networks. Until recently it appeared to be easier to get CPU time funding on supercomputers than to find large-scale trivially parallel clusters, unless one was willing to rely on rather unconventional methods as the *Folding@Home* project [Shirts and Pande (2000); Snow *et al.* (2002)] did.

Next in line are MC simulations of single systems which last (very) long in real time. Parallelization is still trivial, but there is some overhead due to multiple equilibration runs. Typically the CPU time consumption of such a simulation is divided into

$$t_{tot} = t_{eq} + t_{meas} \tag{6.1}$$

where t_{eq} should be large compared with the autocorrelation time τ. We assume $t_{eq} < t_{meas}$. For parallelization of such a computation on n independent nodes the single processor CPU time scales as

$$t_{tot}^1 = t_{eq} + t_{meas}/n . \tag{6.2}$$

How many processors should one choose (assuming a sufficiently large supply)? A good rule of thumb appears to be

$$n = \text{nint} \left[\frac{t_{meas}}{t_{eq}} \right] . \tag{6.3}$$

This means the total processor time used becomes $t_{tot}^n = n\, t_{eq} + t_{meas} \approx 2\, t_{meas}$, which limits the CPU time loss due to multiple equilibrations to less than two

$$R_{loss} = \frac{t_{tot}^n}{t_{tot}} \approx \frac{2\, t_{meas}}{t_{eq} + t_{meas}} < 2 , \tag{6.4}$$

whereas the gain is a reduction of the run time by the factor

$$R_{gain}^{-1} = \frac{t_{tot}^1}{t_{tot}} \approx \frac{2\, t_{eq}}{t_{eq} + n\, t_{eq}} = \frac{2}{1 + n} . \tag{6.5}$$

The worst case of the choice (6.3) is $n = 2$, for which the improvement factor in CPU time is 2/3. Obviously this parallelization makes most sense for $t_{meas} \gg t_{eq}$.

6.2   Message Passing Interface (MPI)

Real parallel computing needs at runtime some exchange of information between the nodes. MPI [MPI Forum (1995); Snir *et al.* (1996)] is a message passing standard which gained some popularity due to its efficient communication, its portability and ease of use, which appear to surpass other solutions. MPI allows convenient C and Fortran bindings. In the following the processor on which the compilation is done and the processes are submitted, is called **local**, whereas the other involved processors are called **non-local**.

In a typical Unix installation a MPI Fortran 77 program will compile with the command

$$...\texttt{pathname/mpif77 fln.f} \qquad (6.6)$$

and generates a executable file, a.out, which runs with

$$...\texttt{pathname/mpirun -nolocal -np n a.out} \qquad (6.7)$$

or .... /a.out in some installations. Here the option -np defines the number n of processes, $n = 1$ if the -np option is omitted. In another implementation the command (6.7) may read

$$...\texttt{pathname/mpirun -machinefile machines.txt -nolocal -np n a.out} \qquad (6.8)$$

where the file machines.txt contain the names (addresses) of the participating non-local processors. The number of processors (CPUs) can be smaller than the number of processes. MPI will then run several, or even all, processes on one CPU. **Processes and processors are not the same for MPI!** To avoid large CPU time losses, care has to be taken that the load is evenly balanced between the processors. The option -nolocal means that the local processor is **not** included in the list of MPI nodes, what is desirable when the local processor is some kind of front end machine. If the -nolocal option is omitted, the local processor will be included.

We limit our presentation of MPI to basic features, which are needed to implement the parallel tempering algorithm of the next subsection. For detailed introductions see [Gropp *et al.* (1999)] or [Pacheco (1996)]. Manuals as well as tutorials are also available on the Web, for instance at the MPI Forum homepage www.mpi-forum.org or at the Argonne National Lab MPI Home page www-unix.mcs.anl.gov/mpi. Type MPI into the google.com search machine to find additional websites.

In our typical applications the same Fortran program runs with differently set variables for all n processes, *i.e.*, they use the Single Program Multiple Data (SPMD) paradigm[2]. The effect of different processes running different instructions is obtained by taking different branches within a single program on the basis of the process rank. **Our first, simple MPI program is listed in table 6.1** and we use it to explain a number of basic MPI concepts. In essence, this program initializes the Potts model heat bath algorithm (3.95) for n processes, performs nequi sweeps for each process and prints out some results. Let us discuss this short program line by line.

As in our previous Fortran programs, we include the implicit declaration (1.21) for our real and logical variables (1 is declared logical) and use the Fortran default declaration for the remaining integers. This line is now followed by the MPI preprocessor directive, which is brought into the routines through the

$$\text{include 'mpi.h'} \tag{6.9}$$

statement. In a properly implemented MPI environment the compiler will find the file mpi.h which contains definitions, macros and function prototypes necessary for compiling the MPI program. The order of the statements (1.21) and (6.9) matters, because mpi.h overwrites some of our implicit declarations with explicit ones. All definitions which MPI makes have names of the form

$$\text{MPI\_ANY\_NAME} . \tag{6.10}$$

Therefore, the user is strongly advised **not to introduce any own variables with names of the form** MPI_... in his or her Fortran routines. In our code we use instead ANY_NAME_MPI..., when we like to emphasize a relationship to MPI.

The lines following include 'mpi.h' are our usual include statements for parameter files and common blocks. For the MPI run documented here our lat.par and potts.par parameter values are the same as those in our previous illustration (3.108) of the heat bath code, *i.e.*, the 10-state Potts model on a 20×20 lattice. Two modifications of our previous parameter and common block structure are notable:

(1) The mc.par parameter file (table 3.5 of chapter 3) has been replaced

---

[2]The MPI interface is also suitable for use by Multiple Instruction Multiple Data (MIMD) or Multiple Program Multiple Data (MPMD) applications.

Table 6.1 The MPI program p_hbtest_mpi.f of ForProg/MPI_Potts is listed. It initializes *n* processes for Potts model heat bath simulations and performs **nequi** equilibration sweeps for each process.

```
      program p_hbtest_mpi
C Copyright, Bernd Berg, Jan 9 2002.
C MPI test for the Potts model Heat Bath (HB) MC algorithm.
      include '../../ForLib/implicit.sta'
      character cd*1,cq*2,cl*3,cmy_id*2,cnp*2
      include 'mpif.h'
      include '../../ForLib/constants.par'
      include 'lat.par'
      include 'mc_pt_mpi.par'
      include 'potts.par'
      include '../../ForLib/lat.com'
      include '../../ForLib/potts_hb.com'
      include '../../ForLib/p_pt_mpi.com'
      include 'lat.dat'
C
      CALL MPI_INIT(IERR)
      CALL MPI_COMM_RANK(MPI_COMM_WORLD,MY_ID,IERR)
      CALL MPI_COMM_SIZE(MPI_COMM_WORLD,N_PROC,IERR)
      IF(MY_ID.EQ.0) WRITE(IUO,'(/," MPI: N_PROC =",I5)') N_PROC
      MY_B=MY_ID
      NEIGH(1)=MY_ID-1
      NEIGH(2)=MY_ID+1
      NSEND(1)=MY_B
      NSEND(2)=NEIGH(1)
      NSEND(3)=NEIGH(2)
C
      call p_inithb_mpi(cd,cq,cl,cmy_id,cnp,.false.) ! Initialize Potts HB MC.
      NSEND(4)=iact
C
      do iequi=1,nequi            ! Sweeps for reaching equilibrium.
        call potts_mchb
      end do
      NSEND(5)=iact
C
C Print some results in a specified order:
      IF(MY_ID.EQ.0) WRITE(IUO,'(/,30X,"     iact            ")')
      IF(MY_ID.EQ.0) WRITE(IUO,
     & '(/,"   MY_ID MY_B      NEIGH   (start) (equilibrium) ")')
      itag=0
      CALL WRITE_MPI_I(NSEND,NRECV,5,IUO,itag)
      CALL MPI_FINALIZE(IERR)
C
      STOP
      END
```

by mc_pt_mpi.par, which is listed in table 6.2. This is distinct from our design of the multicanonical algorithm, where we simply added new parameters in an additional file muca.par. Now the needs of parallel processing make a redesign of mc.par unavoidable. Most notably, the temperature parameter beta is replaced by beta_min and beta_max. A number of parameters have been changed or added: iseed2 $\rightarrow$ isd2_0 (see equation (6.16) below), nequi $\rightarrow$ nequ1, nmeas $\rightarrow$ nmea1, and MPN1, NSR, n_mpi have been added. These choices are suitable for the parallel tempering algorithm discussed in the next subsection (therefore the _pt extension of the filename).

(2) A new common block p_pt_mpi.com, as always located in ForLib, has been introduced. It is listed in table 6.3 and contains variables and arrays needed for our MPI implementation of parallel tempering. This is quite similar to the introduction of the common block potts_muca.com of table 5.3, which we use for the multicanonical method. None of the previously used common blocks needs to be redesigned.

The next three lines of our program are calls to predefined MPI routines. Before calling MPI communication routines, MPI needs to be initialized and it is mandatory to call

$$\text{MPI\_INIT(IERROR)} \qquad (6.11)$$

first. MPI_INIT must be called only once (subsequent calls are erroneous). The integer variable IERROR can be ignored as long as everything proceeds smoothly. It returns MPI error codes, which can be looked up in the MPI manual. Next, each process calls

$$\text{MPI\_COMM\_RANK(COMM, RANK, IERROR)} \qquad (6.12)$$

to find out its rank. The rank is an integer, called MY_ID (for my identity) in our program. The first argument of MPI_COMM_RANK is a **communicator**. A communicator is a collection of processes that can send messages to each other. For basic programs the only communicator needed is the one used here

$$\text{MPI\_COMM\_WORLD .} \qquad (6.13)$$

It is predefined in MPI and consists of all processes running when the program execution begins. The second argument is the integer rank of the process, which takes the values $0, \ldots, n - 1$ if there are n processes. In the program the integer variable MY_ID (which denotes the rank) is used to

Table 6.2 The parameter file `mc_pt_mpi.par` as used in our illustrations of table 6.5 and 6.8.

```
c
c mc_pt_mpi.par: Parameters for parallel tempering (PT) MC using MPI.
c
c Output units, random generator seeds and MPI send-receive:
      parameter(iuo=6,iud=10,iud1=11,iud2=12,iseed1=1,isd2_0=0)
c MPM1: Maximum number of MPI processes - 1.
C NSR:  Dimension for MPI integer arrays NSEND and NRECV.
      parameter(MPM1=31,NSR=10) ! MPM1=31: Up to 32 MPI processes.
c
c Parameters for equilibrium and data production sweeps:
c beta_min and beta_max: Defines the extreme temperatures.
c nequi:   Number of equilibrium sweeps before PT recursion.
c NPT_REC: Number of PT recursion steps.
c nequ2:   Number of equilibrium sweeps of the PT recursion loop.
c nequ1:   Number of equilibrium sweeps inside the PT recursion loop.
c nequ3:   Number of equilibrium sweeps after the PT recursion.
c
c nrpt:   >=1: Number of measurement repetitions in one job.
c nmea2:  Measurement sweeps of the parallel tempering loop.
c nmea1:  Measurement sweeps inside the parallel tempering loop.
      parameter(beta_min=0.65d00,beta_max=0.75d00,H0=ZERO,nhit=0)
C
      parameter(nequi=1000, NPT_REC=05, nequ2=08 000, nequ1=08)
      parameter(nequ3=200, nrpt=32, nmea2=060, nmea1=0200)
```

Table 6.3 The common block `p_pt_mpi.com` of ForLib.

```
COMMON /PT/ BA(0:MPM1),BASUM(0:MPM1),hab(0:mlink,0:MPM1),
& NACPT_PT(0:MPM1),IP_B(0:MPM1),IPI_B(0:MPM1),NEIGH(2),NSEND(NSR),
& NRECV(NSR),MY_BA(nmea2),MY_B,IACPT_PT
```

branch the program by assigning different random numbers and temperatures to the processes.

Some constructs in our program depend on the total number of processes executing the program. MPI provides the routine

$$\text{MPI\_COMM\_SIZE(COMM, ISIZE, IERROR)} \tag{6.14}$$

which returns the number of processes of the communicator COMM in its integer argument ISIZE. In the program of table 6.1 we use the integer variable N_PROC (number of processes) to denote the size. We print N_PROC with the command of the next line. Note that we execute this line only for the process with MY_ID = 0. Otherwise the write statement would be executed by each process, resulting in N_PROC printouts.

In subsequent lines each process performs some initializations for its common block p_pt_mpi.com of table 6.3. The index MY_BETA defines later on the temperature. The array NEIGH(2) keeps track of neighbor processes, which are initially ordered by the process rank (for MY_ID = 0 the value NEIG(1) = −1 means that there is no other process to the left and for MY_ID = N_PROC − 1 the value NEIG(2) = N_PROC means the same to the right). These initializations are only listed for illustrative purposes in the main program of table 6.1 and repeated in the subroutine

$$p\_inithb\_mpi.f\ , \qquad (6.15)$$

where they belong according to the organizational structure of our programs. To print the initial values of MY_BETA and NEIG(2) later in the order of their processes, they are copied into the first three locations of the array NESND(NSR) of p_pt_mpi.com (we will elaborate on the printing problem below).

Next the subroutine p_inithb_mpi.f is called. It is the MPI variant of heat bath initialization routine (3.96) of chapter 3. Important alterations are the following features.

(1) For each process the seed of the Marsaglia random number generator is differently set. Namely, by the two numbers

$$iseed1\ and\ iseed2 = isd2\_0 + MY\_ID\ , \qquad (6.16)$$

where iseed2 and isd2_0 are defined in mc_pt_mpi.par of table 6.2. Assignment 6 of chapter 2.6.4 gives a justification for this initialization. Other implementations of the pseudo random number generators could rely on the SPRNG library [Mascagni and Srinivasan (2000)].

(2) For N_PROC ≥ 2, each process gets its own temperature assigned

$$beta = B\_ARRAY(MY\_ID) = beta\_min + MY\_ID \left(\frac{beta\_max - beta\_min}{N\_PROC - 1} \right)$$
$$(6.17)$$

where **beta_min** and **beta_max** are defined in the parameter file of table 6.2. For reasons explained in the next subsection, the array **BA** of **p_pt_mpi.com** contains, identically for all n processes, all the β values.

For each process a disordered ($\beta = 0$) start is performed, as in our single processor MC simulations of figure 3.5. On exit from **p_inithb_mpi.f** the action variable **iact** of this random start is stored in the array element **NSEND(4)**.

Afterwards **nequi = 1000** equilibration sweeps are performed and the resulting **iact** value is stored in **NSEND(5)**. Now we would like to print the first five elements of the array **NSEND(NSR)**, where **NSR** $= 10 > 5$ is set in **mc_pt_mpi.par** of table 6.2, so that the **NSEND** array allows to print additional numbers. **Printing output** from the processes in a desired order is under MPI a problem which requires some thought. Using the Fortran commands **print** or **write** on each node gives the output in an uncontrolled random order, first come, first serve, for the processes. A solution is to let a single process handle all or most of the input-output issues. This is acceptable as long as the input-output processing time is small compared with the MC simulation time per process. Our implementation is given by the subroutine **write_mpi_i.f** which is listed in table 6.4. It is designed for printing up to ten integer variables in the order of the processes. Let us discuss the details, because new important MPI commands are introduced along the way. Note, that all the effort of our routine **write_mpi_i.f** consists in getting the parallel results into a sequential order, a highly inefficient use of parallel processors.

In the subroutine **write_mpi_i.f** the statement following **include** `'mpif.h'` uses one of the predefined MPI parameters, **MPI_STATUS_SIZE**, to dimension the later needed array **ISTATUS**. We already know the MPI routines (6.12) and (6.14) which are called next. Basically, the routine **write_mpi_i.f** relies then on point to point communication between the processes. Two important MPI routines for this purpose are

$$\text{MPI\_SEND(BUF, ICOUNT, DATATYPE, IDEST, ITAG, COMM, IERROR)} \qquad (6.18)$$

and

$$\text{MPI\_RECV(BUF, ICOUNT, DATATYPE, ISOURCE, ITAG, COMM, ISTATUS, IERROR)} \,.$$
$$(6.19)$$

Differences in the arguments are **DEST** in **MPI_SEND** versus **SOURCE** in **MPI_RECV** and the additional **STATUS** array in **MPI_RECV**. All arguments are

Table 6.4 The MPI subroutine **write_mpi_i.f** of **ForLib** is listed. It writes integer output from *n* processes in a desired order.

```
      SUBROUTINE WRITE_MPI_I(NSEND,NRECV,NSR,IUO,itag)
C Copyright, Bernd Berg, Oct 31 2001.
C Purpose: Prints integer results (up to 10 integers) in a desired order.
      INCLUDE 'implicit.sta'
      INCLUDE 'mpif.h'
      DIMENSION ISTATUS(MPI_STATUS_SIZE)
      DIMENSION NSEND(NSR),NRECV(NSR)
      CALL MPI_COMM_RANK(MPI_COMM_WORLD,MY_ID,IERR)
      CALL MPI_COMM_SIZE(MPI_COMM_WORLD,N_PROC,IERR)
      IF(MY_ID.EQ.0) WRITE(IUO,*) "  "
      DO I=0,(N_PROC-1)
        IF(MY_ID.EQ.I) CALL MPI_SEND(NSEND,NSR,MPI_INTEGER,0,itag,
     &                               MPI_COMM_WORLD,IERR)
        IF(MY_ID.EQ.0) THEN
          CALL MPI_RECV(NRECV,NSR,MPI_INTEGER,I,itag,
     &                  MPI_COMM_WORLD,ISTATUS,IERR)
          WRITE(IUO,'(10I7)') I,NRECV
        END IF
      END DO
      RETURN
      END
```

explained in the following.

(1) BUF: The initial address of the send or receive buffer. In our write routine (table 6.4) this is in the MPI_SEND call the first element of the integer array NSEND and in the MPI_RECV call the first element of the integer array NRECV. Data types other than integer are possible for the buffer.

(2) ICOUNT: An integer variable which gives the number of entries to be send, this is NSR (number send receive) in our calls.

(3) DATATYPE: The datatype of each entry. It is MPI_INTEGER in our example, which agrees with the Fortran datatype integer. Other MPI datatypes which agree with the corresponding Fortran datatypes are MPI_REAL, MPI_DOUBLE_PRECISION, MPI_COMPLEX, MPI_LOGICAL and MPI_CHARACTER. Additional MPI data types (MPI_BYTE and MPI_PACKED) exist.

(4) IDEST: The rank of the destination process. This is 0 in our MPI_SEND call, because each process sends its NSEND array to the process MY_ID =

Table 6.5 The printout of the program p_hbtest_mpi.f when it is run with the parameter values of table 6.2 and the mpirun option -np 4, see assignment 1.

MPI: N_PROC =   4

| | | | | iact | |
MY_ID	MY_B	NEIGH		(start)	(equilibrium)
0	0	-1	1	82	230
1	1	0	2	87	313
2	2	1	3	97	547
3	3	2	4	91	698

0, which handles all the printing.

(5) ISOURCE: The rank of the source. This is $I = 0, \ldots, N\_PROC - 1$ in our routine. Note that the sender and receiver can be identical, what happens for $I = 0$. The receiving processor avoids overwriting its NSEND array by using the NRECV array, which is subsequently printed. The do loop on the receiving process forces the printing to occur in the order of the processes.

(6) ITAG: The message tag, an integer in the range $0, \ldots, 32767$ (many MPI implementations allow for even larger numbers than 32767). The purpose of the tag is that the receiver can figure out the message order when one process sends two or more messages (the receive can be delayed due to other processing). In our example we pass the tag as the write_mpi_i subroutine argument itag.

(7) COMM: The communicator. As before, we use MPI_COMM_WORLD.

(8) ISTATUS: The return status of MPI_RECV. We use the properly dimensioned array ISTATUS and ignore it in the following. Consult the MPI manual, if peculiar completions are encountered.

(9) IERROR: The error code, an integer with which we deal in the same way as with the return status.

The fields **source, destination, tag** and **communicator** are collectively called the **message envelope**. The receiver may specify the wildcard value MPI_ANY_SOURCE for SOURCE, and/or the wildcard value MPI_ANY_TAG for TAG, to indicate that any source and/or tag value is acceptable. In our code we do not do so, because of the danger to loose control about what it transferred by whom to who.

Our thus obtained printout for a run (6.7) with -np 4 processors is listed in table 6.5. The printout for the MY_ID, MY_BETA and the neighbor array values NEIG are the numbers of the initialization in table 6.1. The action variable iact values stored in NSEND(4) are those after the disordered starts with differently seeded (6.16) random numbers. As expected, they are reasonably close to the $\beta = 0$ mean value $80 = 2 \times 20 \times 20/10$. After the disordered start $1\,000$ equilibration sweeps are done at the β values given by (6.17) with beta_min $= 0.6$ and beta_ max$=0.8$. The resulting iact values are stored in NSEND(5). From their printout we notice that they have become ordered, the larger iact values corresponding to the larger MY_BETA values. The MY_BETA $= 0, 1$ iact values stay on the disordered side of the first order transition signal, the MY_BETA $= 2$ result is right on the cross-over and the MY_BETA $= 3$ iact value has crossed to the ordered side (compare figure 5.3, 547/800=0.684 etc.).

Let us return to the main program of table 6.1. After calling write_mpi_i.f it concludes with a call to

$$\text{MPI\_FINALIZE(IERROR)} \qquad (6.20)$$

which tells its process that no further MPI instructions follow. This call is mandatory for an error free termination of MPI.

6.3 Parallel Tempering

The **method of multiple Markov chains** was introduced by [Geyer (1991)] and, independently, by [Hukusima and Nemoto (1996)] under the name **replica exchange method**. The latter work was influenced by the *simulated tempering* [3] method of [Marinari and Parisi (1992)]. This coined the name **parallel tempering** for an exchange of temperatures in multiple Markov chains. A precursor of such ideas is the paper by [Swendsen and Wang (1986)], where actually a more general replica exchange method is formulated. For parallel tempering the amount of exchanged information is small. Therefore, it is particularly well suited for parallel processing on clusters of workstations, which lack the expensive fast communication hardware of the dedicated parallel computers.

Parallel tempering performs n canonical MC simulations at different β-values with Boltzmann weight factors

$$w_{B,i}(E^{(k)}) = e^{-\beta_i E^{(k)}} = e^{-H} , \; i = 0, \ldots, n-1 \qquad (6.21)$$

[3]In turn, simulated tempering is a special case of the *method of expanded ensembles* [Lyubartsev *et al.* (1992)].

where $\beta_0 < \beta_1 < \ldots < \beta_{n-2} < \beta_{n-1}$. It allows the exchange of neighboring β-values

$$\beta_{i-1} \longleftrightarrow \beta_i \quad \text{for} \quad i = 1, \ldots, n-1 . \tag{6.22}$$

These transitions lead to the change

$$-\Delta H = \left(-\beta_{i-1} E_i^{(k)} - \beta_i E_{i-1}^{(k')}\right) - \left(-\beta_i E_i^{(k)} - \beta_{i-1} E_{i-1}^{(k')}\right) \tag{6.23}$$

$$= (\beta_i - \beta_{i-1}) \left(E_i^{(k)} - E_{i-1}^{(k')}\right)$$

which is accepted or rejected according to the Metropolis algorithm, *i.e.*, with probability one for $\Delta H \leq 0$ and with probability $\exp(-\Delta H)$ for $\Delta H > 0$.

The CPU time spent on the β exchange (6.22) should be less than 50% of the total updating time by reasons similar to those discussed at the beginning of chapter 4. In practice much less than 50% of the CPU time (about a few percent) is normally sufficient. Otherwise, one may waste computer time by simply jumping back and forth on strongly correlated configurations. Within this restriction the β_i-values should to be determined so that a reasonably large acceptance rate is obtained for **each** β exchange. This can be done by a straightforward recursion, which is a modification of the method of [Kerler and Rehberg (1994)]. Let us denote by a_i^{pt} the **acceptance rate of the β exchange (6.22)**. We assume that $a_i^{pt} > 0$ holds for all $i = 1, \ldots, n-1$. Iteration m is performed with the β values β_i^m, $i = 0, \ldots, n-1$. We define the β_i^{m+1} values of the next iteration by

$$\beta_0^{m+1} = \beta_0^m \quad \text{and} \quad \beta_i^{m+1} = \beta_{i-1}^{m+1} + a_i^m \left(\beta_i^m - \beta_{i-1}^m\right) \text{ for } i = 1, \ldots, n-1, \tag{6.24}$$

where

$$a_i^m = \lambda^m a_i^{pt,m} \quad \text{with} \quad \lambda^m = \frac{\beta_{n-1}^m - \beta_0^m}{\sum_{i=1}^{n-1} a_i^{pt,m} \left(\beta_i^m - \beta_{i-1}^m\right)} . \tag{6.25}$$

For large acceptance rates a_i^{pt} the distance $\beta_i^m - \beta_{i-1}^m$ is increased, whereas it shrinks for small acceptance rates. The definition (6.25) of the a_i coefficients guarantees $\beta_{n-1}^{m+1} = \beta_{n-1}^m$, *i.e.* the initial $\beta_{n-1} - \beta_0$ distance is kept.

A problem of the recursion (6.24) is that the statistical noise stays similar in each iteration step, assuming that the number of sweeps per iteration is chosen to be constant. It may be preferable to combine all the obtained estimates with suitable weight factors w^k. The final estimate after m iterations reads then

$$\beta_i = \frac{\sum_{k=1}^{m} w^k \beta_i^k}{\sum_{k=1}^{m} w^k} \, .$$
(6.26)

Two reasonable choices for the weights are

$$w^k = \min_i \{a_i^{pt,k}\}$$
(6.27)

and

$$w^k = \frac{1}{\sqrt{\sigma^2}} \quad \text{with} \quad \sigma^2 = \sum_{i=1}^{n} \left(\frac{1}{a_i^{pt,k}}\right)^2 \, .$$
(6.28)

Equation (6.27) determines the weight from the worst exchange acceptance rate and suppresses its run in that way, whereas equation (6.28) relies on what would be the correct statistical weighting of the acceptance rates, if they were statistically independent (what they are not).

6.3.1 *Computer implementation*

We like to program the β exchange according to equations (6.22) and (6.23) using MPI. Initially the indices of the β values are chosen to agree with the rank MY_ID of the process. The exchanges (6.22) lead then to the situation where the β indices become an arbitrary permutation of the numbers $0, \ldots, n-1$. For the MPI program two approaches are possible:

(1) All the n processes send the needed information (the energies of their configuration) to a master processor. The master processor handles the exchange and sends each process its new β value back (which may agree with the old value).
(2) The involved processes handle the exchange among themselves through point to point communication.

We consider only the second approach, because for it the number of sweeps per time unit has a chance to scale linearly with the number of processors, whereas for the first approach the master processor will become the bottleneck of the simulation when many processors are employed[4]. For the second approach each process has to know the identities of the processes

---

[4]When only few processors are used, it may be more efficient to let one master processor handle the β exchange.

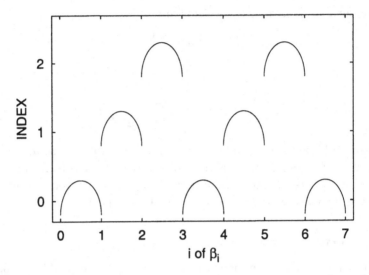

Fig. 6.1 INDEX $= 0, 1, 2$ labels the three interaction groups for the exchange (6.22) of β values through MPI point to point communication. The figure is drawn for eight processes $(i = 0, \ldots, 7)$.

where the neighboring β values reside. They are stored in a neighbor array NEIGH(2) and the initial values are those defined in the MPI test program of table 6.1. With each β exchange the nearest neighbor and next nearest neighbor connections get modified and a correct bookkeeping is the main difficulty of implementing the parallel tempering algorithm. We cannot deal with all pairs simultaneously, but need one buffer process between each pair. This leads to three distinct interaction groups, starting either with β_0, β_1 or β_2, which we label by INDEX $= 0, 1, 2$ as indicated for eight processes in figure 6.1. E.g., for INDEX $= 0$ the buffer processes are $i = 2$ and $i = 5$.

Our MPI Fortran subroutine p_pt_mpi.f which implements the β exchange (6.22) for Potts models is reproduced in table 6.6. As in figure 6.1 the variable INDEX takes on the values 0, 1 or 2 and identifies the lowest β value for which the exchange is tried, the higher β values follow according

Table 6.6 The MPI subroutine p_pt_mpi.f which implements the β exchange according to equations (6.22) and (6.23).

```fortran
      SUBROUTINE P_PT_MPI(N_PROC,itag)
C Copyright, Bernd Berg, December 20, 2001.
C Parallel tempering for the Potts models heat bath code.
C MPI implementation of the beta exchange.
      include '../../ForLib/implicit.sta'
      INCLUDE 'mpif.h'
      include '../../ForLib/constants.par'
      include 'lat.par'
      include 'potts.par'
      include 'mc_pt_mpi.par'
      include '../../ForLib/potts_hb.com'
      include '../../ForLib/p_pt_mpi.com'
      DIMENSION ISTATUS(MPI_STATUS_SIZE)
      DATA INDEX/-1/
      SAVE INDEX
C
      itag=mod(itag+1,32768)
      INDEX=MOD(INDEX+1,3)
      MY_IND1=MOD(MY_B,3)
      MY_IND2=MOD(MY_B+2,3) ! MY_B-1+3
      MY_IND3=MOD(MY_B+1,3) ! MY_B-2+3
C
      IF(MY_IND1.EQ.INDEX) THEN ! Processes MY_IND1.
        NDEST3L=NEIGH(1)
        NSEND3L=-2
        IF(NEIGH(2).LT.N_PROC) THEN
          NSEND(1)=NEIGH(1)
          NSEND(2)=iact
          NSEND(3)=IACPT_PT
          CALL MPI_SEND(NSEND,3,MPI_INTEGER,NEIGH(2),itag,
     &                  MPI_COMM_WORLD,IERR)
          CALL MPI_RECV(NRECV,2,MPI_INTEGER,NEIGH(2),itag,
     &                  MPI_COMM_WORLD,ISTATUS,IERR)
          IF(NRECV(1).NE.-2) THEN
            NSEND3L=NEIGH(2)
            NEIGH(1)=NEIGH(2)
            NEIGH(2)=NRECV(1)
            IACPT_PT=NRECV(2)
            MY_B=MY_B+1
            call potts_wghb(whb_tab,BA(MY_B),H0,nd,nqm1) ! beta=BA(MY_B)
          END IF
        END IF
        IF(NDEST3L.GE.0) CALL MPI_SEND(NSEND3L,1,MPI_INTEGER,NDEST3L,
     &                         itag,MPI_COMM_WORLD,IERR)
      END IF
C
      IF(MY_IND2.EQ.INDEX) THEN ! Processes MY_IND2.
        NDEST3R=NEIGH(2)
        NSEND3R=-2
```

```
      IF(NEIGH(1).GE.0) THEN
        CALL MPI_RECV(NRECV,3,MPI_INTEGER,NEIGH(1),itag,
     &                MPI_COMM_WORLD,ISTATUS,IERR)
        NDEST1=NEIGH(1)
        IF(LPT_EX_IA(BA(MY_B),BA(MY_B-1),iact,NRECV(2))) THEN
          NSEND(1)=NEIGH(2)
          NSEND(2)=IACPT_PT+1
          IACPT_PT=NRECV(3)
          NSEND3R=NEIGH(1)
          NEIGH(2)=NEIGH(1)
          NEIGH(1)=NRECV(1)
          MY_B=MY_B-1
          call potts_wghb(whb_tab,BA(MY_B),HO,nd,nqm1) ! beta=BA(MY_B)
        ELSE
          NSEND(1)=-2
        END IF
        CALL MPI_SEND(NSEND,2,MPI_INTEGER,NDEST1,itag,
     &                MPI_COMM_WORLD,IERR)
      END IF
      IF(NDEST3R.LT.N_PROC) CALL MPI_SEND(NSEND3R,1,MPI_INTEGER,
     &                           NDEST3R,itag,MPI_COMM_WORLD,IERR)
    END IF
C
    IF(MY_IND3.EQ.INDEX) THEN ! Processes MY_IND3.
      IF(NEIGH(1).GE.0) THEN
        CALL MPI_RECV(NRECV3R,1,MPI_INTEGER,NEIGH(1),
     &                itag,MPI_COMM_WORLD,ISTATUS,IERR)
        IF(NRECV3R.NE.-2) NEIGH(1)=NRECV3R
      END IF
      IF(NEIGH(2).LT.N_PROC) THEN
        CALL MPI_RECV(NRECV3L,1,MPI_INTEGER,NEIGH(2),
     &                itag,MPI_COMM_WORLD,ISTATUS,IERR)
        IF(NRECV3L.NE.-2) NEIGH(2)=NRECV3L
      END IF
    END IF
C
    RETURN
    END
```

to the scheme of the figure. The thus identified processes send their iact, IACPT_PT (for a_i^{pt}) and NEIGH(1) values to their NEIGH(2) neighbors, *i.e.* to the processes which accommodate their upper, neighboring β values. These neighbor processes decide whether the β exchange is accepted or rejected. For this purpose they rely on a call to the logical valued function

$$\text{LPT\_EX\_IA(BETA1, BETA2, IACT1, IACT2)} \qquad (6.29)$$

of ForLib, which implements the Metropolis acceptance or rejection according to equation (6.23). For either outcome the processes send two integers back to their lower neighbors NDEST1 (the originally initiating processes) and one integer to their upper neighbors NDEST3L. If the exchange is rejected the first element of each MPI_SEND is set to -2, thus signaling to the receiver that nothing changed. Otherwise, the receivers store the addresses of their new neighbors in the NEIGH array. In addition, two more actions are performed: First, the initiating processes inform their left neighbors about the change by sending a new neighbor address to NDEST3R. This send is performed in any case, where the message -2 informs NDEST3R that nothing changed. Second, the processes whose β values changed re-calculate their heat bath weights and interchange also their parallel tempering acceptance rates IACPT_PT.

After a certain number of β exchanges, the β values are updated along one of the alternatives of equations (6.24) to (6.28). They are implemented by our subroutines

$$\text{pt\_rec0\_mpi.f, pt\_rec1\_mpi.f and pt\_rec2\_mpi.f of ForLib,} \quad (6.30)$$

called in correspondingly labeled main programs of ForProg/MPI_Potts

$$\text{p\_ptmpi0.f, p\_ptmpi1.f and p\_ptmpi2.f .} \quad (6.31)$$

In practice it may be unavoidable that one (or several) of the acceptance rates a_i^{pt} are zero. Applying equation (6.24) blindly gives the undesirable result $\beta'_{i-1} = \beta_i$. To avoid this, the routines (6.30)

$$\text{replace all} \quad a_i^{pt} = 0 \quad \text{values by} \quad a_{\min}^{pt} = 0.5/N_{\text{update}} \quad (6.32)$$

where N_{update} is the number of updates done. This will work when there are enough β_i values to bridge the $[\beta_1, \beta_{n-1}]$ interval. Otherwise, the only solution to the problem of $a_i^{pt} = 0$ values is to increase the number of processes.

To keep MPI instructions out of the routines (6.30), we gather the needed permutation of β values over the processes, as well as their corresponding exchange acceptance rates, in the corresponding main program (6.31). This could be coded using the already introduced point to point MPI communication. However, it is shorter to rely on the

$$\text{MPI\_ALLGATHER (SENDBUF, ISENDCOUNT, SENDTYPE, RECVBUF,}$$
$$\text{IRECVCOUNT, RECVTYPE, COMM, IERROR)} \quad (6.33)$$

instruction. The arguments of MPI_ALLGATHER are: SENDBUF, the starting element of the send buffer; ISENDCOUNT (integer), the number of elements in the send buffer; SENDTYPE, the data type of the send buffer elements; RECVBUF, the first element of the receive buffer; IRECVCOUNT (integer), the number of elements in the receive buffer; RECVTYPE, the data type of the receive buffer; COMM, the communicator; IERROR, the MPI error code.

6.3.2   *Illustration for the 2d 10-state Potts model*

We will bridge the pseudo-transition point with parallel tempering simulations and comment on a number of important differences between the parallel tempering and the multicanonical approach. It has to be stressed that, unlike the multicanonical approach, parallel tempering is not tailored for the study of interface tensions.

Table 6.7 Parallel tempering using the recursion program pt_rec0.f with the parameter file of table 6.2. The rows after BA: give the initial and final β values and the rows in-between accepted exchanges (6.22) of β values, which define the acceptance rates a_i^{pt}. The upper part of the table shows the results obtained with four processes -np 4 and the lower part those obtained with eight processes (-np 8), see assignment 2.

BA:	0.65000	0.68333	0.71667	0.75000				
1.	1140	36	852					
2.	190	2393	83					
3.	114	402	1799					
4.	2152	1827	0					
5.	557	0	2667					
BA:	0.65000	0.74932	0.74943	0.75000				

BA:	0.65000	0.66429	0.67857	0.69286	0.70714	0.72143	0.73571	0.75000
1.	2065	2009	1915	1598	137	1883	2057	
2.	1906	1787	1640	118	2572	1595	1839	
3.	1733	1608	702	2396	2231	1133	1678	
4.	1426	1359	1500	2300	1980	1076	1361	
5.	1475	1315	948	2073	2008	1836	1440	
BA:	0.65000	0.67746	0.69721	0.70321	0.70734	0.71332	0.72619	0.75000

As in section 5.1 we perform our simulations on an 20×20 lattice. We choose

$$\beta_{min} = \beta_0 = 0.65 \quad \text{and} \quad \beta_{max} = \beta_{n-1} = 0.75 \qquad (6.34)$$

for the (fixed) minimum and maximum β values of the recursion. This range includes the double peak situation of figure 5.4 at $\beta = 0.71$. First, we perform parallel tempering runs with $n = 4$ and notice that this number of processes is too small to bridge the desired β range. The upper part of table 6.7 depicts the result of the -np 4 run with the parameter file of table 6.2. The first row shows the initial, equally spaced, β values. The subsequent rows, labeled 1 to 5, give the numbers of the accepted exchanges, $N_{\text{update}}a_i^{pt}$, between the corresponding β values (β_{i-1}, β_i). After each of these rows new β values are calculated according to equations (6.24) and (6.25) with the modification of equation (6.32). Of those only the final estimate (with pt_rec0.f) is shown in the last row. The rows before show that the recursion get gradually unstable, so that in rows 4 and 5 we find zero transitions between certain β values. This behavior has to be attributed to the fact that the recursion (6.24) relies on a linear approximation around a fixed point. This approximation is rendered invalid when no stable fixed point exists, or the starting values are too far away from it.

Next, we double the number of processes to $n = 8$ and obtain a satisfactory coverage over our temperature range. The results (again obtained with the parameter values of table 6.2) are depicted in the lower part of table 6.7. All the $N_{\text{update}}a_i^{pt}$ values are now non-zero. With the recursion progressing, they become equal up to the statistical noise of the iteration run.

In the following we compare the efficiencies of temperature distributions for the subsequent MC production runs. For this purpose we consider the temperature movements for each process. Similar as for multicanonical simulations (5.30) we define a tunneling (or random walk cycle) event, now relying on the smallest and largest temperature instead of energy bounds[5]. For parallel tempering we define a **tunneling event (random walk cycles)** to be a Markov chain of a fixed process which touches the largest temperature, then the smallest temperature, β_{max}, and again the largest temperature,

$$\beta_{\text{min}} \to \beta_{\text{max}} \to \beta_{\text{min}} , \qquad (6.35)$$

where β_{min} and β_{max} are set in the parameter file of table 6.2. In this way one obtains in a production MC run with fixed β_i values a certain number of tunneling events for each process. The **total number of tunneling**

---

[5]If desired, one can also in multicanonical simulations rely on maximum and minimum temperatures instead of energies.

Table 6.8 Total numbers of tunneling events ntun as obtained from MC data production after running either of the recursion subroutines (6.30). We use np=4 or np=8 processes and nrpt*nmea2*nmea1 sweeps with measurements. The nequ2 and nequ1 parameters used for the recursion are given in the first two rows of the table. All other parameters values are as listed in table 6.2. It is assignment 3 to reproduce the tunneling values.

run		1	2	3	4	5	6	7	8
nequ2		64000	32000	16000	8000	4000	2000	1000	500
nequ1		1	2	4	8	16	32	64	128
pt_rec0.f	4	0	0	0	0	0	0	0	0
pt_rec1.f	4	9	13	21	10	17	1	7	1
pt_rec2.f	4	9	6	26	9	12	1	0	1
pt_rec0.f	8	80	69	70	70	70	75	82	79
pt_rec1.f	8	77	61	81	72	61	77	71	70
pt_rec2.f	8	67	68	75	74	66	79	70	76

events is obtained by summing over all processes.

In table 6.8 we collect the total number of tunneling events ntun, which are obtained during the production part of our simulation after running either of our three different recursion procedures (6.24), (6.27) or (6.28). The production parameter values nrpt, nmea2 and nmea1 are those of table 6.2. We use four as well as eight processes and vary the recursion parameter values nequ2 and nequ1, while keeping nequ2*nequ1 constant and NPT_REC=5. With nequ2*nequ1 constant, a low value of nequ2 is preferable, because it saves on the processing time of β exchanges.

The analysis of the produced data files is done with the program

$$\text{ana\_pt.f} \quad \text{of} \quad \text{ForProg/MPI\_Potts}, \quad\quad (6.36)$$

which runs in the single process (often called "scalar") mode. Note that the number of processes used in the data production has to be set as a parameter (N_PROC) inside ana_pt.f.

For eight processes we find that the number of tunneling events depends (within the statistical noise) neither on the distribution of equilibrium sweeps in the range nequ1=1 to nequ1=128, nor on the recursion subroutine used. This is different when using only four processes, so that the parallel tempering method is pushed to its edge. In that case the simplest form of the recursion (6.24) (implemented by pt_rec0.f) fails entirely, while the version (6.26) with either weights (6.27) or (6.28) still works, but only up to nequ1=16 after which this approach collapses too, apparently because

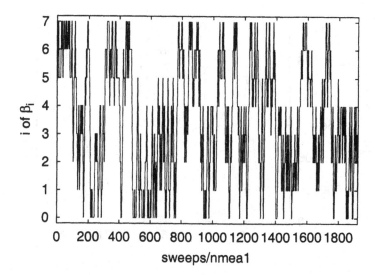

Fig. 6.2 Time series of β values on process zero (MY_ID=0) for the -np 8 parallel tempering run of table 6.7, see assignment 4.

a statistics of less than nequ2=4000 attempted β exchanges is no longer sufficiently large. Even when the recursion with four processes works, the increase in tunneling performance by going from four to eight processes is far larger than the factor of two, which is the increase in CPU time spent. Let us mention that with the initial, uniformly distributed β values there are, with our production statistics, ntun=2 tunneling events when we use four processes and ntun=36 tunneling events for eight processes, *i.e.*, we gain considerably by applying one of the recursions (with the exception of using (6.24) in the case of four processes).

For the eight processes run of table 6.7 (run 4 of table 6.8) figure 6.2 depicts the MC production time series of β values for process zero (MY_ID=0). The β values are identified by their labels $i = 0, \ldots 7$ and the lines between the values are only drawn to guide the eye. As a β exchange is offered every nmea1 sweeps, the time series is given in units of sweeps/nmea1 and it runs from 1 to nrpt*nmea2. Careful counting reveals ten tunneling events. At

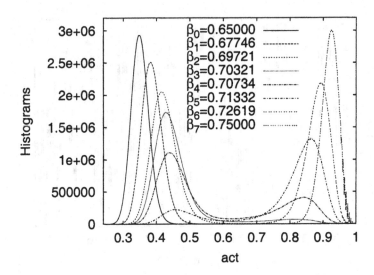

Fig. 6.3 Action histograms for the final β values of the **-np 8** parallel tempering run of of table 6.7. From up to down the β values label the histograms from left to right. It is part of assignment 4 to reproduce this figure.

a second look the reader may think to see some metastability at $i = 4$ and perhaps also at $i = 5$. This is no illusion, but can be made quantitative by calculating the "bounce back" probabilities, that is the probability that after a successful β exchange $\beta_i \rightarrow \beta_{i+1}$ the next accepted change for the β_{i+1} branch is $\beta_{i+1} \rightarrow \beta_i$ (correspondingly $\beta_i \rightarrow \beta_{i-1}$ followed by $\beta_{i-1} \rightarrow \beta_i$). Obviously, this probability is 1.0 at the endpoints, *i.e.* for $i = 0$ and $i = 8$. It turns out to be 0.51 for $i = 4$, 0.48 for $i = 5$ and ≤ 0.31 for the other i values, see assignment 4.

The canonical action variable histograms for this run are depicted in figure 6.3. Each histogram corresponds to a fixed temperature, given by the β values in the figure which agree with the final β values listed in table 6.7. It turns out that the β_4 and β_5 values, for which the "metastability" is encountered in figure 6.2, are precisely those which produce the clearest double peak histograms. For β_4 the higher peak is in the disordered region and for β_5 it is in the ordered region. Let us compare figure 6.3 with the multicanonical results of figure 5.4. Both figures cover the range, which is relevant for the double peak structure at $\beta = 0.71$. The multicanonical method does this by re-weighting the multicanonical distribution, whereas parallel tempering covers the region using β exchanges to jump over the transition. Note that in figure 5.4 the histograms are plotted on a logarithmic scale to emphasize the good precision in the minimum in between the two peaks, whereas a linear scale is used in figure 6.3. Plotting the data of figure 6.3 on a logarithmic scale reveals their rather limited precision in the minimum, see assignment 4. This distinction becomes quickly more pronounced when larger lattices are used. The multicanonical method is well suited for calculations of the interface tensions from energy histograms [Berg and Neuhaus (1992)]. For parallel tempering this is not the case, because the sampling of each histogram is still canonical. Both methods allow to bridge the transition, but in rather different ways.

In this section we have confined our analysis of parallel tempering data to histograms at fixed temperatures and to studying the temperature exchanges between processes. To estimate the free energy and the entropy, one may want to combine overlapping histograms from several β values into one data set. This can be done with **multi-histogram** methods (the author thinks that the approach of [Alves *et al.* (1990)] is conceptually simpler than the one of [Ferrenberg and Swendsen (1989a)]).

6.3.3   *Gaussian Multiple Markov chains*

The parallel tempering method does not allow to explore canonically rare configurations, because each member of the set of simulations samples still a Boltzmann distribution. This can be overcome by applying the parallel tempering idea to Gaussian distributions

$$\sim \exp\left(-\frac{(E - E_i)^2}{\sigma_i^2}\right) . \tag{6.37}$$

Now two parameters (E_i and σ_i) instead of one (β_i) have to be adjusted to get overlapping probability densities and one can focus on canonically rare configurations. For first order phase transitions tests of parallel simulations with Gaussian distributions have given satisfactory results (T. Neuhaus, unpublished).

6.3.4   *Assignments for section 6.3*

These assignments require MPI.

(1) Run the program p_hbtest_mpi.f of ForProg/MPI_Potts, as depicted in table 6.1, with the parameter file of table 6.2 and four processes. The result on the IUO output file should be the one shown in table 6.5

(2) Use the program p_ptmpi0.f of ForProb/MPI_Potts with the parameter file of table 6.2 to reproduce the results of table 6.7.

(3) Use the programs p_ptmpi0.f, p_ptmpi1.f, p_ptmpi2.f and the analysis program ana_pt.f of ForProb/MPI_Potts to reproduce the results of table 6.8.

(4) Run the analysis program ana_pt.f on the data of assignment 2. Plot the appropriate data files to reproduce figures 6.2 and 6.3. The analysis program will also give you the "bounce back" probabilities discussed in connection with figure 6.2. Tell the author if you can find a explanation for the over-all "forward" trend in the replica exchange. Identify the β_4 and β_5 histograms in figure 6.3. Plot the histograms of figure 6.3 also on a logarithmic scale, similar to the one used in the multicanonical figure 5.4. In particular, focus on the histograms for $\beta = 0.70734$ and $\beta = 0.71332$.

6.4   Checkerboard algorithms

The checkerboard labeling of the lattice sites allows for parallel Metropolis or heat bath updating of all sites of the same color of a checkerboard. Lattice sites are labeled by two colors, ic=1,2, so that moving one step in any direction changes the color. For nearest neighbor interactions this allows to update half of the spins (or other dynamical variables) in parallel.

We employ a similar notation as in chapter 3. The ForLib common block lat.com (3.25) becomes replaced by latc.com

$$\text{common/latc/ns}, \text{nla(nd)}, \text{ias(2,msc)}, \text{iac(ms)}, \text{iasc(ms)},$$
$$\&\qquad\qquad \text{icpf(nd,2,msc)}, \text{icpb(nd,2,msc)} \qquad\qquad (6.38)$$

and the parameter statement lat.par (3.26) by latc.par

$$\text{parameter(nd} = 3, \text{ml} = 04, \text{ms} = \text{ml} * *\text{nd}, \text{msc} = \text{ms}/2, \text{mlinkc} = \text{nd} * \text{msc})\,.$$
$$(6.39)$$

To initialize the bookkeeping arrays, the calls to the routines lat_init.f (3.23) and ipointer.f (3.22) have to be replaced by calls to the ForLib routines

$$\text{latc\_init}(ns, nsc, nd, nla, ias, iac, iasc, icpf, icpb, nlink) \qquad (6.40)$$

and

$$\text{icpointer}(ns, nsc, nd, ias, iac, iasc, icpf, icpb, nla, ix, is) \,. \qquad (6.41)$$

New are the arrays ias(2,msc), iac(ms) and iasc(ms). Given the lattice site number is, the color and the site number in the checkerboard algorithm are

$$\text{ic} = \text{iac(is)} \quad \text{and} \quad \text{isc} = \text{iasc(is)} \,. \qquad (6.42)$$

The formula which gives the color assignment of a lattice site is

$$\text{ic} = 1 + \text{mod}\left[\sum_{id=1}^{nd} \text{ix(id)}, 2\right], \qquad (6.43)$$

where ix(id) are the coordinates as in equation (3.19). The other way round, with ic and isc given the lattice site number is

$$\text{is} = \text{ias(ic, isc)} \,. \qquad (6.44)$$

The other arrays have the same meaning as in chapter 3.1.1, adjusted for the checkerboard situation. Table 6.9 illustrates the checkerboard labeling for the same $2^2 4$ lattice for which table 3.1 gave the normal labeling. More details are explored in assignment 1.

The checkerboard parallelization is somewhat academic for our Potts model programs in their present form, because the updating of a single q-state is so fast that the communication overhead of the checkerboard parallelization may swallow all or a substantial part of the gains. However, the situation is quite different for models where the updating of a single site is CPU time extensive. For instance, in $4d$ lattice gauge theory the update of a SU(3) link variable involves a lot of number crunching and a checkerboard parallelization [Barkai and Moriarty (1982)] of the links is efficient[6]. Also for an Ising model with multispin coding [Jacobs and Rebbi (1981); Zorn *et al.* (1981)], where one uses the bits of a computer

---

[6]At the time of the work by [Barkai and Moriarty (1982)] one was actually interested in vectorization. The arguments made for vectorization carry immediately over to parallelization.

Table 6.9 Checkerboard bookkeeping for a $2^2 4$ lattice: Listed are the site number is, the corresponding checkerboard color ic, the checkerboard site number isc, the checkerboard forward pointers icpf and backward pointers icpb. These results are obtained with the program lat3dckb.f of the folder ForProg/lat, see assignment 1.

is	ic	isc	icpf(1,isc,ic)	(2,)	(3,)	icpb(1,isc,ic)	(2,)	(3,)
1	1	1	1	2	3	1	2	7
2	2	1	1	2	3	1	2	7
3	2	2	2	1	4	2	1	8
4	1	2	2	1	4	2	1	8
5	2	3	3	4	5	3	4	1
6	1	3	3	4	5	3	4	1
7	1	4	4	3	6	4	3	2
8	2	4	4	3	6	4	3	2
9	1	5	5	6	7	5	6	3
10	2	5	5	6	7	5	6	3
11	2	6	6	5	8	6	5	4
12	1	6	6	5	8	6	5	4
13	2	7	7	8	1	7	8	5
14	1	7	7	8	1	7	8	5
15	1	8	8	7	2	8	7	6
16	2	8	8	7	2	8	7	6

word to store many lattices in one word, the simultaneous updating of one site of all systems becomes CPU time intensive and the implementation of (additional) checkerboard parallelization may be worthwhile a thought.

6.4.1   *Assignment for section 6.4*

(1) Use the program lat3d_ckb.f of ForProb/lat to reproduce table 6.9. Then, calculate from the checkerboard pointer arrays of table 6.9 back to the pointer arrays of table 3.1.

Chapter 7

Conclusions, History and Outlook

In the first two chapters of this book a self-contained introduction to statistical methods was given with a strict emphasize on sampling, preparing the ground for understanding MC methods as well as for the analysis of MC generated data. To proceed fast to Markov chain MC simulations, a number of elementary topics, which are usually covered early in statistics books, were bypassed. For instance, Kolmogorov axioms, Bayes theorem and a detailed discussion of the binomial and other elementary discrete distributions. It is the intention to present them in a sequel to this book, together with other – more advanced – statistical method, including Bayesian statistics [Lee (1997)], bootstrap data analysis [Efron (1982)], maximum entropy methods [Jaynes (1982)], and extreme value statistics [Gumbel (1958); Galambos (1987)]. On the other hand we already covered a number of statistical methods, which are of central importance for the analysis of MC data. This includes the Student distribution, the F-test and, in particular, jackknife data analysis.

As is emphasized in the preface, one of our goals is to make the numerical results of this book reproducible. By now, the reader should have a good understanding about what is meant by that. Hopefully, this helps to teach the students, and to convince the doubtful, that we are really dealing with quantitatively precise methods. It seems that some of the current numerical research could well benefit by imposing similar standards, at least for computational papers which report surprising results. Disagreements with previous literature or large gains in algorithmic performance could then be verified rather easily, possibly (and hopefully) already on the referee level.

Markov chain MC simulations started in earnest with the 1953 paper by Nicholas Metropolis, Arianna Rosenbluth, Marshal Rosenbluth, Augusta Teller and Edward Teller [Metropolis *et al.* (1953)], although there were previous suggestions by von Neumann and Ulam. The Metropolis (MRRTT may be more appropriate) paper was followed by a number of investigations, which anticipated to some extent modern methods. These early developments are reviewed in articles [Gubernatis (2003); M.N. Rosenbluth (2003); Wood (2003)] of the proceedings of the 2003 Los Alamos conference *The Monte Carlo Method in Physical Sciences: Celebrating the 50th Anniversary of the Metropolis Algorithm.*

In the 1950s and 1960s the time was not ready for a wide acceptance of Markov chain MC simulations as a serious scientific tool. The reason was not only that computer mainframes were expensive and clumsy to handle, but also a certain arrogance of the mainstream analytical researchers. Numerical analysis was considered to be some kind of inferior engineering approach. Instead of simulating the real problems, one preferred analytical insights into (more or less) related questions. So it remained to the 1970s that the use and development of Markov chain MC methods picked up speed. Applications to critical phenomena flourished first and were collected in a review by [Binder (1976)]. Technical progress followed. The acceptance ratio estimator of [Bennett (1976)], the umbrella sampling method of [Torrie and Valleau (1977)] and the emergence of MC renormalization group techniques [Ma (1976); Swendsen (1978)] are just a few examples which stand for this time. In the 1980s the spread of MC simulations accelerated tremendously.

In chapter 3 methods and algorithms are compiled which the author learned and applied during the early 1980s. *E.g.*, the Metropolis and the heat bath algorithms. In chapter 4 we dealt with the autocorrelations and the statistical analysis of the Markov chain MC data. Our focus in chapters 3 and 4 was on techniques and our physical analysis remained limited to estimating expectation values of the internal energy and the magnetization on finite lattices. In chapter 5 we also dealt with the specific heat, the entropy and the free energy. This provides the basis for a thorough physical analysis of statistical physics systems. The number of available analysis techniques is then too numerous to them to be covered within this book. The appropriate choice depends on the questions asked. FSS techniques [Fisher (1971); Brézin (1982); Blöte and Nightingale (1982); Privman and Fisher (1983); Privman and Fisher (1983); Fisher and Privman (1985)] are most useful in the opinion of the author. Selected references are compiled in the following.

Binder cumulants enable accurate estimates of phase transition points [Binder (1981); Binder *et al.* (1986)]. Re-weighting [Ferrenberg and Swendsen (1988)] allows to follow the behavior of the cumulants and other observables more accurately than by simply evaluating them at the (canonical) simulation points. In particular the peaks of the specific heat on finite lattices can be located precisely. The importance of partition function zeros was first recognized by [Yang and Lee (1952)]. Their FSS behavior is linked to that of the correlation length ξ, which is the central observable when one wants to analyze the nature of a phase transition. MC calculations [Bhanot *et al.* (1987); Alves *et al.* (1990); Alves *et al.* (1991)] of partition function zeros allow thus to estimate the critical exponent ν of the correlation length. Direct estimates of the correlation length from calculations of correlation functions are also feasible. They are best carried out on elongated lattices for the zero momentum mass gap $m \sim 1/\xi$ [Alves *et al.* (1990); Alves *et al.* (1991)], a technique which originated in lattice gauge theory glueball calculations. Estimates of critical exponents by a variety of methods are reviewed in [Pelissetto and Vicari (2002)].

In chapter 5 we proceeded by including more advanced algorithmic and data analysis techniques.

Multicanonical sampling techniques were introduced by [Berg and Neuhaus (1992)] for the study of first order phase transitions. With respect to this application they have found applications ranging from chemistry [Wilding and Müller (1995)] to the study of the electroweak transition [Csikor *et al.* (1995)] in elementary particle physics. In addition, it was realized [Berg and Celik (1992); Hansmann and Okamoto (1993)] that the method is well suited for the study of complex systems, such as spin glasses or small proteins (peptides). It appears that nowadays the most fruitful applications of the approach are in this direction. Many variations are possible, so that *Generalized Ensemble Simulations*, became a generic name for such methods, see [Berg (2002); Hansmann and Okamoto (1999); Mitsutake *et al.* (2001)] for reviews. Generalized ensembles rely on modifications of the Gibbs-Boltzmann weights. One may proceed even further and play around with the elements of the transition matrix. This leads to *Transition Matrix Methods*, of which Wang's random walk algorithm is an example, see the article by [Wang and Swendsen (2002)] and references given therein.

At low enough temperatures event driven simulations (EDS) are the method of choice [Bortz *et al.* (1975)]. Extensions of the original EDS approach are given by [Novotny (2001); Korniss *et al.* (2003)] and [Munoz *et al.* (2003)].

The strength of the cluster algorithm [Swendsen and Wang (1987); Wolff (1989)] lies in the study of second order phase transitions for which the critical slowing down is practically overcome. For some measurements improved estimators (which we did not consider) lead to another substantial reduction of statistical errors [Hasenbusch (1995)]. Applications of the cluster algorithm to the study of spin glasses [Swendsen and Wang (1986)] were occasionally tried and success has mainly been achieved in two dimensions [Houdayer (2001)]. Generalization to gauge theories failed [Caracciola *et al.* (1993)]. Still, there is an increasing number of important systems for which cluster algorithms have become available, see [Blöte *et al.* (2002)].

In the final section of chapter 5 we discussed issues which are of importance when performing large scale simulations, defined as taking longer than the average up-time of the computer.

A brief introduction to issues of parallel computing was given in chapter 6. Beginning with the often neglected (but important) issue of trivially parallel computing, we proceeded to give start-up information for using MPI (Message Passing Interface) and chose the parallel tempering MC algorithm [Swendsen and Wang (1986); Geyer (1991); Hukusima and Nemoto (1996)] to illustrate non-trivial parallel computing. As multicanonical simulations parallel tempering can significantly accelerate the dynamics of a simulation, but it does not allow to generate canonically rare configurations. The latter point can be improved by using Gaussian multiple Markov chains. Finally in chapter 6, we dealt with checkerboard updating. It is efficient for models where the updating of a single site is CPU time extensive, like for instance in lattice gauge theory [Barkai and Moriarty (1982)] or when multispin coding [Jacobs and Rebbi (1981); Zorn *et al.* (1981)] is used.

The scope of our treatment has stayed limited to Markov chain MC simulations of equilibrium statistical ensembles and the analysis of the thus produced computer data. *Coupling from the past* [Propp and Wilson (1998)] is a method, which allows in some situations to control the approach to equilibrium rigorously. In practice its applicability has been limited to special situations. Recent interest has also led beyond equilibrium simulations, see [Ódor (2002)] for a review of non-equilibrium results. We briefly discussed the dynamics of importance sampling after equation (3.51) of chapter 3.

There are other, valuable numerical methods, which are often competing with Markov chain MC simulations. Most notable are approaches, which are based on the real dynamics of the statistical system. Methods which fall into that category are the *Langevin algorithm* [Parisi and Wu (1981)] and *molecular dynamics* [Alder and Wainwright (1957-1960)]. Certainly it is an advantage when the real time dynamics can be followed, but in practice there are many technical complications, which may prevent precisely that. Nowadays molecular dynamics simulations tend to use a Nose-Hoover thermostat to introduce a temperature and thermal fluctuations. The textbooks by [Allen and Tildesley (1990); Frenkel and Smit (1996)] provide good introduction to molecular dynamics. *Hybrid Monte Carlo* [Duane *et al.* (1987); Mehlig *et al.* (1992)] is in a combination of the stochastic Markov chain MC approach with molecular dynamics.

In another quite different endeavor, configurations are generated by biased, direct statistical sampling instead of relying on a Markov chain approach. Originating from random walk studies [Rosenbluth and Rosenbluth (1955)], the pruned enriched *Rosenbluth method* (PERM) has been used with success [Grassberger (2002)].

Quantum MC plays a major role in solid state physics as well as in chemistry, see [Foulkes *et al.* (2001)] for a review.

Many of the introduced simulation methods can be combined with one another. Some of such combinations have already been explored in the literature:

(1) Instead of the energy other physical variables can be used for the multi-canonical weighting, for instance the magnetic field [Berg *et al.* (1993)]. The method is then called *multimagnetical*. In spin glass simulation the Parisi overlap parameter was used [Berg *et al.* (2000)], leading to the *multi-overlap algorithm*, and so on.

(2) A cluster variant of the multicanonical algorithm was introduced by [Janke and Kappler (1995)]. Surprisingly, there appear to be no further applications of this method in the literature, although some systems would certainly be worth a try.

(3) Multicanonical and parallel tempering methods were, in the context of peptide folding, studied by [Hansmann (1997)] and others [Sugita and Okamoto (2000)].

More combinations are obviously possible. For instance, one can combine the multicanonical method with EDS updates or use parallel tempering together with cluster updates. Other combinations are with methods that lie outside the scope of this book. Examples are:

(1) *Multigrid* and multicanonical methods [Janke and Sauer (1994)].
(2) Multicanonical variants of Langevin, molecular dynamics and hybrid MC simulation methods have been introduced in the context of peptide folding [Hansmann *et al.* (1996)] and for hybrid MC also in the context of lattice gauge theory [Arnold *et lal.* (1991)].
(3) Parallel tempering and molecular dynamics [Sugita and Okamoto (1999); Sanbonmatsu and García (2002); Rhee and Pande (2003)] go well together when a thermostat is included in the molecular dynamics simulation.
(4) PERM has been combined with multicanonical weighting [Bachmann and Janke (2003)].

The Monte Carlo methods introduced here have many physics applications. To learn more the reader has to consult the literature. Here we compile a few references, which may pave the way.

In the 1980s *lattice gauge theory* introduced large-scale Markov chain MC simulations to particle physics. See the books by [Münster and Montvay (1993)] and [Rothe (1998)] for an overview. Simple MC sampling plays a major role in the simulation of *event generation* in high energy physics, *e.g.* [Marchesini *et al.* (1992)]. A book edited by [Binder (1995)] focuses on *condensed matter physics*, [Binder and Herrmann (2002)] review MC simulations in *statistical physics*. The text by [Allen and Tildesley (1990)] is a classical reference for computer simulations of *liquids*. For MC methods in *chemical physics* [Ferguson *et al.* (1998)] may be consulted. The book by [Gilks *et al.* (1995)] includes, mainly under a Bayesian perspective, *medical, biological* and *financial* applications. There are even stochastic methods to solve partial differential equations (PDEs) [Sabelfeld (1991)].

For the most interesting applications the limits encountered are often those set by the available computer resources. As long as Moore's law is intact, this frontier moves automatically forward by an order of magnitude every five years. At this point it is worthwhile to note that MANIAC, the supercomputer used 1953 by Metropolis et al., was operating at a speed of about one kiloflop (10^3 floating point operations per second). Presently, in the year 2003, the world's leading supercomputer is the Japanese Earth Simulator, operating at a peak speed of about 40 teraflop (40×10^{12} floating point operations per second). The ratio of these two numbers reflects Moore's law quite well. While MC simulations occupy a fair share, the primary applications run on the Earth Simulator are global climate models, which are mainly based on PDEs like, *e.g.*, our daily weather forecast is too.

It is not only Moore's law which moves the pioneering edge of MC simulations forward. Equally (or even more) important has been progress due to algorithmic improvements, new methods and ideas. This kind of progress is by its nature essentially unpredictable. When it happened it has occasionally implied immediate gains by many by orders of magnitude.

In summary, Markov chain MC simulations occupy nowadays a well-established place in the arsenal of scientific tools. They are well positioned to maintain their relative importance. The mayor competition comes no longer from analytical methods, but mainly from other computational techniques.

Appendix A

Computational Supplements

Appendix A.1 deals with the numerical evaluation of the Gamma function $\Gamma(z)$, the incomplete Gamma function $P(a,x)$, the error function $\text{erf}(x)$, the beta function $B(x,y)$ and the incomplete beta function $B_I(x;a,b)$. In appendix A.2 we consider linear linear algebraic equations.

A.1 Calculation of Special Functions

The gamma function is defined by the integral

$$\Gamma(z) = \int_0^\infty t^{z-1} e^{-t} dt \tag{A.1}$$

and satisfies the recursion relation

$$\Gamma(z+1) = z \, \Gamma(z) . \tag{A.2}$$

This implies $\Gamma(k+1) = k!$ for integer k. For $Re\, z > 1$ [Lanczos (1964)] has derived an improved Sterling approximation. A special case of it is

$$\Gamma(z) = \left(z + \frac{9}{2}\right)^{z-\frac{1}{2}} e^{-\left(z+\frac{9}{2}\right)} \sqrt{2\pi} \left(c_0 + \frac{c_1}{z} + \frac{c_2}{z+1} + \cdots + \frac{c_6}{z+5} + \epsilon\right) \tag{A.3}$$

with $Re\, z > 1$, where the c_i, $(i = 1, \ldots, 6)$ are known constants. The error of the approximation is parameterized by ϵ and

$$|\epsilon| < 2 \cdot 10^{-10} .$$

Lanczos equation (A.3) can be used to evaluate the gamma function for complex arguments, but here we limit ourselves to real arguments x.

From (A.3) as well as already from (A.2) it is obvious that the gamma function grows rapidly for large x. Hence, its logarithm and not the function itself has to be evaluated. Our implementation of $\ln\Gamma(x)$ is the Fortran function gamma_ln.f. It relies on Lanczos coefficients, but works for all $x > 0$ by using for $0 < x < 1$ equation (A.2) to yield $\ln\Gamma(x) = \ln\Gamma(x+1) - \ln x$. If results for negative arguments of $\Gamma(x)$ are needed, they can be obtained from those with $x > 1$ by using the reflection formula

$$\Gamma(1-z) = \frac{\pi}{\Gamma(z)\,\sin(\pi z)} \, . \tag{A.4}$$

The definition, equations (6.5.1) and (6.5.2) of [Abramowitz and Segun (1970)], of the incomplete gamma function is

$$P(a, x) = \frac{\gamma(a, z)}{\Gamma(a)} = \frac{1}{\Gamma(a)} \int_0^x e^{-t} t^{a-1} \, dt, \quad (Re\,a > 0). \tag{A.5}$$

For its numerical evaluation the series expansion

$$\gamma(a, x) = e^{-x} x^a \, \Gamma(a) \sum_{k=0}^{\infty} \frac{x^n}{\Gamma(a+1+k)}, \tag{A.6}$$

and the continued fraction expansion

$$\Gamma(a, x) = e^{-x} x^a \left(\frac{1}{x+} \frac{1-a}{1+} \frac{1}{x+} \frac{2-a}{1+} \frac{2}{x+} \cdots \right), \quad (x > 0) \tag{A.7}$$

are suitable. Compare [Abramowitz and Segun (1970)] equations (6.5.4), (6.5.29) and (6.5.31). Our Fortran implementation gamma_p.f of the incomplete gamma function tries, in essence, both until satisfactory results are achieved – or it aborts.

Our implementation of the error function (1.32) simply relies on reducing it to the incomplete gamma function

$$\operatorname{erf}(x) = \operatorname{sgn}(x) \, P\left(\frac{1}{2}, x^2\right) \quad \text{with} \quad \operatorname{sgn}(x) = \begin{cases} +1 \text{ for } x > 0, \\ -1 \text{ for } x < 0, \end{cases} \tag{A.8}$$

which follows immediately from the substitution $t = y^2$ in the defining integral (1.32) of the error function. In our Fortran library we use the long name error_f instead of erf for the Fortran function, because erf is on some platforms already defined as an intrinsic Fortran function.

The beta function is the Eulerian integral of first type

$$B(a,b) = B(b,a) = \int_0^1 dt\, t^{a-1}\,(1-t)^{b-1} \qquad (A.9)$$

which is related to the gamma function by

$$B(a,b) = \frac{\Gamma(a)\,\Gamma(b)}{\Gamma(a+b)}\,, \qquad (A.10)$$

see section 12.4 of [Whittaker and Watson (1927)]. The incomplete beta function is then defined by

$$B_I(x;a,b) = \frac{B(x;a,b)}{B(a,b)} = \frac{1}{B(a,b)} \int_0^x dt\, t^{a-1}\,(1-t)^{b-1} \quad \text{for } a,b > 0\,. \qquad (A.11)$$

In older tables [Pearson (1968)] $B(x;a,b)$ instead of $B_I(x;a,b)$ is called incomplete beta function. For the numerical evaluation of (A.11) we follow the advise of *Numerical Recipes* [Press et al. (1992)] and use the continued fraction representation (equation (26.58) of [Abramowitz and Segun (1970)])

$$B_I(x;a,b) = \frac{x^a\,(1-x)^b}{a\,B(a,b)} \left[\frac{1}{1+} \frac{c_1}{1+} \frac{c_2}{1+} \cdots \right]$$

where

$$c_{2m+1} = -\frac{(a+m)\,(a+b+m)\,x}{(a+2m)\,(a+2m+1)} \quad \text{and} \quad c_{2m} = -\frac{m\,(b-m)\,x}{(a+2m-1)\,(a+2m)}\,.$$

This expansion converges rapidly for $x < (a+1)/(a+b+2)$ and $x > (a+1)/(a+b+2)$. Other values of x can be transformed into this range by using the symmetry relation (26.5.2) of [Abramowitz and Segun (1970)]

$$B_I(x;a,b) = 1 - B_I(1-x;a,b)\,.$$

In our Fortran library ForLib the functions (A.9) and (A.11) are implemented by the routines beta.f and beta_i.f.

A.2   Linear Algebraic Equations

Occasionally, we have to solve linear algebraic equations of the type

$$A\,\vec{x}_1 = \vec{b}_1, \dots, A\,\vec{x}_M = \vec{b}_M \quad \text{and} \quad A\,A^{-1} = 1\,, \qquad (A.12)$$

where the $N \times N$ matrix A and the (N-dimensional) column vectors $\vec{b}_i$, $i = 1, \ldots, M$ are given. We want to calculate the M column vectors $\vec{x}_i$, $i = 1, \ldots, M$ and the inverse matrix A^{-1}. According to standard theorems of linear algebra this is possible when the columns as well as the rows are linearly independent. This means, no column can be presented as a linear combination of the other columns and the same holds for the rows, see for instance [Lipschutz and Lipson (2000)].

Here we adapt Gauss-Jordan[1] elimination as our method of choice. It is straightforward and robust. However, other methods are faster, in particular when one has only the vector equations to solve and the inverse A^{-1} is not needed. Compare for instance the decomposition methods discussed in *Numerical Recipes* [Press *et al.* (1992)].

Elementary facts about the equations (A.12) are:

(1) Interchanging any two rows of A and the corresponding rows of the vectors $\vec{b}_i$ and the unit matrix 1, does not change the solutions $\vec{x}_i$ and A^{-1}, because we only write the linear equations in a different order.
(2) If we replace any row in A by a linear combination of itself and any other row, together with the same linear combination of rows of the matrix $\vec{b}_M$ and the unit matrix 1, the solutions $\vec{x}_i$ and A^{-1} do not change.
(3) Interchanging any two columns of A gives the same solution set, when we interchange simultaneously the corresponding rows of the vectors $\vec{x}_i$ and the inverse matrix A^{-1}.

Gauss-Jordan elimination uses the above operations to reduce the matrix A to the identity matrix. Once this is accomplished, the right-hand side of (A.12) has become the solution set. Assume a_{11}, the first element of the matrix A, is non-zero. The basic idea of the Gauss-Jordan elimination is then easily illustrated: We divide the first row by a_{11} and subtract from the other, $i \geq 2$, rows the right amount of the first row to make all a_{i1}, $i \geq 2$ matrix elements zero. The first column agrees then with that of the identity matrix. Next, we proceed in the same way with a_{22}, so that the second column agrees then with that of the identity matrix, and so on. The procedure runs into trouble when we hit a zero element or a small diagonal

---

[1]For three linear equations the method was already explained in ancient Chinese manuscripts. Gauss developed it to solve least square problems and the first appearance in print was in a handbook on geodesy by Wilhelm Jordan [Athloen and McLaughlin (1987)].

element a_{ii} (in the latter case due to roundoff errors are the problem). This difficulty is overcome by **pivoting**: Rows and columns are interchanged to put desirable large elements into the diagonal position from which the next quotient is being selected. In practice picking the in magnitude largest available element as pivot is a good choice, assuming that the equations in (A.12) have all been scaled to a similar order of magnitude. Details can be found in many books (some references are listed below).

These ideas are implemented in the subroutine mat_gau.f of ForLib, which follows closely appendix B of [Bevington and Robinson (2003)]. A call to

$$\text{MAT\_GAU(AMAT, N, BMAT, M)} \tag{A.13}$$

needs on input the matrices A and $\vec{b}_M$ of equation (A.12), AMAT(N,N) and BMAT(N,M). It returns on output the inverse matrix of A, A^{-1} in AMAT and the solution vectors of the linear equations (A.12) in BMAT. An illustration for a random matrix acting on a random vector is given by the program lran.f in ForProg/Tests.

For the general treatment of linear algebraic equations the book by [Wilkinson (1965)] is a classical text. See also the textbooks by [Ralston and Rabinowitz (1978); Westlake (1968); Bevington and Robinson (2003)]. On the web excellent software is freely available, for instance LINPACK and EISPACK at www.netlib.org. Further useful routines can be found in the CERNLIB.

Appendix B

More Exercises and some Solutions

Exercises in addition to the assignments of the main text are collected in part B.1 of this appendix. Solutions to some of them as well as to some assignments of the main text are given in part B.2.

B.1 Exercises

For section 1.1:

(1) Use Marsaglia's first 600 random numbers for `iseed1=1` and `iseed2=0` to throw a dice: `i=1+int(six*xr)`. Find the frequencies `nfreq(i)`, *i.e.* the number of occurrences of $i = 1, 2, 3, 4, 5$ and 6.

(2) Calculate the probability density $f(t)$ for the random variable $t^r = (x^r)^2$ where x^r is uniformly distributed in the range $[0, 1)$.

(3) Estimate π using 10^6 random numbers, which are uniformly distributed in the range $[0, 1)$.

For section 1.8:

(1) Use the residue theorem, see for instance [Whittaker and Watson (1927)] p.111,

$$\frac{1}{2\pi i} \oint dz' \frac{f(z')}{z' - z} = f(z) \qquad (B.1)$$

(where the contour closes counter-clockwise) to calculate

$$\phi(t) = \int_{-\infty}^{+\infty} dx \frac{e^{itx}}{\pi (1 + x^2)} \ .$$

For section 2.1:

(1) Proof equation (2.21).

(2) Sample from two Gaussian distributions with identical variances $\sigma_1 = \sigma_2 = \sigma$ and their expectation values separated by two standard deviations: $|\hat{x}_1 - \hat{x}_2| = 2\sigma$. We call a pair of data, one from each distribution, an event. Let us perform the Gaussian difference test for each event. For which choice of Q_{cut} will, in the average, half the events be rejected and half be accepted?

(3) Assume you perform the Gaussian difference test with $Q_{\text{cut}} = 0.05$ on pairs of data points, respectively corresponding to Gaussian distributions with identical standard deviations σ, but different means $\hat{x}_1$, $\hat{x}_2$. How large has $|\hat{x}_1 - \hat{x}_2|/\sigma$ to be, so that the data points are rejected with (a) 50% likelihood and (b) 95% likelihood?

For section 3.2:

(1) Consider the Metropolis algorithm, but instead of choosing sites randomly, select them in the systematic order $i = 1, \ldots, N$. Show that, although detailed balance is now violated, balance still holds.

For section 3.3:

(1) Verify equation (3.70) for the Potts magnetization in an external magnetic field at infinite temperature.

For section 3.4:

(1) Assume, you change the Metropolis updating routine **xy_met.f** so that the acceptance rate is (via changing $\Delta\Phi$) adjusted after every sweep, not just during equilibration, but also during the production part. Will this algorithm still generate configurations with the correct Boltzmann weights?

For section 4.1:

(1) Consider the Metropolis generation of Gaussian random numbers (4.32). Assume that the desired probability density is

$$P_\alpha(x) = \sqrt{\frac{\alpha}{\pi}} e^{-\alpha x^2}, \quad \alpha > 0 \tag{B.2}$$

and that the change proposals are now

$$x' = x + a$$

where a is drawn from the Gaussian probability density

$$P_\beta(a) = \sqrt{\frac{\beta}{\pi}}\, e^{-\beta a^2}, \quad \beta > 0. \tag{B.3}$$

Calculate the acceptance rate as function of α and β.

For section 5.2:

(1) Find a way to reduce the dimension of the array IAcase in table 5.10 from nconf to

$$\frac{\text{nconf}}{q\,(2d)!}\,.$$

B.2 Solutions

Solutions for some of the assignments in the main text and some of the exercises in the previous section B.1 are given.

For section 1.1, exercises:

(1) nfreq: 96, 88, 118, 108, 90, 100 in the order i = 1, 2, 3, 4, 5, 6.
(2) The events are obtained by squaring a uniformly distributed random variable. According to equation (1.10) the distribution function is therefore

$$F(t) = \begin{cases} 0 & \text{for} \quad t \le 0, \\ \sqrt{t} & \text{for} \quad 0 \le t \le 1, \\ 1 & \text{for} \quad t \ge 1. \end{cases}$$

The probability density is the derivative of $F(t)$, *i.e.*

$$f(t) = \begin{cases} t^{-\frac{1}{2}} & \text{for} \quad 0 \le t \le 1, \\ 0 & \text{elsewhere.} \end{cases}$$

For section 1.7, assignments:

(1) For a non-zero expectation value a the Gaussian probability density

and distribution function are

$$g_a(x) = \frac{1}{\sigma\sqrt{2\pi}}\, e^{-(x-a)^2/(2\sigma^2)} \quad \text{and} \quad G_a(x) = \frac{1}{2} + \frac{1}{2}\operatorname{erf}\left(\frac{x-a}{\sigma\sqrt{2}}\right).$$

(B.4)

The reduced moments (1.57) are unchanged.

For section 1.8, assignments:

(1) The results are

$$\phi(t) = \exp(ita)\,\exp\left(-\frac{1}{2}\sigma^2 t^2\right),$$

(B.5)

where a is the mean value $\hat{x}$,

$$\phi_y(t) = \exp(it2a)\,\exp\left(-\frac{1}{2}2\sigma^2 t^2\right),$$

(B.6)

where the mean value is now $\hat{y} = 2a$, and

$$\phi_{\bar{x}}(t) = \exp(ita)\,\exp\left(-\frac{1}{2}\frac{\sigma^2}{2}t^2\right),$$

(B.7)

with the mean value back to $\hat{x} = a$.

For section 1.8, exercises:

(1) Write

$$\phi(t) = \lim_{a\to\infty}\int_{-a}^{+a} dx\,\frac{1}{2}\left\{\frac{e^{itx}}{1+ix} + \frac{e^{itx}}{1-ix}\right\}$$

and introduce $z = x + iy$. Close the contour on a square in the upper or lower complex plane, chosen so that $\exp(-|ty|)$ holds. For $a\to\infty$ the staple outside the real axis does not contribute and the residue theorem yields $\phi(t) = \exp(-|t|)$.

For section 2.1, exercises:

(1) Let

$$M = \sum_{i=1}^{N} N_i,\quad \text{then}\quad w_i = \frac{N_i}{M} \quad \text{and}\quad \sigma_{\bar{x}}^2 = w_i\,\sigma_{\bar{x}_i}^2 \quad \text{holds.}$$

Equation (2.11) implies

$$\sigma_{\bar{x}}^2 = \langle (\bar{x})^2 \rangle - \hat{x}^2 \quad \text{and} \quad \sigma_{\bar{x}_i}^2 = \langle (\bar{x}_i)^2 \rangle - \hat{x}^2 .$$

Then

$$\langle (\bar{x}_i - \bar{x})^2 \rangle = \langle (\bar{x}_i)^2 \rangle - 2 \langle \bar{x}_i \, \bar{x} \rangle + \langle (\bar{x})^2 \rangle$$

and

$$\langle \bar{x}_i \, \bar{x} \rangle = w_i \langle (\bar{x}_i)^2 \rangle + \sum_{j, j \neq i} w_j \langle \bar{x}_i \, \bar{x}_j \rangle = w_i \langle (\bar{x}_i)^2 \rangle + (1 - w_i) \, \hat{x}^2 .$$

Putting these equations together,

$$\langle (\bar{x}_i - \bar{x})^2 \rangle = (1 - 2w_i) \langle (\bar{x}_i)^2 \rangle - (1 - 2\,w_i) \, \hat{x}^2 - \hat{x}^2 + \langle (\bar{x})^2 \rangle$$

$$= (1 - 2w_i) \, \sigma_{\bar{x}_i}^2 + \sigma_{\bar{x}}^2 = \sigma_{\bar{x}_i}^2 - w_i \, \sigma_{\bar{x}_i}^2$$

and, hence,

$$\frac{1}{N-1} \sum_{i=1}^{N} w_i \langle (\bar{x}_i - \bar{x})^2 \rangle = \frac{1}{N-1} \sum_{i=1}^{N} \left[w_i \, \sigma_{\bar{x}_i}^2 - (w_i)^2 \, \sigma_{\bar{x}_i}^2 \right]$$

$$= \frac{1}{N-1} \sum_{i=1}^{N} \left[\sigma_{\bar{x}}^2 - w_i \, \sigma_{\bar{x}}^2 \right] = \frac{1}{N-1} \, [N - 1] \, \sigma_{\bar{x}}^2 = \sigma_{\bar{x}}^2 .$$

(2) 0.1556.

(3) (a) 2.77165 and (b) 5.09798.

For section 3.2, exercises:

(1) It is sufficient to consider two states which are connected by the update of a single spin. Assume the states have energies E. The Boltzmann vector of the two states is

$$P_B = c \begin{pmatrix} e^{-\beta E} \\ e^{-\beta E'} \end{pmatrix}$$

and the Metropolis update matrix is of the form

$$W_0 = \begin{pmatrix} 1 - q & p \\ q & 1 - p \end{pmatrix} .$$

Without restriction of generality we can assume $E \leq E'$, so that

$$q = \exp[-\beta (E' - E)] \leq 1 \quad \text{and} \quad p = 1$$

holds. Therefore,

$$W_0 \, P_B \;=\; c \left(\frac{e^{-\beta E} - e^{-\beta E'} + e^{-\beta E'}}{e^{-\beta E'}} \right) \;=\; P_B \,.$$

The Boltzmann vector is indeed an eigenstate with eigenvalue one. The transition matrix (3.37) is given by the product

$$W \;=\; \prod_{is=1}^{N} W_{0,is} \tag{B.8}$$

where is is the site number and $W_{0,is}$ the transition matrix for the spin at is, suitably embedded in a $K \times K$ matrix which acts on the configurations.

For section 3.4, exercises:

(1) It follows from exercise 1 to section 3.3 that the balance (not detailed balance) still holds after every change of the acceptance rates. Nevertheless problems emerge when these changes are correlated with the configuration of the previous sweep.

For section 4.1, exercises:

(1) The transition is accepted with probability

$$W_a(x;\alpha) = W(x \to x') = \begin{cases} 1 & \text{if } (x+a)^2 \leq x^2 \\ e^{-\alpha(x+a)^2 + \alpha x^2} & \text{if } (x+a)^2 \geq x^2. \end{cases}$$

Assuming that we are in equilibrium, the acceptance rate for a given $a \geq 0$ is

$$a_c(a;\alpha) = \int_{-\infty}^{+\infty} W_a(x;\alpha)\, P_\alpha(x)\, dx = 1 - \text{erf}\left(\frac{a}{2\sqrt{\alpha}}\right)$$

and $a_c(-a) = a_c(a)$, where erf(x) is the error function (1.32). The acceptance rate for the Metropolis procedure which proposes a with a Gaussian probability distribution $P_\beta(a)$ becomes

$$A_c(\beta,\alpha) = \int_{-\infty}^{+\infty} a_c(a) P_\beta(a)\, da = 1 - \int_{-\infty}^{+\infty} \text{erf}\left(\frac{a}{2\sqrt{\alpha}}\right) P_\beta(a)\, da \,.$$

Using polar coordinates the constraint of the error function integral becomes a constraint of the polar angle. The integrations are elementary and yield

$$A_c(\beta, \alpha) = 1 - \frac{2}{\pi} \arctan \sqrt{\frac{\alpha}{\beta}} \, .$$

Limit are

$$\lim_{\beta \to \infty} A_c(\beta, \alpha) = 1 \quad \text{and} \quad \lim_{\beta \to 0} A_c(\beta, \alpha) = 0 \, .$$

For section 5.2, exercises:

(1) First, we may cyclically permute the central spin together with its nearest neighbor spin, so that the central spin always assumes the same value iq_in. This gives the factor q. Next, the local energy distribution does not depend on the permutations of the nearest neighbor spins. Hence, we can reduce the nearest neighbors to their sorted order and this gives the factor $(2d)!$.

Appendix C

More Fortran Routines

There are additional routines in ForLib, which are not introduced in the main text. These are

(1) heap_iper.f: A version of the heapsort routine heap_per.f, see section 1.6, which sorts an integer data array.

(2) heapisort.f: A version of the heapsort routine heapsort.f, see section 1.6, which sorts an integer data array.

(3) f_interpol.f: Simple, linear interpolation of a function given by an array of function and argument values.

(4) nsum.f: Integer function, which returns the sum of the elements of an integer array.

(5) parabola.f: Three point fit to a parabola.

(6) tun_cnt2.f: Another version of the tunneling (random cycle) count. This routine increments the count not only at the minimum, but also at the maximum.

(7) rmasave_a.f: Append version of rmasave.f. Writes the state of the Marsaglia random number generator to the end of an open file.

(8) rmaset_a.f: A version of rmaset.f, which reads the state of the Marsaglia random number generator from an open file (for use together with rmasave_a.f).

(9) rwght_ts_zln.f: Ferrenberg-Swendsen re-weighting for a simple action-variable time series. This routine uses logarithmic coding and prepares for jackknife binning with the subroutine rwght_ts_zlnj.f (quite similar to the multicanonical time series re-weighting discussed in chapter 5.1).

(10) rwght_ts_zlnj.f: Jackknife binning for data calculated with the routine rwght_ts_zln.f.

Bibliography

Abramowitz, M. and Segun, I.A. (1970). *Handbook of Mathematical Functions*, Dover Publication, New York, fifth Dover edition.

Alder, B.J. and Wainwright, T.E. (1957). *Phase Transition for a Hard Sphere System*, J. Chem. Phys. **27**: 1208–1209. Alder, B.J. and Wainwright, T.E. (1960). *Studies in Molecular Dynamics. I. General Method*, J. Chem. Phys. **31**: 459–466. (1960). *Studies in Molecular Dynamics. II. Behavior of a Small Number of Elastic Spheres*, J. Chem. Phys. **33**: 1439–1451.

Allen, M.D. and Tildesley, D.J. (1993). *Computer Simulations of Liquids*, Oxford University Press, New York.

Alves, N.A., Berg, B.A. and Villanova, R. (1990). *Ising-Model Monte Carlo Simulations: Density of States and Mass Gap*, Phys. Rev. B **41**: 383–394.

Alves, N.A., Berg, B.A. and R. Villanova, R. (1991). *Potts Models: Density of States and Mass Gap from Monte Carlo Calculations*, Phys. Rev. B **43**: 5846–5856.

Arnold, G., Lippert, Th. and Schilling, K. (1999). *Multicanonical Hybrid Monte Carlo: Boosting Simulations of Compact QED*, Phys. Rev. D **59**: 054509.

Athloen, S. and McLaughlin, R. (1987). *Gauss-Jordan Reduction: A Brief History*, Am. Math. Monthly **94**: 130–142.

Bachmann, B. and Janke, W. (2003). *Multicanonical Chain-Growth Algorithm*, Phys. Rev. Lett. **91**: 208105.

Barkai, D. and Moriarty, K.J.M. (1982). *Can the Monte Carlo Method for Lattice Gauge Theory Calculations be effectively vectorized?*, Comp. Phys. Commun. **27**: 105–111.

Baxter, R.J. (1973). *Potts Models at the Critical Temperature*, J. Phys. C **8**: L445–L448.

Beale, P.D. (1996). *Exact Distribution of Energies in the Two-Dimensional Ising Model*, Phys. Rev. Lett. **76**: 78–81.

Belavin, B.A. and Polyakov, A.M. (1975). *Metastable States of Two-Dimensional Isotropic Ferromagnets*, JETP Lett. **22**: 245–247.

Bennett, C.H. (1976). *Efficient Estimation of Free Energy Differences from Monte Carlo Data*, J. Comp. Phys. **22**: 245–268.

Berg, B.A. and Lüscher, M. (1981). *Definition and Statistical Distributions of a*

Topological Number in the Lattice $O(3)$ σ-Model, Nucl. Phys. B [FS] **190**: 412–424.

Berg, B.A. (1992a). *Double Jackknife Bias-Corrected Estimators*, Comp. Phys. Commun. **69**: 7–15.

Berg, B.A. (1992b). *Monte Carlo Calculations of Confidence Levels for Realistic Least Square Fitting*, Comp. Phys. Commun. **69**: 65–75.

Berg, B.A. and Neuhaus, T. (1992). *Multicanonical Ensemble: A New Approach to Simulate First-Order Phase Transitions*, Phys. Rev. Lett. **68**: 9–12.

Berg, B.A. and Celik, T. (1992). *A New Approach to Spin Glass Simulations*, Phys. Rev. Lett. **69**: 2292–2295.

Berg, B.A., Hansmann, U.H. and Neuhaus, T. (1993). *Simulation of an Ensemble with Varying Magnetic Field: A Numerical Determination of the Order-Order Interface Tension of the $D = 2$ Ising Model*, Phys. Rev. B **47**: 497–500; *Properties of Interfaces in the Two and Three Dimensional Ising Model*, Z. Phys. B **90**: 229–239.

Berg, B.A., Billoire, A. and Janke, W. (2000). *Spin Glass Overlap Barriers in Three and Four Dimensions*, Phys. Rev. B **61**: 12143–12150.

Berg, B.A. (2002). *Generalized Ensemble Simulations for Complex Systems*, Comp. Phys. Commun. **147**: 52–57.

Berg, B.A. (2003). *Multicanonical Simulations Step by Step*, Comp. Phys. Commun. **153**: 397–406.

Bevington, P.R. and Robinson, D.K. (2003). *Data Reduction and Error Analysis for the Physical Sciences*, McGraw-Hill, New York, 3rd edition.

Bhanot, G., Salvador, R., Black, S., Carter, P. and Toral, R. (1987). *Accurate Estimate of ν for the Three-Dimensional Ising Model from a Numerical Measurement of its Partition Function*, Phys. Rev. Lett. **59**: 803–806.

Binder, K. (1976). *Monte Carlo Investigations of Phase Transitions and Critical Phenomena*, in *Phase Transitions and Critical Phenomena*, C. Domb and M.S. Green (editors), Vol. 5B, Academic Press, New York.

Binder, K. (1981). *Critical Properties from Monte Carlo Coarse Graining and Renormalization*, Phys. Rev. Lett. **47**: 693–696.

Binder, K. (1982). *The Monte Carlo Calculation of the Surface Tensions for Two- and Three-Dimensional Lattice-Gas Models*, Phys. Rev. A **25**: 1699–1709.

Binder, K., Challa, M.S.S. and Landau, D.P. (1986). *Finite-Size Effects at Temperature-Driven First-Order Phase Transitions*, Phys. Rev. B **34**: 1841–1852.

Binder, K. (editor), (1995). *The Monte Carlo Method in Condensed Matter Physics*, Topics in Applied Physics, Vol. 71, Springer, Berlin.

Binder, K. (1997). *Quadrupolar Glasses and Random Fields*, in *Spin Glasses and Random Fields*, edited by Young, A.P., World Scientific, Singapore.

Binder, K. and Herrmann, D.W. (2002). *Monte Carlo Simulation in Statistical Physics – An Introduction*, Springer Series in Solid-State Sciences 80, 4th edition, Springer, Berlin.

Birnbaum, Z.W. and Tingey, F.H. (1951). *One-sided confidence contours for distribution functions*, Ann. Math. Stat. **22**: 592–596. Birnbaum, Z.W. (1953). *On the power of a one-sided test of fit for continuous probability functions,*

Ann. Math. Stat. **24**: 484-489.

Blöte, H.W.J. and Nightingale, M.P. (1982). *Critical Behavior of the Two-Dimensional Potts Model with a Continuous Number of States: A Finite Size Scaling Analysis*, Physica A **112**: 405–465.

Blöte, H.W.J. and Nightingale, M.P. (2000). *Monte Carlo Computation of Correlation Times of Independent Relaxation Modes at Criticality*, Phys. Rev. B **62**: 1089–1101.

Blöte, H.W.J., Heringa, J.R. and Luijten, E. (2002). *Cluster Monte Carlo: Extending the Range*, Comp. Phys. Commun. **147**: 58–63.

Borgs, C. and Janke, W. (1992). *An Explicit Formula for the Interface Tension of the 2D Potts Model*, J. Phys. I France **2**: 2011–2018.

Bortz, A.B., Kalos, M.H. and Lebowitz, J.L. (1975). *A New Algorithm for Monte Carlo Simulation of Ising Spin Systems*, J. Comp. Phys. **17**: 10–18.

Brandt, S. (1983). *Statistical and Computational Methods in Data Analysis*, North-Holland Publishing, Amsterdam.

Brézin, E. and J. Zinn-Justin, J. (1976). *Spontaneous Breakdown of Continuous Symmetries near Two Dimensions*, Phys. Rev. B **14**: 3110–3120.

Brézin, E. (1982). *An Investigation of Finite Size Scaling*, J. Phys. (Paris) **43**: 15–22.

Bouzida, D., Kumar, S. and Swendsen, R.H. (1992). *Efficient Monte Carlo methods for computer simulations of biological molecules*, Phys. Rev. A **45**: 8894–8901.

Cabibbo, N. and Marinari, E. (1982). *A New Method for Updating SU(N) Matrices in Computer Simulations of Gauge Theories*, Phys. Lett. B **119**: 387–390.

Caracciola, S., Edwards, R.G., Pelissetto, A. and Sokal, A.D. (1993). *Wolff-Type Embedding Algorithms for General Nonlinear σ-Models*, Nucl. Phys. B **403**: 475–541.

Caselle, M., Fiore, R., Gliozzi, F., Hasenbusch, M., Pinn, K. and Vinti, S. (1994). *Rough Interfaces Beyond the Gaussian Approximation*, Nucl. Phys. B **432**: 590–628.

Caselle, M., Tateo, R. and Vinti, S. (1999). *Universal Amplitude Ratios in the 2D four state Potts model*, Nucl. Phys. B **562**: 549-566.

Chaikin, P.M. and Lubensky, T.C. (1997). *Principles of condensed matter physics*, Cambridge University Press, table 8.6.1, p.467.

Coleman, S. (1973). *There are no Goldstone Bosons in Two Dimensions*, Comm. Math. Phys. **31**: 259–264.

Creutz, M. (1980). *Monte Carlo Study of Quantized SU(2) Gauge Theory*, Phys. Rev. D **21**: 2308–2315.

Csikor, F., Fodor, Z., Heim, J. and Heitger, J. (1995). *Interface Tension of the Electroweak Phase Transition*, Phys. Lett. B **357**: 156–162.

Duane, S., Kennedy, A.D., Pendleton, B.J. and Roweth, D. (1987). *Hybrid Monte Carlo*, Phys. Lett. B **195**: 216–222.

Edwards, S.F. and Anderson, P.W. (1975). *Theory of Spin Glasses*, J. Phys. F **5**: 965–974.

Eisenmenger, F., Hansmann, U.H., Hayryan, S. and Hu, C.-K. (2001). *[SMMP] A Modern Package for Simulation of Proteins*, Comp. Phys. Commun. **138**:

192–212.

Efron, B. (1982). *The Jackknife, the Bootstrap and other Resampling Plans*, SIAM, Philadelphia.

Ferdinand, A.E. and Fisher, M.E. (1969). *Bounded and Inhomogeneous Ising Models. I. Specific-Heat Anomaly of a Finite Lattice*, Phys. Rev. **185**: 832–846.

Ferguson, D.M., Siepmann, J.I. and Truhlar, D.G. (editors), (1998). *Monte Carlo Methods in Chemical Physics*, Advances in Chemical Physics, Vol. 105, John Wiley & Sons, New York.

Ferrenberg, A.M. and Swendsen, R.H. (1988). *New Monte Carlo Technique for Studying Phase Transitions*, Phys. Rev. Lett. **61**: 2635–2638.

Ferrenberg, A.M. and Swendsen, R.H. (1989a). *Optimized Monte Carlo Data Analysis*, Phys. Rev. Lett. **63**: 1195–1198.

Ferrenberg, A.M. and Swendsen, R.H. (1989b). *New Monte Carlo Technique for Studying Phase Transitions (Errata)*, Phys. Rev. Lett. **63**: 1658.

Fisher, M.E. (1971). *The theory of critical point singularities*, in *Critical Phenomena*, Proceedings of the 1970 Enrico Fermi International School of Physics, Vol. 51, M.S. Green (editor), Academic Press, New York, pp. 1–99.

Fisher, M.E. and Privman, V. (1985). *First-Order Transitions Breaking O(n) Symmetry: Finite-Size Scaling*, Phys. Rev. B **32**: 447–464.

Fisher, R.A. (1930). *Inverse Probability*, Proc. Camb. Phil. Soc. **26**: 528–535.

Fisher, R.A. (1933). *The Concepts of Inverse Probability and Fiducial Probability Referring to Unknown Parameters*, Proc. Roy. Soc. A **139**: 343–348.

Fisher, R.A. (1926). *Applications of Student's Distribution*, Metron **5**: 90.

Foulkes, W.M.C., Mita, L., Needs, R.J. and Rajagopal, G. (2001). *Quantum Monte Carlo simulations of solids*, Rev. Mod. Phys. **73**: 33–83.

Frenkel, D. and Smit, B. (1996). *Understanding Molecular Simulation*, Academic Press, San Diego.

Flyvberg, H. and Peterson, H.G. (1989). *Error Estimates on Averages of Correlated Data*, J. Chem. Phys. **91**: 461–466.

Fortuin, C.M. and Kasteleyn, P.W. (1972). *On the Random-Cluster Model*, Physica (Utrecht) **57**: 536–564.

Galambos, J. (1987). *The Asymptotic Theory of Extreme Order Statistics*, Krieger Publishing Co., Malibar, Florida.

Gell-Mann, M. and Neeman, Y. (1964). *The Eightfold Way*, Benjamin, New York.

Gilks, W.R., Richardson, S. and Spiegelhalter, D.J. (editors), (1995). *Markov Chain Monte Carlo in Practice*, CRC Press.

Glauber, R.J. (1963). *Time-Dependent Statistics of the Ising Model*, J. Math. Phys. **4**: 294–307;

Geman, S. and Geman, D. (1984). *Stochastic Relaxation, Gibbs Distributions, and the Bayesian Restoration of Images*, IEEE Transactions of Pattern Analysis and Machine Intelligence **6**: 721–741.

Gould, H. and Tobochnik, J. (1996). *An Introduction to Computer Simulation Methods: Applications to Physical Systems*, second edition, Addison-Wesley, Reading, MA.

Grassberger, P. (2002). *Go with the Winners: a General Monte Carlo Strategy*, Comp. Phys. Commun. **147**: 64–70.

Geyer, G.J. (1991). *Markov Chain Monte Carlo Maximum Likelihood*, in *Computing Science and Statistics*, Proceedings of the 23rd Symposium on the Interface, Keramidas, E.M. (editor), Interface Foundation, Fairfax, Virginia, pp. 156–163.

Gropp, W., Lusk, E. and Skjellum, A. (1999). *Using MPI*, MIT Press.

Gubernatis, J. (2003). *The Heritage*, in Proceeding of the Los Alamos 2003 conference *The Monte Carlo Method in the Physical Sciences: Celebrating the 50th Anniversary of the Metropolis Algorithm*, Gubernatis, J. (editor), AIP Conference Proceedings, Volume 690, Melville, NY, pp. 3–21.

Gumbel, E.J. (1958). *Statistics of Extremes*, Columbia University Press, New York.

Hansmann, U.H. and Okamoto, Y. (1993). *Prediction of Peptide Conformation by Multicanonical Algorithm: New Approach to the Multiple-Minima Problem*, J. Comp. Chem. **14**: 1333-1338.

Hansmann, U.H. Okamoto, Y. and Eisenmenger, E. (1996). *Molecular Dynamics, Langevin and Hybrid Monte Carlo Simulations in a Multicanonical Ensemble*, Chem. Phys. Lett. **259**: 321–330.

Hansmann, U.H. and Okamoto, Y. (1999). *The Generalized-Ensemble Approach for Protein Folding Simulations*, Ann. Rev. Comp. Phys. **6**: 129–157.

Hansmann, U.H. (1997). *Parallel tempering algorithm for conformational studies of biological molecules*, Chem. Phys. Lett. **281**: 140–150.

Hasenbusch, H. (1995). *An Improved Estimator for the Correlation Function of 2D Nonlinear Sigma Models*, Nucl. Phys. B. [Proc. Suppl.] **41**: 764–766.

Hastings, W.K. (1970). *Monte-Carlo Sampling Methods Using Markov Chains and Their Applications*, Biometrika **57**: 97–109.

Hoshen, J. and Kopelman, R. (1976). *Percolation and Cluster Distribution. I. Cluster multiple labeling technique and critical concentration algorithm.* Phys. Rev. B **14**: 3438-3445.

Houdayer, J. (2001). *A Cluster Monte Carlo Algorithm for 2-Dimensional Spin Glasses*, Eur. Phys. J. B **22**: 479–484.

Huang, K. (1987). *Statistical Mechanics*, John Wiley & Sons, New York.

Hukusima, K. and Nemoto, K. (1996). *Exchange Monte Carlo Method and Applications to Spin Glass Simulations*, J. Phys. Soc. Japan **65**: 1604–1608.

Ising, E. (1925). *Beitrag zur Theorie des Ferromagnetizmus*, Z. Phys. **31**: 253–258.

Jacobs, L. and Rebbi, C. (1981). *Multi-spin Coding: A Very Efficient Technique for Monte Carlo Simulations of Spin Systems*, J. Comp. Phys. **41**: 203–210.

Janke, W. and Kappler, S. (1995). *Multibondic Cluster Algorithm for Monte Carlo Simulations of First-Order Phase Transitions*, Phys. Rev. Lett. **74**: 212-215.

Janke, W. and Sauer, T. (1994). *Multicanonical Multigrid Monte Carlo Method*, Phys. Rev. E **49**: 3475–3479.

Jaynes, E.T. (1982). *On the rationale of maximum-entropy methods*, Proc. IEEE **70**: 939-952.

Kalos, M.H. and Whitlock, P.A. (1986). *Monte Carlo Methods*, John Wiley & Sons, New York.

Kaufmann, B. (1949). *Crystal Statistics II. Partition Function Evaluated by Spinor Analysis*, Phys. Rev. **76**: 1232–1243.

Kawasaki, K. (1972). *Kinetics of Ising Models*, in *Phase Transitions and Critical Phenomena*, Domb, C. and Green, M.S. (editors), Vol. 2, Academic Press, New York, pp. 443–501.

Kerler, W. and Rehberg, P. (1994). *Simulated Tempering Procedure for Spin-Glass Simulations*, Phys. Rev. E **50**: 4220–4225.

Kirkpatrick, S. Gelatt Jr., C.D. and Vecchi, M.P. (1983). *Optimization by Simulated Annealing*, Science **220**: 671–690.

Knuth, D.E. (1968). *Fundamental Algorithms*, Vol. 1 of *The Art of Computer Programming*, Addison-Wesley, Reading, MA.

Knuth, D.E. (1981). *Seminumerical Algorithms*, Vol. 2 of *The Art of Computer Programming*, Addison-Wesley, Reading, MA.

Kolmogorov, A. (1933). *Determinazione empirica di una legga di distribuzione*, Giornale Istit. Ital. Attuari **4**: 83–91.

Korniss, G., Novotny, M.A., Guclu, H., Toroczkai, Z. and Rikvold, P.A. (2003). *Suppressing Roughness of Virtual Times in Parallel Discrete Event Simulations*, Science **229**: 677–679.

Kosterlitz, M. and Thouless, D.J. (1973). *Ordering, Metastability and Phase Transitions in Two-Dimensional Systems*, J. Phys. C **6**: 1181–1203.

Kosterlitz, M. (1974). *The Critical Properties of the Two-Dimensional XY Model*, J. Phys. C **7**: 1046–1060.

Kuiper, N.H. (1962). *Tests Concerning Random Points on a Circle*, Proceedings of the Koninklijke Nederlandse Akademie van Wetenschappen **63**: 38–47.

Lanczos, C. (1964). *A precision approximation of the Gamma function*, SIAM Numerical Analysis B **1**: 86-96.

Landau, D.P. and Binder, K. (2000). *A Guide to Monte Carlo Simulations in Statistical Physics*, Cambridge University Press, Cambridge.

Landau, L.D. and Lifshitz, E.M. (1999). *Statistical Physics*, Course of Theoretical Physics Vol. 5, Butterworth-Heinemann, Boston.

Levenberg, K. (1944). *A Method for the Solution of Certain Nonlinear Problems in Least Squares*, Qty. Appl. Math. **2**: 164–168.

Lee, P.M. (1997). *Bayesian Statistics: An Introduction*, Arnold, London.

Lipschutz, S. and Lipson, M. (2000). *Schaum's Outlines Linear Algebra*, 3rd edition, McGraw-Hill.

Liu, J.S. (2001). *Monte Carlo Strategies in Scientific Computing*, Springer, New York.

Lüscher, M. (1994). *A portable high-quality random number generator for lattice field theory simulations*, Comp. Phys. Commun. **79**: 100–110.

Lyubartsev, A.P., Martsinovski, A.A., Shevkanov, S.V. and Vorontsov-Velyaminov, P.N. (1992). *New Approach to Monte Carlo Calculation of the Free Energy: Method of Expanded Ensembles*, J. Chem. Phys. **96**: 1776–1783.

Ma, S.-K. (1976). *Renormalization Group by Monte Carlo Methods*, Phys. Rev. Lett. **37**: 461–464.

Marchesini, G., Webber, B.R., Abbiendi, G., Knowles, I.G., Seymour, M.H. and

Sanco, L. (1992). *HERWIG 5.1 - a Monte Carlo event generator for simulating hadron emission reactions with interfering gluons*, Comp. Phys. Commun. **67**: 465–508.

Marinari, E. and Parisi, G. (1992). *Simulated Tempering: A New Monte Carlo Scheme*, Europhys. Lett. **19**: 451–458.

Marquardt, D.W. (1963). *An Algorithm for the Estimation of Non-Linear Parameters*, SIAM **11**: 431–441.

Marsaglia, G., Zaman, A. and Tsang, W.W. (1990). *Toward a Universal Random Number Generator*, Stat. Prob. **8**: 35–39.

Mascagni, M. and Srinivasan, A. (2000). *Algorithm 806: SPRNG: A Scalable Library for Pseudorandom Number Generation*, ACM Transactions on Mathematical Software **26**: 436-461. Available on the web at sprng.cs.fsu.edu.

Mehlig, B., Heermann, D.W. and Forrest, B.M. (1992). *Hybrid Monte Carlo Methods for Condensed-Matter Systems*, Phys. Rev. B **45**: 679–685.

Mermin, N.D. and Wagner, H. (1966). *Absence of Ferromagnetism or Antiferromagnetism in One- or Two-Dimensional Isotropic Heisenberg Models*, Phys. Rev. Lett. **17**: 1133–1136.

Message Passing Interface Forum (1995). *MPI: A Message-Passing Interface Standard*, University of Tennessee, Knoxville, TN.

Metropolis, N. Rosenbluth, A.W., Rosenbluth, M.N., Teller, A.H. and Teller, E. (1953). *Equation of State Calculations by Fast Computing Machines*, J. Chem. Phys. **21**: 1087–1092.

Michael, C. (1994). *Fitting Correlated Data*, Phys. Rev. D **49**: 2616–2619.

Miller, A.J. (2002). *Subset Selection in Regression*, Second Edition, Chapman & Hall (managed by Kluwer Academic Publishers, Norwell, MA).

Miller, R. (1974). *The Jackknife – a Review*, Biometrika **61**: 1–15.

Mitsutake, A., Sugita, Y. and Okamoto, Y. (2001). *Generalized-Ensemble Algorithms for Molecular Simulations of Biopolymers*, Biopolymers (Peptide Science) **60**: 96–123.

Moore, G.E. (1965). *Cramming more components onto integrated circuits*, Electronics **38**: 114–117.

Munoz, J.D., Mitchel, S.J. and Novotny, M.A. (2003). *Rejection-free Monte Carlo algorithms for models with continuous degrees of freedom*, Phys. Rev. E **67**: 026101.

Münster, G. and Montvay, I. (1993). *Quantum Fields on a Lattice*, Cambridge University Press, New York.

Negele, J.W. (1999). *Instantons, the QCD Vacuum, and Hadronic Physics*, Nucl. Phys. B (Proc. Suppl.) **73**: 92–104.

Neuhaus, T. and Hager, J.S. (2003). *2d Crystal Shapes, Droplet Condensation and Supercritical Slowing Down in Simulations of First Order Phase Transitions*, J. Stat. Phys. **113**: 47–83.

Newman, M.E.J. and Barkena, G.T. (1999). *Monte Carlo Methods in Statistical Physics*, Clarendon Press, Oxford.

Neyman, J. and Pearson, E.S. (1933). *On the Problem of the most Efficient Tests of Statistical Hypothesis*, Phil. Trans. Roy. Soc. A **231**: 289–337.

Novotny, M.A. (2001). *A Tutorial on Advanced Dynamic Monte Carlo Methods for Systems with Discrete State Spaces*, Ann. Rev. Comp. Phys. **9**: 153–210.

Ódor, G. (2002). *Universality classes in nonequilibrium lattice systems*, cond-mat/0205644, to appear in Rev. Mod. Phys..

Onsager, L. (1944). *Crystal Statistics.I. A Two-Dimensional Model with an Order-Disorder Transition*, Phys. Rev. **65**: 117–149.

Parisi, G. and Wu Yongshi, (1981). *Perturbation Theory without Gauge Fixing*, Sci. Sin. **24**: 483–496.

Pathria, R.K. (1972). *Statistical Mechanics*, Pergamon Press, New York.

Pacheco, P. (1996). *Parallel Programming with MPI*, Morgan and Kaufmann Publishers.

Pearson, K. (1968). *Tables of the Incomplete Beta-Function*, Cambridge University Press.

Pelissetto, A. and Vicari, E. (2002). *Critical Phenomena and Renormalization-Group Theory*, Phys. Rep. **368**: 549–727.

Polyakov, A.M. (1975). *Interactions of Goldstone Particles in Two Dimensions. Applications to Ferromagnets and Massive Yang-Mills Fields*, Phys. Lett. B **59**: 79–81.

Potts, R.B. (1952). *Some Generalized Order-Disorder Transformations*, Proc. Cambridge Philos. Soc. **48**: 106–109.

Privman, V. and Fisher, M.E. (1983). *Finite-Size Effects at First-Order Transitions*, J. Stat. Phys. **33**: 385–417.

Propp, J. and Wilson, D. (1998). *Coupling from the Past User's Guide*, DIMACS Series in Discrete Mathematics and Theoretical Computer Science (AMS) **41**: 181–192.

Press, W.H., Flannery, B.P., Teukolsky, S.A. and Vetterling, W.T. (1992). *Numerical Recipes in Fortran*, Second Edition, Cambridge University Press, Cambridge.

Quenouille, M.H. (1956). *Notes on Bias in Estimation*, Biometrika **43**: 353–360.

Ralston, A. and Rabinowitz, P. (1978). *A First Course in Numerical Analysis*, McGraw-Hill, New York, 2nd edition.

Rhee, Y.M. and Pande, V.S. (2003). *Multiplexed-Replica Exchange Molecular Dynamics Method for Protein Folding Simulation*, Biophys. J. **84**: 775–786.

Rosenbluth, M.N. and Rosenbluth, A.W. (1955). *Monte Carlo Calculation of the Average Extension of Molecular Chains*, J. Chem. Phys. **23**: 356–359.

Rosenbluth, M.N. (2003). *Genesis of the Metropolis Algorithm for Statistical Mechanics*, in Proceeding of the Los Alamos 2003 conference *The Monte Carlo Method in the Physical Sciences: Celebrating the 50th Anniversary of the Metropolis Algorithm*, Gubernatis, J. (editor), AIP Conference Proceedings, Vol. 690, Melville, NY, pp. 22–30.

Rothe, H.J. (1998). *Lattice Gauge Theories: An Introduction*, World Scientific, Singapore.

Sabelfeld, K.K. (1991). *Monte Carlo Methods in Boundary Value Problems*, Springer Series in Computational Physics, Springer, Berlin.

Sanbonmatsu, K.Y. and Garcia, A.E. (2002). *Structure of Met-enkephalin in explicit aqueous solution using replica exchange molecular dynamics*, Proteins **46**: 225–234.

Shirts, M. and Pande, W.J. (2000). *Screen savers of the world unite*, Science **290**: 1093–1904.

Smirnov, N. (1939). *Sur les écarts de la courbe de distribution empirique*, Mat. Sbornik **48**: 3.

Snir, M., Otto, S., Huss-Lederman, S., Walker, D. and Dongarra, J. (1996). *MPI: The Complete Reference*, MIT Press.

Stauffer, D. and Aharony, A. (1994). *Introduction to Percolation Theory*, Second Edition, Taylor & Francis.

Snow, C.D., Nguyen, H., Pande, W.J. and Gruebele, M. (2002). *Absolute comparison of simulated and experimental protein-folding dynamics*, Nature **420**: 102–106.

Sokal, A. (1997). *Monte Carlo Methods in Statistical Mechanics: Foundations and New Algorithms*, in *Functional Integration: Basics and Applications*, Cargése Summer School 1996, DeWitt-Morette, C., Cartier, P. and Folacci, A. (editors), Plenum Press, New York.

Stephens, M.A. (1970). *Use of the Kolmogorov-Smirnov, Cramer-Von Mises and Related Statistics Without Extensive Tables*, J. Royal Stat. Soc. B **32**: 115–122.

Student (1908). *The Probable Error of a Mean*, Biometrika **6**: 1–25.

Sugita, Y. and Okamoto, Y. (1999). *Replica-exchange molecular dynamics method for protein folding*, Chem. Phys. Lett. **314**: 141–151.

Sugita, Y. and Okamoto, Y. (2000). *Replica-exchange multicanonical algorithm and multicanonical replica exchange method for simulating systems with a rough energy landscape*, Chem. Lett. **329**: 261–270.

Swendsen, R.H. (1978). *Monte Carlo Renormalization Group*, Phys. Rev. Lett. **42**: 859–861.

Swendsen, R.H. and Wang, J.-S. (1986). *Replica Monte Carlo Simulations of Spin Glasses*, Phys. Rev. Lett. **57**: 2607–2609.

Swendsen, R.H. and Wang, J.-S. (1987). *Nonuniversal Critical Dynamics in Monte Carlo Simulations*, Phys. Rev. Lett. **58**: 86–88.

't Hooft, G. (1976). *Computation of the Quantum Effects due to a Four-Dimensional Pseudoparticle*, Phys. Rev. D **14**: 3432–3450.

Tobochnik, J. and Chester, G.V. (1979). *Monte Carlo Study of the Planar Spin Model*, Phys. Rev. B **20**: 3761–3769.

Torrie, G.M. and Valleau, J.P. (1977). *Nonphysical Sampling Distributions in Monte Carlo Free-energy Estimation: Umbrella Sampling*, J. Comp. Phys. **23**: 187–199.

Tukey, J.W. (1958). *Bias and Confidence in not-quite large Samples (Abstract)*, Ann. Math. Stat. **29**: 614–615.

Van der Waerden, B.L. (1969). *Mathematical Statistics*, Springer, New York.

Vattulainen, I., Ala-Nissila, T. and Kankaala, K. (1995). *Physical Models as Tests for Randomness*, Phys. Rev. E **52**: 3205–3214.

Verbeke, J. and Cools, R. (1995). *The Newton-Raphson Method*, Int. J. Math. Educ. Sci. Technol. **26**: 177–193.

Villain, J. (1977). *Spin Glass with Non-Random Interactions*, J. Phys. C **10**: 1717–1734.

Wang, F. and Landau, D.P. *Efficient, Multiple-Range Random Walk Algorithm to Calculate the Density of States*, Phys. Rev. Lett. **86**: 2050–2053.

Wang, J.-S. and Swendsen, R.H. (2002). *Transition Matrix Monte Carlo Method*, J. Stat. Phys. **106**: 245–285.

Westlake, J.R. (1968). *A Handbook of Numerical Matrix Inversion and Solution of Linear Equations*, John Wiley & Sons, New York.

Weinberg, S. (1972). *Gravitation and Cosmology*, John Wiley & Sons, New York, pp. 193–194.

Whittaker, E.T. and Watson, G.N. (1927). *A Course of Modern Analysis*, Cambridge University Press, New York.

Wilding, N.B. and Müller, M. (1995). *Liquid-Vapor Asymmetry in pure Fluids: A Monte Carlo Simulation Study*, J. Chem. Phys. **102**: 2562–2573.

Wilkinson, J.H. (1965). *The Algebraic Eigenvalue Problem*, Clarendon Press, Oxford.

Wolff, U. (1989). *Collective Monte Carlo Updating for Spin Systems*, Phys. Rev. Lett. **62**: 361–363.

Wood, W. (2003). *A Brief History of the Use of the Metropolis Method at LANL in the 1950s*, in Proceeding of the Los Alamos 2003 conference *The Monte Carlo Method in the Physical Sciences: Celebrating the 50th Anniversary of the Metropolis Algorithm*, Gubernatis, J. (editor), AIP Conference Proceedings, Volume 690, Melville, NY, 2003, pp. 39–44.

Wu, F.Y. (1982). *The Potts Model*, Rev. Mod. Phys. **54**: 235–268.

Yang, C.N. and Lee, T.D. (1952). *Statistical Theory of Equations of State and Phase Transitions. I. Theory of Condensation*, Phys. Rev. **87**: 404–409.

Yang, C.N. and Lee, T.D. (1952). *Statistical Theory of Equations of State and Phase Transitions. II. Lattice Gas and Ising Model*, Phys Rev. **87**: 410–419.

Zorn, R., Herrmann, H.J. and Rebbi, C. (1981). *Tests of the multi-spin-coding technique in Monte Carlo simulations of statistical systems*, Comp. Phys. Commun. **23**: 337–342.

Index